Protein Structure

The Practical Approach Series

SERIES EDITOR

B. D. HAMES
Department of Biochemistry and Molecular Biology
University of Leeds, Leeds LS2 9JT, UK

★ **indicates new and forthcoming titles**

Affinity Chromatography

Anaerobic Microbiology

Animal Cell Culture
(2nd edition)

Animal Virus Pathogenesis

Antibodies I and II

★ Antibody Engineering

Basic Cell Culture

Behavioural Neuroscience

Biochemical Toxicology

Bioenergetics

Biological Data Analysis

Biological Membranes

Biomechanics—Materials

Biomechanics—Structures and
Systems

Biosensors

Carbohydrate Analysis
(2nd edition)

Cell–Cell Interactions

The Cell Cycle

Cell Growth and Apoptosis

Cellular Calcium

Cellular Interactions in
Development

Cellular Neurobiology

Clinical Immunology

Crystallization of Nucleic
Acids and Proteins

Cytokines (2nd edition)

The Cytoskeleton

Diagnostic Molecular Pathology
I and II

Directed Mutagenesis

★ DNA and Protein Sequence
Analysis

DNA Cloning 1: Core
Techniques (2nd edition)

DNA Cloning 2: Expression
Systems (2nd edition)

★ DNA Cloning 3: Complex
Genomes (2nd edition)

★ DNA Cloning 4: Mammalian
Systems (2nd edition)

Electron Microscopy in
Biology

Electron Microscopy in
Molecular Biology

Electrophysiology

Enzyme Assays

★ Epithelial Cell Culture

Protein Structure

A Practical Approach

Edited by

T. E. CREIGHTON
European Molecular Biology Laboratory,
Heidelberg, Germany

OXFORD UNIVERSITY PRESS
Oxford New York Tokyo

Oxford University Press, Great Clarendon Street, Oxford OX2 6DP

Oxford   New York

Athens   Auckland   Bangkok   Bogota   Bombay   Buenos Aires
Calcutta   Cape Town   Dar es Salaam   Delhi   Florence   Hong Kong
Istanbul   Karachi   Kuala Lumpur   Madras   Madrid   Melbourne
Mexico City   Nairobi   Paris   Singapore   Taipei   Tokyo   Toronto

and associated companies in
Berlin   Ibadan

Oxford is a trade mark of Oxford University Press

Published in the United States
by Oxford University Press Inc., New York

A catalogue record for this book is available from the British Library

Library of Congress Cataloging in Publication Data
(Data available)
ISBN 0 19 963619 2 (Hbk)
ISBN 0 19 963618 4 (Pbk)

Two volume set
ISBN 0 19 963617 6 (Hbk)
ISBN 0 19 963620 6 (Pbk)

Typeset by Footnote Graphics, Warminster, Wilts
Printed in Great Britain by Information Press, Ltd, Eynsham, Oxon.

Preface

The first edition of this volume presented relatively simple electrophoretic, chromatographic, and spectrophotometric techniques that were likely to be practical in an average laboratory. All that was required was some simple electrophoretic equipment, a chromatography column and fraction collector, and a UV spectrophotometer. The present volume attempts to maintain that spirit, while keeping pace with technical advances. Accordingly, the present volume contains new chapters on the relatively sophisticated techniques of mass spectrometry and ultracentrifugation, even though they require relatively expensive equipment. Nevertheless, the emphasis is not on the technical aspects of these complex instruments, but on the more practical aspects of sample preparation and the possibilities of experimental strategy and design that would be most useful to a general protein chemist about to collaborate with an expert in either of these techniques.

Mass spectrometry has suddenly become a very powerful method for protein characterization and verification. It is almost indispensable in working with proteins. In the ideal world, a protein chemist would be able to verify the molecular mass of every protein or peptide sample generated; this would avoid many unpleasant surprises (or artefactual results) arising from unsuspected reactions of the protein. Chapter 2 describes what is possible with this technique and gives practical protocols for sample preparation for the various types of mass analysis of both intact proteins and peptide digests.

Mass spectrometry is extremely useful in identifying the products of post-translational modification of proteins. The long list of these modifications keeps growing, and very many have probably been missed thus far. The most distinctive result of most such modifications is the change in mass of the protein; nearly all modifications can be detected in this way, with no need for the modification to have been suspected beforehand. General strategies to identify and characterize protein post-translational modifications are given in Chapter 4.

Very many polypeptide chains function as part of non-covalent complexes with other polypeptide chains or other macromolecules. Often, the complexes are readily dissociable. They are best characterized by measuring the apparent size of the molecule in solution under normal physiological conditions, using hydrodynamic methods; these are described in Chapter 9. In some cases, an ultracentrifuge or light scattering instrument is required, but much can still be done with a simple gel filtration column.

There is a new description, in Chapter 7, of how to determine which cysteine residues are paired in disulfide bonds. The previous edition described the very elegant technique devised originally by J. R. Brown and B. S. Hartley (1966) (*Biochem. J.*, **101**, 214). It used paper electrophoresis under identical conditions in two dimensions, with an intervening breakage of all disulfide bonds by performic acid vapour. No superior, more sophisticated technique has appeared, but paper electrophoresis is not practical these days, and modern separation technologies are not suited to two-dimensional separations. The best modern methods are described in Chapter 7. A method for counting the number of disulfide bonds present, as well as cysteine and lysine residues, is given in Chapter 6.

Electrophoretic techniques for determining the molecular weight of a polypeptide chain and for characterizing its conformational properties are presented in Chapters 1 and 8, respectively. Chapter 1 provides useful illustrations of the effects of varying the parameters of SDS–PAGE, which retains its central importance in molecular biology. Chapter 8 describes the power of simple gel electrophoresis in characterizing the conformational properties of proteins. Other electrophoretic techniques useful with proteins are found in the related volume '*Gel electrophoresis of proteins: a practical approach*'.

Immunological approaches to identifying and characterizing proteins, which are both simple and powerful, are described in Chapters 3 and 13. Peptide mapping procedures are described in Chapter 5.

Spectral methods are very useful in characterizing protein structure and function, as described in Chapter 11. UV absorbance is also the most accurate method for determining the concentration of a pure protein, but only if its molar absorbance (extinction) coefficient is known. Fortunately, it has been realized that the UV absorbance of a simple protein (without other prosthetic groups) is determined largely by its content of aromatic amino acid residues and disulfide bonds, especially if the protein is unfolded. With the techniques described in Chapter 10, it is possible to calculate or determine experimentally the absorbance coefficient of a protein, so that its concentration can subsequently be determined very accurately. Surely there is no longer any reason or excuse for using with a purified protein an assay like the classic Lowry folin phenol method 'calibrated' with another protein, such as BSA.

The folded conformations of proteins are essential for their biological functions. How to measure and to increase the stability of that conformation are described in Chapters 12 and 14, respectively.

Comparable simple techniques for characterizing protein function are to be found in the companion volume '*Protein function: a practical approach*'. For a comprehensive description of the properties of proteins, see the second edition of my volume '*Proteins: structures and molecular properties*' (W. H. Freeman, New York, 1993).

Heidelberg   T. E. C.
December 1996

Contents

Contents

Contents

13. Immunochemical analysis of protein conformation 323

Bertrand Friguet, Lisa Djavadi-Ohaniance, and Michel E. Goldberg

Contents

Contributors

TSUTOMU ARAKAWA
Graduate Department of Biochemistry, Brandeis University, Waltham, MA 02254–9110, USA.

ELIZABETH A. CARREY
Department of Biochemistry, University of Dundee, Dundee DD1 4HN, UK.

LISA DJAVADI-OHANIANCE
Unité de Biochimie Cellulaire, Institut Pasteur, 28, rue du Docteur Roux, 75724 Paris Cedex 15, France.

BERTRAND FRIGUET
Unité de Biochimie Cellulaire, Institut Pasteur, 28, rue du Docteur Roux, 75724 Paris Cedex 15, France.

MICHEL E. GOLDBERG
Unité de Biochimie Cellulaire, Institut Pasteur, 28, rue du Docteur Roux, 75724 Paris Cedex 15, France.

DAVID P. GOLDENBERG
Department of Biology, University of Utah, Salt Lake City, UT 84112, USA.

WILLIAM R. GRAY
Department of Biology, University of Utah, Salt Lake City, UT 84112, USA.

STEPHEN E. HARDING
National Centre for Macromolecular Hydrodynamics, Nottingham Laboratory, University of Nottingham, Department of Applied Biochemistry and Food Science, Sutton Bonington LE12 5RD, UK.

MICHELLE HOLLECKER
Centre de Biophysique Moléculaire, Rue Charles-Sadron, 45071 Orléans Cedex 2, France.

OLE NØRREGAARD JENSEN
European Molecular Biology Laboratory, Meyerhofstr. 1, D-69012 Heidelberg, Germany.

RADHA G. KRISHNA
Department of Biochemistry and Molecular Biology, Medical School, The University of Texas, Houston Health Science Center, 6431 Fannin Street, Houston, TX 77225, USA.

Contributors

GREGORY S. MAKOWSKI
Department of Laboratory Medicine, University of Connecticut Health Center, Farmington, CT 06030, USA.

MATTHIAS MANN
European Molecular Biology Laboratory, Meyerhofstr. 1, D-69012 Heidelberg, Germany.

NICK A. MORRICE
Department of Biochemistry, University of Dundee, Dundee DD1 4HN, UK.

C. NICK PACE
Department of Medical Biochemistry, 440 Reynolds Medical Building, Texas A&M University, College Station, TX 77843–1114, USA.

MELINDA L. RAMSBY
School of Medicine, University of Connecticut Health Center, Farmington, CT 06030, USA.

SILKE REINARTZ
Poliklinik für Geburtshilfe und Gynäkologie, Universitäts-Frauenklinik, Sigmund-Freud-Str. 25, D-53105 Bonn, Germany.

KARL HEINZ SCHEIDTMANN
Institut für Genetik, Universität Bonn, Römerstr. 164, D-53117 Bonn, Germany.

HARALD SCHLEBUSCH
Poliklinik für Geburtshilfe und Gynäkologie, Universitäts-Frauenklinik, Sigmund-Freud-Str. 25, D-53105 Bonn, Germany.

FRANZ X. SCHMID
Biochemie, Universität Bayreuth, D-95440 Bayreuth, Germany.

J. MARTIN SCHOLTZ
Department of Medical Biochemistry, 440 Reynolds Medical Building, Texas A&M University, College Station, TX 77843–1114, USA.

ANDREJ SHEVCHENKO
European Molecular Biology Laboratory, Meyerhofstr. 1, D-69012 Heidelberg, Germany.

SERGE N. TIMASHEFF
Graduate Department of Biochemistry, Brandeis University, Waltham, MA 02254–9110, USA.

FINN WOLD
Department of Biochemistry and Molecular Biology, Medical School, The University of Texas, Houston Health Science Center, 6431 Fannin Street, Houston, TX 77225, USA.

Abbreviations

A_x	absorbance at wavelength x nm
Ac	acetyl group
ACN	acetonitrile
AEC	3-amino-9-ethylcarbazole
Ala	alanine residue
amu	atomic mass unit
Arg	arginine residue
Asn	asparagine residue
Asp	aspartic acid residue
ATZ	anilinothiazolinone
BDB	*bis*-diazobenzidine
bisacrylamide	*N,N'*-methylene-*bis*-acrylamide
BPTI	bovine pancreatic trypsin inhibitor (aprotinin, Trasylol™)
BrdU	bromodeoxyuridine
BSA	bovine serum albumin
Caps	3-(cyclohexylamino)-1-propanesulfonic acid
CD	circular dichroism
CHAPS	3-[(3-cholamidopropyl)dimethylammonio]-1-propanesulfonate
Cys	cysteine residue
Cys(cam)	*S*-carboxamidomethylcysteine residue
CZE	capillary zone electrophoresis
D	diffusion coefficient
DEAE	diethylaminoethyl
DHB	2,5-dihydroxybenzoic acid
DLS	dynamic light scattering
DMSO	dimethyl sulfoxide
DTT	dithiothreitol
ε_x	molar absorbance at wavelength x, when specified
ECL	enhanced chemiluminescence
EDTA	ethylenediaminetetraacetic acid
ELISA	enzyme-linked immunosorbent assay
Epps	4-(2-hydroxyethyl)-1-piperazinepropanesulfonic acid
ES	electrospray
EtOH	ethanol
FAB	fast atom bombardment
Fab	immunoglobulin fragment, antigen binding
FCS	fetal calf serum
FITC	fluorescein isothiocyanate
Gaba	γ-amino-*n*-butyric acid

Abbreviations

Gdm$^+$	guanidinium ion
GdmCl	guanidinium chloride, guanidine hydrochloride
GlcNAc	*N*-acetylglucosamine residue
Gln	glutamine residue
Glu	glutamic acid residue
Gly	glycine residue
GSH	glutathione, thiol form
GSSG	glutathione, disulfide form
GST	glutathione-*S*-transferase
HCCA	4-hydroxy-α-cyanocinnamic acid
Hepes	*N*-2-hydroxyethylpiperazine-*N'*-2-ethanesulfonic acid
HFIP	hexafluoroisopropanol
His	histidine residue
HPLC	high-pressure liquid chromatography
IAEDANS	5-((2-iodoacetyl)aminoethyl)aminonaphthalene-1-sulfonic acid
IEF	isoelectic focusing
Ig	immunoglobulin
Ile	isoleucine residue
IPTG	isopropyl-β-D-thiogalactoside
k_B	Boltzmann's constant
KLH	keyhole limpet haemocyanin
α-LA	α-lactalbumin
LC	liquid chromatography
Leu	leucine residue
LSIMS	liquid secondary ionization mass spectrometry
Lys	lysine residue
m	molecular mass
mAb	monoclonal antibody
MALDI	matrix-assisted laser desorption/ionization
MALLS	multi-angle laser light scattering
Man	mannose residue
MeOH	methanol
Mes	2-(*N*-morpholino)ethanesulfonic acid
Met	methionine residue
Mol. wt	molecular weight (dimensionless)
Mops	3-(*N*-morpholino)propanesulfonic acid
MPD	2-methyl-2,4-pentanediol
M_r	relative molecular mass (dimensionless)
MRW	mean residue weight
MS	mass spectrometry
MUG	methyl-umbelliferyl-β-galactopyranoside
NMR	nuclear magnetic resonance
ONPG	*o*-nitrophenyl-β-D-galactopyranoside

ORD	optical rotatory dispersion
P	phosphate group
PAGE	polyacrylamide gel electrophoresis
PBS	phosphate-buffered saline
PC	personal computer
PCR	polymerase chain reaction
PD	plasma desorption
PEG	polyethylene glycol
p.f.u.	plaque-forming units
Phe	phenylalanine residue
P_i	inorganic phosphate
PMSF	phenylmethylsulfonyl fluoride
PNPP	*p*-nitrophenyl phosphate
Pro	proline residue
PTH	phenylthiohydantoin
PVDF	polyvinylidene difluoride
R	electrical resistance
RIA	radioimmunoassay
RNase	ribonuclease
RP	reverse phase
scFv	single chain variable fragment of immunoglobulin
SDS	sodium dodecyl sulfate
SEC	size exclusion chromatography
Ser	serine residue
SPDP	*N*-succinimidyl-3-(2-pyridyl dithio)-propionate
TBS	tris-buffered saline
TCA	trichloroacetic acid
TCEP	*tris*-(2-carboxyethyl)phosphine
TEMED	*N,N,N',N'* tetramethylethylenediamine
TFA	trifluoroacetic acid
THAP	2,4,6-trihydroxyacetophenone
Thr	threonine residue
T_m	midpoint of temperature-induced transition
TMAO	trimethylamine *N*-oxide
TOF	time-of-flight
Trp	tryptophan residue
Tyr	tyrosine residue
Val	valine residue
z	net charge of a molecule

Protein molecular weight determination by sodium dodecyl sulfate polyacrylamide gel electrophoresis

GREGORY S. MAKOWSKI and MELINDA L. RAMSBY

1. Introduction

Sodium dodecyl sulfate polyacrylamide gel electrophoresis (SDS–PAGE) is a widely used analytical technique for the separation and characterization of complex mixtures of proteins and peptides and for the estimation of the relative molecular weight (M_r) of a protein. Despite numerous variations (1–3 and references therein), the discontinuous SDS–PAGE system described by Laemmli in 1970 (4) remains the most popular. Using this classical method as a paradigm, this article will review the principles and theory of electrophoresis, describe the essential components of SDS–PAGE, and demonstrate relevant applications including the use of SDS–PAGE for detection of enzymatic activity.

2. Principles and theory

Electrophoresis refers to the movement of charged particles under the influence of an electric field (5). In essence, charged particles migrate in an electric field towards the electrode of opposite sign. Proteins, although amphoteric polyvalent macromolecules, migrate in an electric field in a fashion consistent with their net charge. Because net charge is a conformation-dependent characteristic, protein electrophoresis is markedly affected by the pH, temperature, and ionic composition of the electrophoretic medium. Independent of charge, electrophoretic mobility is also affected by macromolecular size and shape (Chapter 8). Within gel-based systems, electrophoretic migration is further dependent upon the physical properties of the gel, especially with respect to pore size. Pore size determines sieving capacity and is a reflection of the extent of cross-linking within the three-dimensional matrix. Thus, electrophoretic separation of proteins is a function of macro-molecular charge, size, and shape, as well as the physicochemical properties of the electrophoretic medium. Thus, through selection of appropriate conditions, electrophoretic fractionation of complex protein mixtures can be modulated to achieve the desired resolution.

3. Components of SDS–PAGE

SDS–PAGE is an example of zone electrophoresis in a polyacrylamide gel matrix and affords excellent control of pore size. The three components integral to understanding the fundamental principles of SDS–PAGE are polyacrylamide, SDS, and multiphasic buffers.

3.1 Polyacrylamide

Acrylamide and bisacrylamide are synthetic organic molecules that polymerize to generate linear polyacrylamide and its cross-linked forms, respectively, in a three-dimensional matrix. Free radical-induced polymerization of acrylamide is catalysed by ammonium persulfate and an accelerator (*N,N,N′,N′*-tetramethylenediamine, TEMED). Acrylamide solutions are degassed (see *Protocol 1*) prior to catalyst addition because dissolved oxygen is a free radical scavenger and thus inhibits polymerization.

Polyacrylamide gels are transparent and flexible, yet relatively strong and resilient. They are chemically inert and are compatible with numerous buffers, salts, and detergents.

Overall, protein mobility through polyacrylamide gels is proportional to the pore size, which is a function of both the acrylamide concentration ($\%T$) and that of the bisacrylamide cross-linker ($\%C$). In general, the pore size is inversely proportional to $\%T$.

$$\%T = \frac{\text{acrylamide (g)} + \text{bis-acrylamide (g)}}{100 \text{ ml}} \times 100\% \qquad [1]$$

$$\%C = \frac{\text{bis-acrylamide (g)}}{\text{acrylamide (g)} + \text{bis-acrylamide (g)}} \times 100\% \qquad [2]$$

For a 10% Laemmli gel (4), T is equal to 10% and C is equal to 2.6% (see *Protocol 1*). In contrast, a 10%T Dreyfuss gel (6) has larger pore size because it contains only 0.9%C. Depending on the application, many $\%T/\%C$ variations have been reported (1–3). Specifically, lower cross-linker concentrations improve the separation of filamentous proteins and have been used to improve resolution of cytoskeletal proteins (7).

In general, best resolution is achieved when the pore size approximates the protein molecular size (1–3). The $\%T$ of the gel determines the range over which the protein M_r is proportional to the rate of migration (*Table 1*). Gels may also be composed of a gradient of polyacrylamide (typically 5–20%T). Although more difficult to prepare (see Section 6.1 and Chapter 8), gradient gels resolve wider ranges of protein mol. wt.

3.1.1 Slab and tube gels

SDS–PAGE is conducted using either tube or slab gels, depending upon the application. Tube (rod) gels are cylinders of polyacrylamide (1.5–3 mm thickness, 5–15 cm length) formed using a glass tube mould, and only one sample is analysed per gel. Tube gels are convenient for preparative gel electrophoresis, isoelectric focusing, and two-dimensional procedures. Slab gels, in contrast, are flat rectangular sheets of polyacrylamide generated in a mould composed of two glass plates separated slightly (0.75–1.5 mm). Slab gels can contain many application sites. Thus, numerous samples, including mol. wt standards, can be analysed simultaneously and compared directly. Slab gels are easier to handle during staining or destaining, and they dry easily for long-term storage. In addition, slab gels lend well to densitometric analysis, as well as to transblotting techniques (Chapter 3). These advantages have resulted in almost exclusive use of the slab gel format.

3.1.2 Continuous and discontinuous gels

Polyacrylamide gels may be cast with continuous or discontinuous buffers. Continuous gels as initially described by Weber and Osborn (8) contained a constant acrylamide concentration. In contrast, Laemmli gels are discontinuous composite gels containing a short stacking gel of low %T (3–4%) layered on top of a long resolving gel of higher %T (7–25%) (*Figure 1*). The two gels are also discontinuous with respect to buffer composition, and the upper stacking gel functions to concentrate samples prior to entering the lower fractionating gel (see Section 3.3).

3.2 Sodium dodecyl sulfate

SDS is an anionic detergent that is used to denature proteins, giving them all the same conformational properties, and to prevent protein interactions during electrophoresis. SDS also masks the intrinsic protein charge and gives all proteins a similar net negative charge, so that electrophoretic migration is towards the anode. In general, SDS coats proteins with a uniform negative charge and constant charge-to-mass ratio. Hence, in an electric field SDS-coated proteins experience the same field strength and migrate at identical intrinsic rates towards the anode. Proteins of different size are subjected to different degrees of sieving by the gel, however, and the mobility of a protein is inversely proportional to its size (8–11).

Uniform SDS binding is critical for the appropriate electrophoretic migration, and it generally approximates one SDS molecule per two amino acid residues, or 1.4 g SDS/g protein (12–14). To achieve adequate binding, proteins are heated (100°C) in the presence of excess SDS (see *Protocol 4*), and protein disulfide bonds are chemically reduced with 2-mercaptoethanol (or other sulfhydryl reagent) (15). The exact shape of SDS-coated proteins, however, remains unclear (16,17).

Although SDS is effective for most proteins, some SDS-coated proteins have unequal charge density or atypical shapes and therefore demonstrate anomalous electrophoretic behaviour. These include low mol. wt proteins (< 10 kDa), glycosylated proteins, proteins with high proline content, and proteins with very high or low isoelectric points (2).

3.3 Multiphasic buffers

Laemmli gels (4), in addition to being discontinuous with respect to %T, are also discontinuous with respect to the pH and ionic composition of the stacking and resolving gels (multiphasic buffers). This functions to concentrate (stack) proteins prior to fractionation in the resolving gel (*Figure 1*). Multiphasic buffers sandwich the SDS-coated proteins between rapidly (Cl⁻) and slowly (glycine) migrating ions as described below (see Section 3.3.2). Stacking is critical for obtaining optimal electrophoretic resolution but is absent in monophasic (continuous) buffer systems (8). Consequently, monophasic

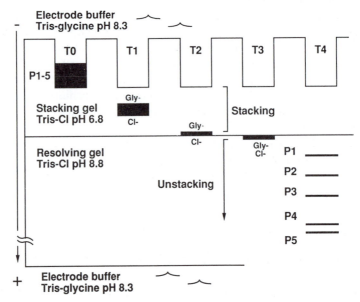

Figure 1. Explanation of Kohlrausch boundary migration under Laemmli multiphasic buffer conditions. Electrophoresis through the stacking (pH 6.8) and resolving (pH 8.8) gels of a hypothetical sample containing five proteins (P1–5) is shown as a function of time (T). T0: sample loaded into well; T1: partial stacking of sample between leading (Cl⁻) and trailing edge (glycine) ions; T2: complete stacking of P1–5; T3: unstacking takes place with increased ionization of glycine in the pH 8.8 resolving gel; T4: resolution of P1–5.

systems are limited by the requirement that samples be both highly concentrated and low in volume. The use of multiphasic systems overcomes both of these limitations and is thus more conducive to analysis of ordinary biological samples.

3.3.1 Kohlrausch boundary

The presence of a discontinuous multiphasic gel–buffer system enables a high voltage gradient to be generated between rapidly migrating and slowly migrating ions during electrophoresis. This voltage gradient serves to concentrate ions of intermediate mobility, such as proteins, and is referred to as a Kohlrausch boundary (18). Ornstein (19) and Davis (20) first described the application of a Kohlrausch boundary to PAGE under non-denaturing conditions, and Laemmli subsequently expanded its application to use under denaturing conditions (4).

3.3.2 Laemmli modification

Laemmli modifications to the method of Ornstein (19) and Davis (20) include the addition of SDS and the adjustment of the stacking and resolving gel buffers to pH 6.8 and 8.8, respectively (see *Figure 1*). Briefly, during electrophoresis through the upper gel, sample proteins, due to their intermediate mobilities, become highly concentrated ('stacked') between the leading edge chloride ions of the upper gel buffer and the trailing edge glycine ions of the electrode reservoir buffer to form a Kohlrausch boundary. Upon encountering the higher pH of the lower gel buffer (pH 8.8), glycine becomes ionized and hence exhibits increased mobility resulting in dissipation of the Kohlrausch boundary. The stacked proteins are thus no longer confined, and their subsequent migration through the resolving gel results in fractionation based upon their size.

Although multiphasic zone electrophoresis theory initially predicted over 4200 combinations of leading and trailing edge ion combinations (21), which were subsequently simplified to 19 (22), the buffer system used by Laemmli (4) remains the most popular.

4. Performing SDS–PAGE

Due to its inherent advantages, this article will focus on performing Laemmli (4) SDS–PAGE in the slab gel format. Identical solutions are, however, used for generating tube gels. A summary procedure for running SDS–PAGE gels is as follows:

- prepare stock solutions (see *Protocol 1*)
- assemble glass plates for gel casting
- mix and pour resolving gel (see *Protocol 1*)
- mix and pour stacking gel (see *Protocol 2*)

- prepare samples (see *Protocol 4*)
- set-up and run electrophoresis system (see *Protocol 5*)
- fix, stain/destain, and dry gel (see *Protocols 6* and *7*)

4.1 Slab SDS–PAGE

Slab gel systems in a variety of dimensions and thickness are available. The apparatus routinely used by the authors (Shadel) (*Figure 2*) generates an 11 cm (height) × 14 cm resolving gel and a 1.5 cm stacking gel, both 0.75 mm thickness. This gel accommodates 16 samples. SDS–PAGE systems can be ordered with equipment for casting gels and performing electrophoresis. A power source that operates in fixed DC mode (i.e. constant voltage or constant current) is also required. Gel size and number will determine power requirements, but at least up to 500 V/200 mA are generally necessary.

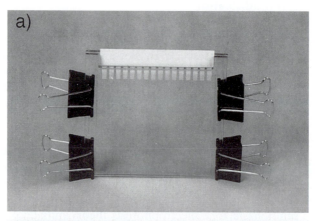

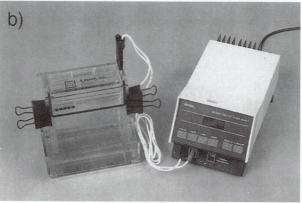

Figure 2. Electrophoresis unit. (a) Slab gel assembly with Teflon comb inserted. (b) Slab gel attached to electrophoresis tank with power supply.

4.2 Preparation of SDS–PAGE gels

For proteins of known approximate size, determine the optimum acrylamide concentration required for SDS–PAGE from *Table 1*. For unknown proteins, a 10%T gel is an excellent choice for initial characterization. The slab gel is poured in two stages: resolving gel (*Protocol 1*), followed by stacking gel (*Protocol 2*). Reagents for SDS–PAGE can be purchased from most commercial chemical suppliers. For most laboratory purposes, electrophoretic grade reagents should be used, and the solutions should be filtered if silver staining is to be used. Unless specified, stock solutions are prepared with ultrapure water (> 18 megaohms resistance) in glass or plastic bottles and are stable for at least six months at 4 °C.

Table 1. Protein mol. wt range resolved by SDS–PAGE

Polyacrylamide gel (%T)	M_r range
6	30 000–200 000
8	20 000–175 000
10	15 000–150 000
12	10 000–100 000
15	6000–50 000
5–20 gradient	6000–250 000

Protocol 1. Resolving gel preparation

Equipment and reagents

- Polyacrylamide slab gel glass plates and casting equipment—follow manufacturers instructions (*Figure 2*)
- Acrylamide/bisacrylamide solution: dissolve 150 g acrylamide and 4 g *N,N'*-methylene-*bis*-acrylamide in 300 ml water. Add water to 500 ml final volume. Solution may be warmed slightly to facilitate solubilization. (Caution: unpolymerized acrylamide is neurotoxic; use gloves and appropriate outerwear.)
- TEMED: use as supplied by the manufacturer (store at 4 °C)
- Ammonium persulfate solution: 100 mg of ammonium persulfate in 5 ml water, freshly prepared

- Resolving gel buffer solution: 90.9 g Tris base in 350 ml water, adjust the solution to pH 8.8 by slow addition of concentrated HCl (note: avoid back titration with NaOH). Add water to 500 ml final volume. Alternatively, a pH 9.2 buffer enhances resolution and shortens run time (see Section 5.3).
- 10% (w/v) SDS solution: dissolve 10 g SDS in 90 ml water (note: wear mask while measuring SDS powder). Add water to 100 ml final volume. Store below 25 °C.
- Water-saturated butanol solution: combine 100 ml *n*-butanol and 5 ml distilled water in a plastic bottle. Shake well. Use top layer for overlaying gels. Store below 25 °C.

Method

1. Assemble glass plates for casting slab gels.

2. Label one plate with an indelible marker at the resolving gel height (11 cm).

Protocol 1. *Continued*

3. Combine reagents for two resolving gels:

Stock solution	Resolving gel (%*T*)		
	7.5%	10%	12.5%
Acrylamide/bisacrylamide (ml)	10	13.3	16.6
Resolving gel buffer (ml)	10	10	10
Ultrapure water (ml)	19.5	17.2	12.9
Degas under vacuum (10 min)			
10% SDS (ml)	0.8	0.8	0.8
Ammonium persulfate (ml)	0.6	0.6	0.6
TEMED (μl)	20	20	20

Caution: use gloves when handling acrylamide reagent. To prevent reagent loss during degassing, the 10% SDS solution is added post-evacuation.

4. Pipette reagents into casting mould. Fill to the 11 cm mark (avoid air bubbles).

5. Overlay the resolving gel with water-saturated butanol (2 mm depth).

6. Allow polymerization to proceed for 30 min at 25°C.

7. Wash top of resolving gel with distilled water. Gently pad dry with paper towel.

8. For storage, see *Protocol 3*.

The stacking gel solution is prepared and poured into the gel apparatus (see *Protocol 2*). A Teflon comb is used to generate wells for sample loading. Depending on the application, analytical (10–16 wells) or preparative scale (one or two wells) combs can be used. The effective stacking gel height is 1.5 cm.

Protocol 2. Stacking gel preparation

Reagents

• Stacking gel buffer solution: dissolve 15.2 g Tris base in 200 ml water, adjust to pH 6.8 by slow addition of concentrated HCl, add water to 250 ml final volume

Method

1. Wash top of resolving gel with distilled water to remove butanol overlay.

2. Remove water. Gently pad the top of resolving gel with paper towel.

3. Combine reagents for two stacking gels:

Stock solution	3%*T* stacking gel
Acrylamide/bisacrylamide	1.0 ml
Stacking gel buffer	2.5 ml
Ultrapure water	6.3 ml
Degas under vacuum (10 min)	
10% SDS	0.2 ml
Ammonium persulfate	0.13 ml
TEMED	10 μl

Caution: use gloves when handling acrylamide reagents.

4. Pipette reagents to within 0.5 cm of top of the glass plate mould.

5. Gently insert the Teflon comb. Avoid trapping air bubbles.

6. Allow polymerization to proceed for 30 min at 25°C.

7. Carefully remove the Teflon comb under a gentle stream of distilled water (in sink).

8. Remove extraneous polymerized acrylamide from top of gel.

9. Attach the polyacrylamide gel to the electrophoresis chamber (see *Protocol 5*).

4.2.1 Storage of the gel

If not used immediately, the resolving gel may be stored (*Protocol 3*) for up to 48 hours, without substantial differences in electrophoretic properties. It is also possible to store the gel following addition of the stacking gel with the Teflon comb inserted. Prolonged storage could, however, result in dissolution of the sharp pH boundary between the lower (pH 8.8) and upper (pH 6.8) buffers.

Protocol 3. Storage of the resolving gel

Reagents

• Overnight overlay solution: combine 1.5 ml 10% SDS, 30 ml resolving gel buffer solution, 45 ml water, and mix well

Method

1. Wash the top of the resolving gel with distilled water to remove butanol overlay.

2. Tilt the gel to remove water.

3. Overlay the gel with overnight overlay solution (height of 1–2 cm).

4. Apply a Parafilm cover to limit evaporation.

5. Store the resolving gel at 25°C (24 h) or 4°C (48 h).

Protocol 3. *Continued*

6. To use, remove the Parafilm and decant overlay solution.

7. Gently blot the top of resolving gel with paper towel to remove excess fluid.

8. Pour the stacking gel (see *Protocol 2*).

4.2.2 Preparation of samples for SDS–PAGE

Samples are prepared for SDS–PAGE as outlined in *Protocol 4*. The amount of protein required will depend upon the detection system (i.e. Coomassie blue or silver staining) (23–25). About 1–2 μg each protein should be applied for Coomassie blue staining, but substantially lower quantities can be detected by silver staining (23–25). Sample volumes of 5–40 μl can be routinely analysed.

Protocol 4. Sample preparation

Equipment and reagents

- Polypropylene Eppendorf (or like) tubes (0.5–1.5 ml) with cap (screw cap preferable)

- 2 × sample buffer solution: combine 5 ml stacking gel buffer solution, 8 ml 10% SDS solution, and 1 ml water. Add 4 ml glycerol, 2 ml 2-mercaptoethanol and mix well (preferably in fume-hood). Add a small amount (1–2 mg) of bromphenol blue indicator dye. Divide the sample buffer in 1 ml portions (plastic screw cap vials are preferable) and store at −70°C (or at −20°C for short periods). Note: the solution may require warming to dissolve crystallized SDS.

Method

1. Add an equal volume of 2 × sample buffer to the sample. Mix well (vortex).

2. Heat at 100°C (boiling water-bath) for 1–2 min (note: excessive heating will cause non-screw cap tubes to pop open resulting in inconsistent sample concentration or loss).

3. Cool to 25°C and load on to the gel (see *Protocol 5*).

4. Samples can be stored at −70°C (or −20°C short-term).

4.2.3 Assembly of the chamber and electrophoresis

The slab gel is attached to the electrophoresis chamber, the buffer reservoirs are filled, and the samples applied. Wires are connected to the power supply, and electrophoresis is performed at constant current (typically 20 mA/gel) (see *Protocol 5* and *Figure 2*).

It should be noted that the buffer resistance decreases as the gel temperature rises due to heating during electrophoresis. This change, however, is insufficient to compensate for the overall increased resistance (R). To ensure constant current (I), a corresponding increase in voltage (V) is automatically made by the power supply (Ohm's Law):

$$V \text{ (volts)} = I \text{ (amperes)} \times R \text{ (ohms)}. \qquad [3]$$

Heat generation is more pronounced with thick gels (1.5 mm, 40 mA/gel). Thus, active cooling (water jacket or gel run in cold room) may be required to prevent anomalous electrophoretic migration or damage to the SDS–PAGE apparatus.

Protocol 5. Chamber assembly and electrophoresis

Equipment and reagents

- Use electrophoresis equipment and a DC power supply as recommended by the manufacturers (*Figure 2*)
- Hamilton microlitre syringe is used for loading samples (rinse the syringe three times with 2 × sample buffer between samples)

- 4 × electrophoresis tank buffer solution: dissolve 48 g Tris base, 230.4 g glycine, and 16 g SDS in 4 litres final volume with distilled water (note: do not titrate the pH with NaOH). Store below 25°C. Mix one part with three parts distilled water for use as 1 × electrophoresis tank buffer.

Method

1. Remove casting attachments from the gels.

2. Fill the lower chamber with 1 × tank buffer.

3. Clamp the gel to the electrophoretic chamber. Avoid trapping of air bubbles along the bottom.

4. Fill the upper chamber with 1 × tank buffer. Note: the buffer must contact the stacking gel. Make certain that there are no leaks.

5. Carefully load samples (*Protocol 4*) into the bottom of the sample wells (note: the density of the sample buffer is greater than that of tank buffer).

6. Attach wires to the electrodes and power supply, with the red, positive wire to the lower buffer tank.

7. Set the power supply to constant current (typically 20 mA for one gel).

8. Perform electrophoresis until the dye front is 0.5 cm from bottom of gel (2–3 h).

9. Turn off power supply, detach wires, and remove gel.

10. Stain the polyacrylamide gel (see *Protocols 6* and *7*).

4.3 Staining the gel

The most commonly used procedure for protein detection is Coomassie blue staining (see *Protocol 6*); an alternative procedure especially appropriate for small proteins is given in Chapter 8, *Protocol 4*. A more sensitive detection method is silver staining (24,25) (see *Protocol 7*). Specific methods for detection of modified proteins (glycosylation, phosphorylation) have been described (26,27). Immunological detection is generally accomplished following electrophoretic transfer to nitrocellulose filters (Western blotting) (28) (see Chapter 3). An enzyme activity stain for detecting proteases is described in Section 7.

Protocol 6. Staining polyacrylamide gels with Coomassie blue

Equipment and reagents

- Plastic tray with cover (approx. dimensions 15 × 20 cm, 4 cm height)
- Destaining solution: combine 1.6 litres methanol, 400 ml glacial acetic acid, and 2 litres distilled water. Mix well and store below 25°C.
- Coomassie blue staining solution: dissolve 2 g Coomassie brilliant blue R-250 in 500 ml methanol, 100 ml glacial acetic acid (17.4 M), and 400 ml distilled water. Mix well, filter through Whatman No. 1 filter paper, and store below 25°C.

Method

1. Gently pry glass plates apart to disassemble the slab gel.
2. Break the stacking/resolving gel interface with a metal or plastic spatula.
3. Remove the stacking gel and discard (note: polymerized acrylamide is non-toxic).
4. Cut a designated corner of the resolving gel for later orientation.
5. Place the gel in 200 ml Coomassie blue staining solution in covered plastic tray.
6. Stain for 1 h at 25°C on a mixing platform.
7. Decant the used stain solution (note: the stain may be used several times).
8. Destain the gel with destaining solution until desired background is achieved.
9. Photograph the wet gel or dry it between porous cellophane sheets.

Protocol 7. Staining polyacrylamide gels with silver

Equipment and reagents

SDS–PAGE reagents may require filtering to lower background silver staining artefacts. Unless specified, reagents are stable for three months when stored below 25 °C.

- Plastic tray with lid
- 50% methanol solution: combine 500 ml methanol, 100 ml glacial acetic acid, 400 ml water, and mix well
- 5% methanol solution: combine 50 ml methanol, 70 ml glacial acetic acid, 880 ml water, and mix well
- 10% glutaraldehyde solution: combine 200 ml 50% glutaraldehyde, 800 ml water, and mix well (note: glutaraldehyde is toxic)
- Reduction solution: just before use, dissolve 5 mg dithiothreitol into 1 litre ultrapure water

- 0.1% silver nitrate solution: dissolve 1 g silver nitrate in 1 litre ultrapure water, mix well, and store in a brown bottle
- Developer solution: dissolve 30 g sodium carbonate in 900 ml ultrapure water, adjust volume to 1 litre—just before use, add 50 ml of 37% formaldehyde per 100 ml and mix well
- 2.3 M citric acid stop solution: dissolve 44 g citric acid in 60 ml water, adjust volume to 100 ml

Method

1. These instructions are for one slab gel; all incubations take place on a rotatory mixer.

2. Incubate the gel in 100 ml 50% methanol solution for 30 min.

3. Transfer the gel to 100 ml 5% methanol solution and incubate for 30 min.

4. Incubate the gel in 50 ml 10% glutaraldehyde solution for 30 min.

5. Wash the gel extensively with water to completely remove the glutaraldehyde.

6. Incubate the gel in 100 ml reduction solution for 30 min.

7. Rinse the gel quickly in 50 ml water (15 sec).

8. Rinse twice in 50 ml developer solution (15 sec each).

9. Incubate in 100 ml developer solution, until the desired level of staining is achieved.

10. Add 5 ml citric acid stop solution. Incubate 15 min.

11. Wash the gel extensively in water. Due to expansion, the gels are extremely fragile. To increase stability, contract the gel by incubation in 40% methanol solution.

4.4 Calibration proteins and mol. wt determination

Calibration proteins ('standards') are essential for determination of the mol. wt of an unknown protein (8–11). Four sets of calibration proteins have been

Table 2. Calibration proteins for SDS–PAGE

Protein	M_r ($\times 10^{-3}$)	1[a]	2[a]	3[a]	4[b]
Myosin	205		+	+	
β-Galactosidase	116		+	+	
Phosphorylase b	97		+	+	+
Fructose-6-phosphate kinase	84		+	+	
Albumin	66	+	+	+	+
Glutamate dehydrogenase	55		+	+	
Ovalbumin	45	+	+	+	+
Glyceraldehyde-3-phosphate dehydrogenase	36	+	+	+	
Carbonic anhydrase	29	+	+		+
Trypsinogen	24	+	+		
Trypsin inhibitor	20	+	+		+
α-Lactalbumin	14.2	+	+		+
Aprotinin (BPTI)	6.5	+	+		
Number of proteins		8	13	8	6

[a] Sigma standards: 1, low; 2, wide; 3, high mol. wt.
[b] Pharmacia LKB low mol. wt standards.

evaluated (*Table 2*). Proteins were prepared for SDS–PAGE as described by the manufacturer or as described above (see *Protocol 4*).

Calibration graphs can be generated in several forms (*Figure 3*). Graphs obtained by plotting mol. wt versus migration distance yield a hyperbolic curve, making mol. wt determination difficult. An almost linear relationship is, however, obtained when logarithm (mol. wt) is plotted versus migration distance (8–11). The mol. wt of the unknown protein can be determined from the regression equation or directly from the graph.

Plotting logarithm (mol. wt) versus relative migration distance (R_f)

$$R_f = \frac{\text{calibrator migration distance (mm)}}{\text{bromophenol blue migration distance (mm)}} \qquad [4]$$

is required for comparing results obtained from separate gels (i.e. tube gels).

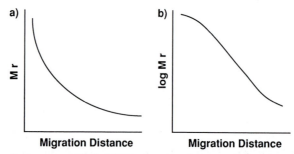

Figure 3. Theoretical graphs of SDS–PAGE migration distance relative to protein M_r. Plotting M_r gives a hyperbolic curve, while plotting logarithm M_r gives an approximately linear curve over the middle range, but sigmoidal at the extremities.

5. Factors affecting SDS–PAGE

The versatility of SDS–PAGE is illustrated below, altering several parameters to demonstrate their function. These included changes in polyacrylamide concentration (%*T*), cross-linking (%*C*), pH, and the effects of sample volume and sample reduction.

5.1 Effect of polyacrylamide concentration

Migration of protein standards was measured at three polyacrylamide concentrations: 8.5, 10, 12.5%*T* (see *Figure 4*). Increasing the acrylamide

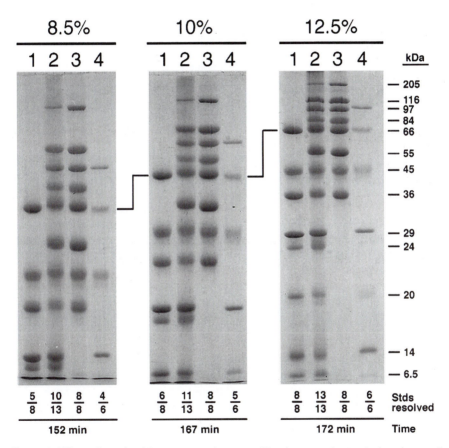

Figure 4. Effect of acrylamide concentration on calibration protein resolution. Laemmli slab gel electrophoresis was performed at 8.5%*T*, 10%*T*, and 12.5%*T*. Protein standards 1–4 are those described in *Table 2*. The *M*ᵣ of each protein is indicated for the gel on the right. The number of protein standards resolved and the time of electrophoresis are shown at the bottom. For orientation, the zig-zag line links the albumin bands. The gels were stained with Coomassie blue.

15

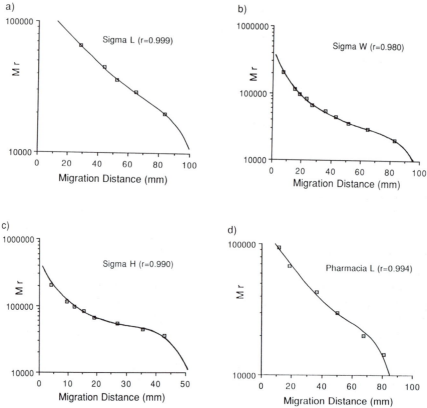

Figure 5. Correlation between the migration distance and the protein M_r in 10%T Laemmli gels. The protein standards 1–4 of *Table 2* are presented in parts (a) to (d), respectively. The correlation coefficient (r) is given in each case.

concentration (i.e. decreasing the pore size) resulted in decreased migration for all four groups of protein standards. The resolution was also influenced.

Using data obtained from 10%T gels, protein migration distance was plotted versus logarithm (mol. wt) (*Figure 5*). Good correlation between mobility and mol. wt was observed for all four groups of protein standards. As expected, the best linearity was achieved over the smallest mol. wt range, especially for low mol. wt standards, and the wide mol. wt standards resulted in the poorest correlation.

The overall effect of %T on the mobility of protein standards for Laemmli gels is shown in *Figure 6*. Increasing the polyacrylamide concentration from 8.5 to 12.5%T resulted in substantially decreased electrophoretic migration for the low mol. wt protein standards. The 8.5% gel resolved only four of the six protein standards.

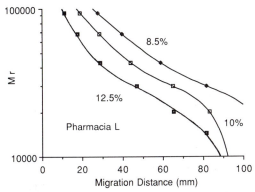

Figure 6. Effect of varying the acrylamide concentration (%*T*). Data were compiled for Pharmacia low mol. wt protein standards using 8.5%*T*, 10%*T*, and 12.5%*T* Laemmli gels.

5.2 Effect of cross-linker concentration

The effect of cross-linker (bisacrylamide) concentration on electrophoretic resolution was evaluated. Using constant 10%*T* gels, the cross-linker was varied from 1.3%*C* and 4%*C* and compared to the control gel, 2.6%*C* (*Figure 7*). Decreased cross-linker (i.e. larger pore size) resulted in a substantial increase in electrophoretic mobility for protein standards (*Figure 8*). Changes in cross-linker concentration resulted in an inverse change in mobility for all protein standards. Increased mobility did not, however, correlate with enhanced resolution (relative to 2.6%*C* and 4%*C* gels). The electrophoretic run time was not appreciably affected (compare with *Figure 4*).

5.3 Effect of resolving gel pH

Gels of 8.5, 10, and 12.5%*T* were assembled as described above, except that the resolving gel buffer was prepared at pH 9.2, rather than 8.8 (see *Protocol 1*) (29). A substantial improvement in the number of standards resolved was noted (*Figure 9*) as compared to normal Laemmli gels (see *Figure 4*). A significant decrease (approximately 40%) in electrophoretic run time was observed at pH 9.2.

Data obtained from 10%*T* gels resulted in good correlation for all four groups of protein standards (*Figure 10*). In agreement with the normal Laemmli system (see *Figure 5*), the best correlation was observed with protein standards that spanned a narrow mol. wt range.

Electrophoretic parameters were compared for pH 8.8 and pH 9.2 gels by monitoring the dye front migration. No substantial difference in migration rate was noted through the 1.5 cm stacking gel. The migration rate through the resolving gel was linear for both the pH 8.8 and pH 9.2 gels. The electrophoretic run time was, however, substantially shorter at pH 9.2, which probably resulted from a pH-dependent increase in glycine ionization that facilitated

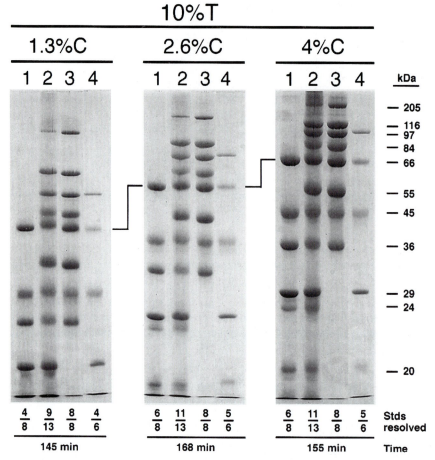

Figure 7. Effect of varying the cross-linker, bisacrylamide concentration. Electrophoresis was performed on 10%*T* Laemmli gels containing bisacrylamide at 1.3%*C*, 2.6%*C* (standard conditions), and 4%*C*. The other conditions were as in *Figure 4*.

unstacking (19,20,29). As expected, the voltage increased for both gels during the electrophoretic run. Higher voltage (by about 60 volts) was consistently required for pH 9.2 gels. This difference is most likely a consequence of the increased resistance (29).

It has been noted (14) that the actual composition of the Laemmli buffer system has not been clearly established, since the temperature at which the temperature-sensitive Tris–HCl buffer was prepared was not cited (4). In our experience (29), slow addition of concentrated HCl to the 1.5 M Tris resolving buffer (see *Protocol 1*) is essential to prevent heat generation and artificially decreased pH (e.g. a pH 8.8 solution at 37°C would be pH 9.1 at 25°C).

1: Protein molecular weight determination

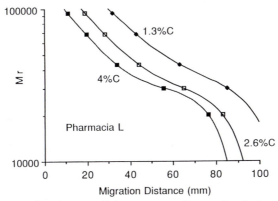

Figure 8. Effect of varying the cross-linker concentration on the electrophoretic mobility. The data are from *Figure 7*.

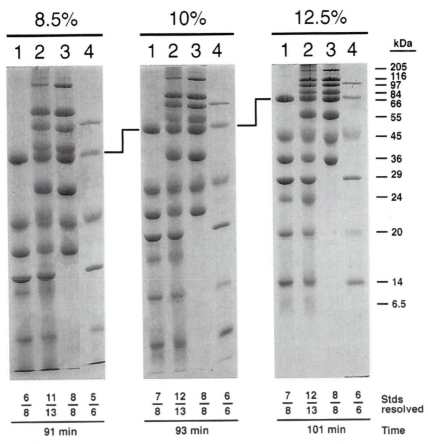

Figure 9. Effect of increased resolving gel pH. The conditions were as in *Figure 4*, except that the pH of the resolving gel was increased from 8.8 to 9.2.

19

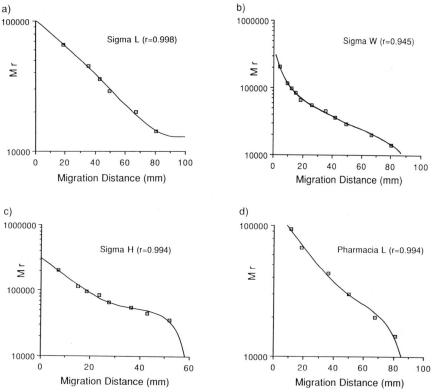

Figure 10. Correlation between electrophoretic mobility and protein M_r in 10%T modified Laemmli slab gel with the resolving gel at pH 9.2. The data are from *Figure 9*.

Despite the distinct advantages of pH 9.2 gels, the buffer system as originally described by Laemmli (4) continues to be most popular.

5.4 Effect of sample volume

Band broadening increases with the application of increasing volumes of samples. Broadening is a consequence of resistance changes/uneven field strength ('frowning', see Section 9.4). In contrast, small differences in salt concentration are tolerated. Samples that contain substantially different salt concentrations (e.g. following ion exchange chromatography) should be normalized prior to SDS–PAGE (30).

5.5 Effect of sample reduction

Reduction of disulfide bonds in a protein prior to electrophoresis is critical for accurate M_r determination. A substantial increase in electrophoretic mobility of disulfide bonded proteins is observed in samples prepared without reduction, which could be erroneously interpreted as a decreased M_r

(10,31). On the other hand, this difference can be utilized to monitor disulfide formation in a protein.

6. Additional SDS–PAGE systems

Depending on the application, numerous SDS–PAGE systems have been described (1–3 and references therein). For brevity we will consider only a few of the more commonly used procedures here.

6.1 Gradient gels

Gradient gels contain a gradient of polyacrylamide. These gels contain a range of pore sizes, and thus are capable of resolving a wider range of protein M_r (see *Table 1*). Calibration curves are generated by plotting logarithm M_r versus $\%T$ (32). Preparation of gradient gels is complex and requires additional apparatus (see *Chapter 8*). Solutions must be cooled (0–2 °C) prior to addition of persulfate to avoid premature polymerization, unless photochemical polymerization is used (*Chapter 8*). These factors make reproducing such gels difficult. In addition, dry storage of gradient gels is compromised by an uneven rate of contraction, which facilitates cracking.

6.2 Other gels

Esoteric SDS–PAGE gels include urea gels for low mol. wt ($< 10^4$) proteins and Tricine gels for low mol. wt basic proteins (e.g. histones) (33–35). Detergents (Triton X-100, sodium deoxycholate) have been successfully incorporated for resolution of membrane proteins (36). Gels have been used for resolution of *in situ* protease generated peptide fragments (Cleveland peptide mapping) (37) (see Chapter 5).

7. Enzyme activity determination in gels

Some intrinsic enzymatic activity may be retained in SDS–PAGE samples if they are prepared without reduction and heating and are allowed to renature after electrophoresis (38). Following removal of SDS, activity can be detected within the gel itself, by diffusion into a substrate-containing overlay, or by elution into free solution. The following protocol describes the detection of certain proteases, but the procedures can be adapted to other uses.

7.1 Gelatin zymography

Proteases that degrade gelatin (denatured collagen) can be detected directly in gels (38). Gelatin is incorporated prior to polymerization (see *Protocol 8*) and remains immobile during electrophoresis. Protease activity is indicated by unstained (i.e. proteolytically degraded) regions against a stained background, in contrast to conventional gels.

7.1.1 Preparation of gelatin zymogram solutions

Protocol 8. Gelatin zymography

Equipment and reagents

- Plastic tray with cover (approx. dimensions 15 × 20 cm, 4 cm height)
- Gelatin solution: add 300 mg gelatin to 20 ml water in a 50 ml polypropylene conical tube (or similar). Heat tube in a boiling water-bath to promote solubilization (cap should be loose). Store at 4°C (solution will gel). Warm solution to 37°C for use.
- Non-reducing 2 × sample buffer: substitute 2 ml water for 2-mercaptoethanol in 2 × sample buffer of *Protocol 4*
- 2.5% (v/v) Triton X-100: dissolve 25 ml Triton X-100 in 975 ml distilled water

- 0.5 M Tris–HCl stock solution: dissolve 60.5 g Tris base in 900 ml water, adjust the solution to pH 7.6 by slow addition of concentrated HCl, make volume to 1 litre with water
- 50 mM calcium chloride solution: dissolve 7.35 g $CaCl_2$ in 1 litre final volume of water
- Zymogram development solution: combine 100 ml 0.5 M Tris–HCl solution, 100 ml 50 mM calcium chloride solution, 800 ml water, and mix well

Method

1. Prepare the gelatin zymogram resolving gel.

Stock solution[a]	Acrylamide concentration		
	7.5%	10%	12.5%
Acrylamide/bisacrylamide (ml)	10	13.3	16.6
Resolving gel buffer (ml)	10	10	10
Ultrapure water (ml)	15.5	13.2	8.9
Gelatin solution (37°C) (ml)	4	4	4
Degas under vacuum (10 min)			
10% SDS (ml)	0.8	0.8	0.8
Ammonium persulfate (ml)	0.6	0.6	0.6
TEMED (µl)	20	20	20

2. Prepare the stacking gel as in *Protocol 2*.

3. Prepare samples as in *Protocol 4*, but use non-reducing 2 × sample buffer and heat the sample only to 37°C.

4. Perform electrophoresis as in *Protocol 5*.

5. Disassemble glass plates and discard stacking gel.

6. Place the resolving gel in a plastic tray containing 100 ml of 2.5% Triton X-100.

7. Wash gel on a rotating platform incubator for 30 min at 25°C; replace solution and repeat a second time.

8. Decant the wash solution and add 100 ml zymogram development solution.

9. Incubate (covered) in a thermostatically controlled water-bath or incubator for 18 h at 37°C.

10. Decant the development solution. Stain the gelatin zymogram with Coomassie blue and destain (see *Protocol 6*).

ª Stock solutions are from *Protocol 1*, except for gelatin solution.

7.1.2 Evaluation of gelatin zymograms

Protease activity is indicated by unstained (cleared) regions against a dark background (see *Figure 11*). As the electrophoretic migration of each protease is proportional to its mol. wt, this technique can be used to determine the apparent masses of unknown proteases (i.e. those migrating between 72 kDa and 92 kDa in human serum and plasma). The size and intensity of the cleared region are proportional to the amount of protease activity.

7.2 Overlay techniques

Overlay techniques have been successfully used for detection of plasminogen activators (39). An agarose indicator gel containing fibrin (plasminogen-rich) prepared on a plastic support sheet (Gelbond, FMC BioProducts) is overlaid on a Triton X-100 washed SDS–PAGE slab gel (see *Figure 12a*). Diffusion of plasminogen activators results in the activation of plasminogen and subsequent fibrin degradation. The indicator gel is stained and bands of activity are

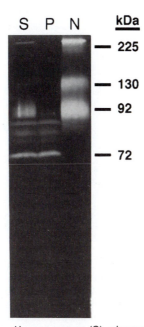

Figure 11. Gelatin zymography. Human serum (S), plasma (P), and polymorphonuclear leucocyte extract (N) were subjected to electrophoresis as in *Protocol 8*.

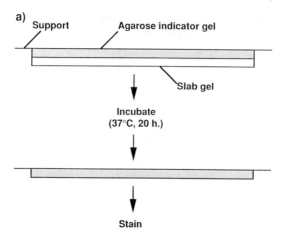

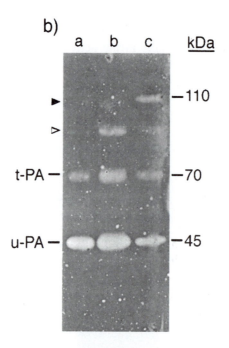

Figure 12. Enzyme overlay technique. (a) Schematic illustration of the technique. An agarose overlay containing fibrinogen (plasminogen-enriched) is incubated on a Triton X-100 washed SDS–PAGE slab gel. (b) Identification of tissue plasminogen activator (t-PA) and urokinase-plasminogen activator (u-PA) in a stained zymogram. Open and closed triangles indicate relative positions of t-PA–t-PA dimer and t-PA–inhibitor complex, respectively. M_r markers are noted on the right. Samples a–c were conditioned media from bovine corneal endothelial cell culture treated in different ways (39).

indicated by cleared regions against a dark background (see *Figure 12b*). By incorporating a plasminogen activator (urokinase) into the indicator gel, it is also possible to detect inhibitors (i.e. reverse zymography) (39).

7.3 Enzyme elution into free solution

Enzymes can be eluted from gels into free solution. The region of interest is excised, placed into a buffered solution, and crushed. This process (i.e. passive diffusion) is slow and generally yields poor recovery. In contrast, electrophoretic elution provides more rapid and consistent means to recover SDS–PAGE proteins (70–80%). Prior to activity determination, eluted proteins may require prolonged dialysis or extraction with solvents to remove SDS (40).

8. Additional SDS–PAGE applications

Several of the most common applications of SDS–PAGE are described below.

8.1 Western blotting

SDS–PAGE protein can be electrotransferred (Western blotted) to membranes (filters) composed of nitrocellulose or polyvinylpyrrolidone (28) (see Chapters 3 and 5). This technique is advantageous because proteins are more accessible for immunological detection (28) and for amino acid sequencing (41). Western blots have also been successfully used for identification of metal-binding proteins (42).

8.2 Preparation of antigens

SDS–PAGE resolved proteins are ideal antigens for antibody production. Substantial quantities of protein in relatively pure form can be generated under preparative scale conditions. Because polyacrylamide acts as an adjuvant, the protein of interest can be excised, the gel crushed, and injected directly. This eliminates the need for exogenous adjuvants.

8.3 Radiolabelled proteins

Radiolabelled proteins can be detected by exposure of the gel to X-ray film (autoradiography), as described in Chapter 5.

9. Troubleshooting

9.1 Gels fail to polymerize

Ensure that the reagents are at room temperature prior to addition of catalyst. Freshly prepare ammonium persulfate, which is hygroscopic. Degas solutions longer. Check catalyst and TEMED addition. Increase catalyst concentrations by 25–50%. Note: ethanol inhibits polymerization; do not wash glass plates with ethanol.

9.2 Electrophoresis time too long

Check power supply current/voltage settings and wire connections. Check pH and concentration of gel and tank buffers. Compare voltage with previous electrophoretic run. Check for tank buffer leaking.

9.3 Poorly resolved/smeared protein bands

Decrease sample load. Increase amount of 2 × sample buffer. Sample may require centrifugation to remove particulates. Check sample well polymerization integrity (trapped bubbles under the comb result in distorted application zone). Extraneous polymerized acrylamide can become trapped in a sample well; remove it from top of gel with gentle stream of water.

9.4 Uneven/diffuse dye front migration

Dye front smiling is caused by uneven heat dissipation; decrease current or provide active cooling. Dye front frowning is caused by bubble trapping (along bottom of gel), current leaks, or high salt concentration in sample (normalize samples). For diffuse dye front migration, check pH of buffers. Compare voltage/current output with previous run.

References

1. Hames, B. B. (1990). In *Gel electrophoresis of proteins: a practical approach* (ed. B. D. Hames and D. Rickwood), p. 1. IRL Press, Oxford.
2. See, Y. P. and Jackowski, G. (1989). In *Protein stucture: a practical approach* (1st edn) (ed. T. E. Creighton), p. 1. IRL Press, Oxford.
3. Dunn, M. J. (1990). In *Protein purification methods: a practical approach* (ed. E. L. V. Harris and S. Angal), p. 18. IRL Press, Oxford.
4. Laemmli, U. K. (1970). *Nature*, **227**, 680.
5. Tselius, A. (1937). *Biochem. J.*, **31**, 1464.
6. Dreyfuss, G., Adam, S. A., and Choi, Y. D. (1984). *Mol. Cell Biol.*, **4**, 415.
7. Alcover, A. and Alvia, J. (1985). *Anal. Lett.*, **18,** 1157.
8. Weber, K. and Osborn, M. (1969). *J. Biol. Chem.*, **244**, 4406.
9. Shapiro, A. L., Vinuela, E., and Maizel, J. V. (1967). *Biochem. Biophys. Res. Commun.*, **28**, 815.
10. Dunker, A. K. and Rueckert, R. R. (1969). *J. Biol. Chem.*, **244**, 5074.
11. Neville, D. M. (1971). *J. Biol. Chem.*, **246**, 6328.
12. Pitt-Rivers, R. and Impiombato, F. S. (1968). *Biochem. J.*, **109**, 825.
13. Reynolds, J. A. and Tanford, C. (1970). *J. Biol. Chem.*, **245**, 5161.
14. Scopes, R. K. (ed.) (1987). *Protein purification: principles and practices* (2nd edn), p. 284. Springer-Verlag Press, New York.

15. Tanford, C. and Reynolds, J. A. (1976). *Biochim. Biophys. Acta*, **457**, 133.
16. Mattice, W. L., Riser, J. M.. and Clark, D. S. (1976). *Biochemistry*, **15**, 4264.
17. Leach, B. S., Collawn, J. F., and Fish, W. W. (1980). *Biochemistry*, **19**, 5734.
18. Kohlrausch, F. (1897). *Ann. Physik. Chem.*, **62**, 209.
19. Orstein, L. (1964). *Ann. N.Y. Acad. Sci.*, **121**, 321.
20. Davis, B. J. (1964). *Ann. N.Y. Acad. Sci.*, **121**, 404.
21. Jovin, T. M. (1973). *Biochemistry*, **12**, 871.
22. Chrambach, A. and Jovin, T. M. (1983). *Electrophoresis*, **4**, 190.
23. Switzer, R. C., Merril, C. R., and Shifrin, S. (1979). *Anal. Biochem.*, **82**, 580.
24. Merril, C. R., Goldman, D., Sedman, S. A., and Ebert, M. H. (1981). *Science*, **211**, 1437.
25. Morrissey, J. H. (1981). *Anal. Biochem.*, **117**, 307.
26. Zaccharias, R. J., Zell, T. E., Morrison, J. H., and Woodlock, J. J. (1969). *Anal. Biochem.*, **31**, 148.
27. Green, M. R., Pastewka, J. V., and Peacock, A. C. (1973). *Anal. Biochem.*, **56**, 43.
28. Towbin, H. M., Staehelin, T., and Gordon, J. (1979). *Proc. Natl. Acad. Sci. USA*, **76**, 4350.
29. Makowski, G. S. and Ramsby, M. L. (1993). *Anal. Biochem.*, **212**, 283.
30. Weber, K. and Osborn, M. (1975). In *The proteins* (3rd edn) (ed. H. Neurath and R. L. Hill), Vol. 1, p. 179. Academic Press, New York.
31. Marshall, T. (1984). *Clin. Chem.*, **30**, 475.
32. Ferguson, K. A. (1964). *Metabolism*, **13**, 21.
33. Swank, R. W. and Munkres, K. D. (1971). *Anal. Biochem.*, **39**, 462.
34. Schagger, H. and von Jagow, G. (1987). *Anal. Biochem.*, **166**, 368.
35. Thomas, J. O. and Kornberg, R. D. (1975). *Proc. Natl. Acad. Sci. USA*, **72**, 2626.
36. Dewald, B., Dulaney, J. T., and Touster, O. (1974). In *Methods in enzymology* (ed. W. B. Jacoby), Vol. 104, p. 305. Academic Press, New York.
37. Cleveland, D. W., Fischer, S. G., Kirschner, M. W., and Laemmli, U. K. (1977). *J. Biol. Chem.*, **252**, 1102.
38. Heussen, C. and Dowdle, E. B. (1980). *Anal. Biochem.*, **102**, 196.
39. Ramsby, M. L. and Kreutzer, D. L. (1993). *Invest. Ophthalmol. Vis. Sci.*, **34**, 3207.
40. Hager, D. A. and Burgess, R. R. (1980). *Anal. Biochem.*, **109**, 76.
41. Matsudaira, P. (1987). *J. Biol. Chem.*, **262**, 10035.
42. Makowski, G. S., Lin, S.-M., Brennan, S. M., Smilowitz, H. M., Hopfer, S. M., and Sunderman, F. W. (1991). *Biol. Trace Elem. Res.*, **29**, 93.

2

Protein analysis by mass spectrometry

OLE NØRREGAARD JENSEN, ANDREJ SHEVCHENKO, and
MATTHIAS MANN

1. Introduction

The molecular weight is a highly specific characteristic of a molecule and is often the first physicochemical property that is measured of a protein. Traditionally, SDS–PAGE (Chapter 1), analytical ultracentrifugation, or gel permeation chromatography (Chapter 9) are used for molecular weight measurements. Recently, mass spectrometric techniques have been developed that permit mass determination of intact proteins with an accuracy far superior to the above mentioned methods. At the same time biological mass spectrometry has changed from being a complex and exotic discipline performed only by specialists into being an invaluable analytical technique used by skilled biochemists at all levels of protein characterization. This change is mainly due to the emergence of two new ionization techniques that allow production of gas phase molecular ions of proteins and peptides, as well as to the availability of relatively simple and easy to operate mass spectrometers. The power of the techniques is apparent from the following performance data. Only a few picomoles of a protein are necessary to obtain its molecular weight with an accuracy of 0.01–0.1%. Even less sample is necessary for peptide analysis: subpicomole amounts can now be analysed with a mass accuracy better than 0.01%, i.e. an error of 0.1 Da in 1000 Da. Although mass spectrometry is a versatile and powerful analytical technique, it has some limitations. This chapter focuses on practical guidelines for sample preparation for mass spectrometric analysis which, if followed, should ensure good results. The principles of operation and applications of mass spectrometry in biological research can be found in recent reviews (1–5). Here, we only provide a brief overview so that the reader can appreciate the utility and limitations of the new mass spectrometric techniques.

Mass spectrometry is typically employed for protein primary structure analysis in the last stages of protein purification and prior to techniques used for higher order structure determination, such as circular dichroism, NMR spectroscopy, and X-ray crystallography. It is often a complement to other microcharacterization techniques, such as reversed-phase HPLC, amino acid analysis, and Edman degradation.

Mass spectrometric molecular weight determination of a purified protein may verify a proposed amino acid sequence or reveal microheterogeneities. However, mass spectrometry also has the unique capability to analyse protein and peptide mixtures without prior separation. Molecular weight information obtained by peptide mixture analysis has many uses, such as rapid identification of a protein by its 'peptide mass fingerprint', verification by mass of specific stretches of amino acid sequence within a protein, or identification of amide bonds that are highly susceptible to proteolytic attack such as those located between folding domains. Detailed primary structure analysis can be performed by amino acid sequencing of individual peptides by tandem mass spectrometry at sample levels too low for Edman degradation.

2. Mass spectrometry

A mass spectrometer generates gas phase ions from a sample, separates the ions according to their mass-to-charge ratio, and records a spectrum of their abundances. An instrument typically consists of three modules:

- ion source
- mass analyser
- detector and data acquisition system

The most relevant techniques for practical biological work are matrix-assisted laser desorption/ionization (6,7) combined with time-of-flight mass analysis (MALDI MS), and electrospray ionization (8–10) combined with quadrupole mass analysis (ES MS). Older methods, such as fast atom bombardment (FAB) MS and ^{252}Cf plasma desorption (PD) MS are still used but they are less sensitive and generally are only applicable when the molecular mass is less than 10 kDa.

2.1 Generating gas phase molecular ions

For mass spectrometric analysis, intact gas phase ions have to be generated from the analyte in the ion source of the mass spectrometer. The critical step in MALDI is sample preparation where analyte and a UV absorbing matrix compound (*matrix*, in short) are mixed and dried to produce protein-doped matrix crystallites on the sample probe. These crystallites are introduced into the mass spectrometer and irradiated by a pulsed UV laser. Rapid sublimation of the analyte/matrix crystallites leads to the formation of gas phase, protonated molecular ions from the solid, crystalline matrix/analyte substrate. Currently used matrices are listed in *Table 2*.

In electrospray mass spectrometry (ES MS) analyte solution is passed through a hypodermic needle held at high potential. Due to the high electric field a spray cone of fine droplets forms at the tip of the needle. These 'electrosprayed' highly charged droplets evaporate rapidly, which causes an additional increase of surface charge density. Charged analyte molecules are thought to desorb from the surface into the gas phase to carry off excess charge. ES ionization of large molecules produces multiply charged ions. Because ions are produced from solution and at atmospheric pressure ES is readily interfaced to chromatographic techniques.

2.2 Mass analysis of molecular ions

The mass analyser separates molecular ions based on their mass-to-charge ratio (m/z) (11). MALDI is usually combined with a time-of-flight (TOF) mass spectrometer. Ions are accelerated to a kinetic energy of approximately 25 keV and subsequently allowed to fly through a fixed distance field-free region of about one metre in length before they are recorded by the electric signal generated upon impact at a detector. The time-of-flight in the field-free region is related to the m/z of a given ion. Having identical kinetic energy, small molecular ions are moving faster than large molecular ions and thus arrive earlier at the detector. The m/z for an ion can be determined from its time-of-flight by comparison to the time-of-flight of known standards. An advantage of the TOF analyser is that the total ion population is detected in one experiment resulting in very high sensitivity. The mass range of the TOF analyser is theoretically unlimited. However, large molecular ions move relatively slowly and are therefore not detected as efficiently as small, fast ions which generate a stronger response upon impact at the detector. The practical mass range for MALDI TOF MS is from 500 Da up to 150 000 Da. Note that high mass analysis is considerably more challenging than low mass analysis.

ES ionization is normally coupled to a quadrupole mass filter, a relatively simple and compact mass spectrometer. The ions, which are generated at atmospheric pressure, are guided into the vacuum of the quadrupole mass spectrometer through an orifice or through a narrow heated capillary. Ions are separated based on their trajectory in an oscillating electric field created between four parallel metal rods, the quadrupole. At a given oscillation frequency and amplitude of the electric potentials applied to the rods, ions of just one m/z value are allowed to pass through to the detector. By scanning the amplitude a mass spectrum is recorded. The quadrupole mass filter has a limited mass range (typically up to m/z 3000) but as ES ionization produces highly charged molecular ions (ion signals at m/z 400–2000) it is well suited as a mass analyser for this ionization method.

2.3 Tandem mass spectrometry

For primary structure analysis (sequencing) of peptides an electrospray tandem mass spectrometer (ES MS/MS) can be employed. A tandem mass spectrometer consists of two mass analysers (e.g. quadrupole mass filters) that are assembled in a linear arrangement separated only by a quadrupole collision cell ('triple quadrupole' mass spectrometer) (12). Selected precursor ions (parent ions) which are filtered through the first quadrupole mass spectrometer collide with inert gas molecules (Ar or He) in the collision cell. This causes the ions to decompose into structurally informative fragment ions (product or daughter ions) which are immediately mass analysed in the second mass spectrometer. It is possible to obtain amino acid sequence information by tandem mass spectrometry, even for N terminally blocked peptides or modified peptides (see Section 8).

3. When and how to use mass spectrometry for protein and peptide characterization

Both MALDI MS and ES MS have their advantages and disadvantages. Some characteristic features of these techniques are compared in *Table 1*. Sample preparation for MALDI MS analysis is fast and simple and consumes very small amounts of sample. Mass spectra can be obtained in a short period of time (success or otherwise of an analysis is known within a few minutes) even on impure or heterogeneous samples. MALDI MS is therefore often used for screening of samples prior to submitting them for more detailed and time-consuming analysis, such as chromatography, Edman degradation, ES MS or -MS/MS, NMR spectroscopy, X-ray diffraction, etc. One disadvantage of MALDI can be its preferential ionization of some components in a crude mixture at the expense of others (suppression) which is one reason why MALDI MS is not a quantitative technique. MALDI MS has a relatively low resolution (due to peak broadening) when analysing proteins of 30 kDa and above. This results in a reduced, although still acceptable, mass accuracy of 0.05–0.1%.

ES MS, on the other hand, is excellent for accurate molecular weight determination of intact proteins up to ~ 150 kDa in size. A mass error within 0.01–0.02% is routinely achieved. The requirements for protein purity and homogeneity are, however, more stringent than for MALDI MS. Desalting prior to ES MS is often necessary. Another feature of ES MS is that the method is semi-quantitative, i.e. the mass spectrometric response is proportional to the protein concentration (but independent of the flow rate), over several orders of magnitude. In some cases, ES MS can be used to estimate the relative abundances of components in simple heterogeneous samples. In general, MALDI MS is not a quantitative technique mainly because the mass spectrometric response depends on the inclusion of analyte into the matrix crystals and on the ionization efficiency of individual analyte species. Only in carefully calibrated and highly reproducible experiments can quantitative information be obtained.

Table 1. Comparison of MALDI MS and ES MS[a]

	MALDI MS	**ES MS**
Ionization principle	Energy sudden	Spray
Mass analyser (typical)	Time-of-flight	Quadrupole mass filter
Mass range (Da)	400 to $> 2 \times 10^5$	50 to 10^5
Mass accuracy (peptide)	0.1–0.001%[b]	0.05–0.008%[c]
Mass accuracy (protein)	0.1%	0.05–0.01%
Sensitivity	Low fmol to pmol	Low fmol to pmol
Quantification	No	Semi-quantitative
Resolution (peptide)	500–10 000[d]	1000–2000
Resolution (protein)	50–500	1000–2000
Mixture analysis	Excellent	Good
Tolerance to impurities	Good	Moderate
Versatility	Excellent	Good
LC-MS capability	Off-line	On-line
MS/MS	Some[e]	Yes
Ease of operation	Simple	Moderately simple
Cost ($)	> 80 000	> 150 000

[a] These are *typical* values for the performance of MALDI and ES MS in a *routine* laboratory setting.
[b] Highest accuracy when using fast evaporation matrix surfaces in combination with delayed extraction and/or reflector TOF mass analyser.
[c] In general, protein mass determination by ES MS is five- to tenfold more accurate than by MALDI MS.
[d] Lower values are for linear TOF, higher values for delayed extraction ion source or reflector TOF analyser.
[e] Post-source decay analysis in reflector time-of-flight instrument.

Mass spectrometry is not an obvious method for studying non-covalent molecular interactions (protein–protein, protein–DNA, protein–ligand complexes) or protein folding although it has potential for this kind of analysis (13–15). Only in special cases of strong interaction can specific non-covalent complexes be observed by ES MS and, very rarely, by MALDI MS. Protein clusters (non-specific interactions) are often observed with both techniques. These clusters are generated during the ionization process.

4. Sample preparation for mass spectrometry

4.1 Sample purity—a mass spectrometrist's point of view

A common misunderstanding between mass spectrometrists and biochemists concerns the purity of a sample. Biochemists think of contamination by other proteins. Mass spectrometrists want to know the concentrations of non-volatile low mol. wt substances, such as salts, buffers, chaotropic agents, detergents, or stabilizers. Involatile non-protein or non-peptide sample constituents may prevent mass analysis because they interfere with sample preparation and molecular ion production or obscure the signals corresponding from the analyte of interest. In MALDI, the protein or peptide analyte may not be included into the matrix crystallites if too much salt or detergent is present. In ES, salts and detergents interfere with desolvation of molecules and may act as charge scavengers, thereby decreasing the yield of analyte molecular ions.

It is of utmost importance that the biochemist, at an early stage of sample work-up, discuss sample preparation methods with the person performing the actual mass analysis. This can save considerable time and labour and usually ensures good results. The use of the word 'purity' in the following sections implies the absence of involatile non-protein substances.

4.2 Solvents used in mass spectrometry

As indicated above, volatile buffers are preferable when preparing a protein sample for mass spectrometric analysis—especially when using ES MS. Most proteins are soluble in 50 mM ammonium bicarbonate buffer pH 7–8. The ammonium bicarbonate can be removed by vacuum centrifugation and the protein redissolved in 0.1% (v/v) TFA, 5% (v/v) formic acid, or some of the solvent systems listed in *Table 3*. If non-volatile buffers or additives have been used during protein purification they should be exchanged with ammonium bicarbonate prior to the sample being submitted for MS analysis. Note that MALDI is more tolerant of common buffers and salts (see *Table 4*) than ES. In cases where the protein concentration is relatively high (greater than 20 μM), five- to tenfold dilution with water may reduce the effects of non-volatile buffer constituents on molecular ion production. Alternatively, lyophilization, HPLC, desalting cartridges, acetone precipitation, centrifugal filtration cartridges, or dialysis are useful methods for the exchange or removal of buffer components. It is not always possible to remove all involatile buffer constituents, so rinsing methods have been developed which allow 'clean-up' of samples immediately prior to MALDI MS analysis (see *Protocols 1* and *2*). In the case of ES MS, chromatographic columns or desalting cartridges can be interfaced directly to the ion source in which case sample purity is not as critical (see below). Detergents and surfactants, such as sodium dodecyl sulfate (SDS), Triton X-100, CHAPS, etc., quench or obscure the mass spectrometric response and can be difficult or impossible to remove from protein samples. They should not be used during protein purification if mass spectrometric analysis is desired. The only exception is SDS used in gel electrophoresis (SDS–PAGE) (Chapter 1). Peptides generated by in-gel digestion of proteins isolated by SDS–PAGE can be analysed by MALDI and ES MS because the SDS is removed by thorough rinsing of the gel pieces prior to digestion (*Protocol 6*). MALDI MS is compatible with or even improved by some non-ionic detergents (for example 1% octylglucoside or 1% MEGA-7) and to 0.1% deoxycholate, an anionic detergent. ES MS is only slightly affected by 0.1% octylglucoside which may produce a higher background level (16). The use of stabilizers such as glycerol and polyethylene glycols (Chapter 14) should be avoided.

Table 2. Commonly used MALDI matrices for analysis of peptides, proteins, carbohydrates, and nucleic acids, using 337 nm or 355 nm UV lasers[a]

Matrix	Peptide or protein	Carbohydrate	Nucleic acid	Reference
HCCA (4-hydroxy-α-cyanocinnamic acid)	X			17
Sinapinic acid (3,5-dimethoxy-4-hydroxycinnamic acid)	X			18
DHB (2,5-dihydroxybenzoic acid)	X	X		19
THAP (2,4,6-trihydroxyacetophenone)	X		X	20
3-Hydroxy-picolinic acid			X	21
Ferulic acid (4-hydroxy-3-methoxycinnamic acid)	X		X	18

[a] The matrix materials can be obtained from Aldrich and Sigma.

Table 3. Commonly used solvent combinations for MS analysis of proteins and peptides[a]

Analyte	MALDI MS[b]	ES MS
Soluble proteins	0.1% TFA/ACN (2:1)	1% Formic acid/MeOH (1:1) 1% Acetic acid/MeOH (1:1)
Membrane proteins	30–99% Formic acid 30–100% HFIP 70% Formic acid/HFIP (2:1, 1:1) 1% Octylglucoside	10–70% Formic acid/MeOH (1:1) 1% Acetic acid/MeOH/chloroform (2:5:2)
Peptide mixtures	0.1% TFA/ACN (2:1) 5% TFA/ACN (4:1) Formic acid/isopropanol/water (1:1:1) 0.1–1% Octylglucoside	1–5% Formic acid/MeOH (1:1) 1% Acetic acid/MeOH (1:1)
Peptide	0.1–2% TFA/ACN (2:1) 5% Formic acid/ACN (2:1)	0.1–1% Formic acid/MeOH (1:1) 1% Acetic acid/MeOH (1:1)

[a] Dissolve analyte and sonicate briefly immediately before analysis. Instead of methanol (MeOH), one can use other organic solvents such as isopropanol and acetonitrile. All proportions are by volume.
[b] Using sinapinic acid or HCCA matrix. A typical solvent for 2,5-DHB matrix is 0.1% TFA/EtOH (2:1 or 1:1); It is mainly used for peptide analysis.

Sample preparation methods for MALDI MS analysis of peptides and proteins are described in *Protocols 1* and *2*. Sample preparation for ES MS is more critical and should be discussed with the operator of the ES mass spectrometer using the guidelines given in *Protocol 3* and in *Table 3*. Again, the use of a desalting/concentration method prior to ES MS analysis is recommended.

Protocol 1. Sample preparation for MALDI MS: dried droplet method[a]

Reagents

- 0.1 % (v/v) TFA
- Acetonitrile
- Matrix[b]

Method

1. Make a saturated solution ($\sim$ 10 g/litre) of matrix by adding 1 vol. acetonitrile and 2 vol. of 0.1% TFA to an aliquot of matrix (dry powder). Vortex for at least 30 sec. The solution should appear slightly turbid. Centrifuge to precipitate insoluble matrix material. Use only the supernatant in subsequent steps. The matrix solution can be used for several days. Alternatively, ready-made matrices can be purchased from some MALDI MS manufacturers.

2. Dilute protein samples to a final concentration of 5–50 μM in 0.1% TFA, if possible. Dilute peptide samples to a final concentration of 0.5–10 μM in 0.1% TFA, if possible.

3. In a microcentrifuge tube, mix 1 μl of sample solution into 9 μl matrix solution and vortex. The mixing ratio of analyte solution to matrix solution can be varied from 1:1 to 1:9 (v/v) depending on the amount of analyte available. Use a 1:1 ratio for less abundant samples.

4. Deposit 0.5–1 μl of the analyte/matrix solution on to the mass spectrometer probe.

5. Dry the droplet in a stream of air or nitrogen at ambient temperature (do NOT use heat!).

6. If interfering substances are expected to be present, the sample deposit can be cleaned by gently 'etching' the surface of the crystals with cold water or 0.1% TFA (22). Dip the probe tip into cold water or add a drop of cold water on top of the sample deposit, followed by wiping or shaking off the water. Very impure samples can be prepared by using a modified dried droplet method, where a matrix seed layer is deposited prior to application of the analyte/matrix solution (23).

7. Insert the probe into the MALDI mass spectrometer and acquire a mass spectrum.

---

[a] This is the original sample preparation method for MALDI MS. The sample (peptide or protein solution, in this case) is mixed with the matrix solution, deposited on the MS probe, and dried.
[b] This protocol works well for sinapinic acid and HCCA matrices. The matrix can be chosen from *Table 2* and the solvents mentioned here may be substituted by the solvent combinations listed in *Table 3*. We use HCCA and sinapinic acid as MALDI matrices almost exclusively for peptide and protein work.

Protocol 2. Sample preparation for MALDI MS: fast evaporation method[a]

Reagents
- Matrix
- Acetone
- 0.1% (v/v) TFA
- 2.0% (v/v) TFA

Method

1. Make a saturated solution (~ 40 g/litre) of matrix in acetone/0.1% TFA (99:1, v/v) in a 1.5 ml microcentrifuge tube. Vortex briefly. Centrifuge to remove insoluble matrix material. Note: the matrix has to be soluble in acetone or some other fast evaporating solvent. Sinapinic acid and HCCA (see *Table 2*) are routinely used, but ferulic acid and THAP can also be employed.

2. Deposit 0.5–1 μl matrix solution by rapid transfer from the micro-centrifuge tube to the MS probe. It is important that the droplet can spread out (no confining edges on the probe) and dry in a matter of 1–2 sec. The resulting thin matrix film should appear homogeneous and, in the case of HCCA yellowish, and almost transparent. No crystals should be visible. Depending on the particular MS probe design, it may be necessary to optimize conditions by varying the concentration of matrix in acetone or the volume of liquid applied on to the MS probe.

3. Prepare the sample as in *Protocol 1*. Further dilution is possible to obtain a 0.01–0.1 μM polypeptide solution, which can help to reduce the concentration of contaminating species. Refer to *Table 3* for selecting a suitable solvent combination for the analyte. Make sure that the sample solution is acidic (pH < 4) and contains less than 30% organic solvent so that the matrix film is not completely redissolved.

4. Deposit 0.5–1 μl sample solution on top of the thin film and allow it to dry at ambient temperature. If the sample contains more than 30% organic solvent, first deposit 0.5 μl of 2% TFA on the film followed by injection of 0.5 μl of the sample solution into this droplet.

5. Rinse the sample by depositing 10–20 μl water or 0.1% TFA on the probe followed by immediate wiping or shaking of the probe. This rinsing procedure can be repeated several times as long as the matrix film is not damaged.

6. Insert into MALDI mass spectrometer and acquire a spectrum.

[a] This procedure allows the matrix and the sample to be applied on to the MS probe in independent steps, i.e. the sample solution can be optimized independent of the matrix solution. This simplifies sample preparation and also improves sensitivity, resolution, and mass accuracy attainable in peptide analysis (24,25). This method is very tolerant of salts and buffers present in protein and peptide samples, as long as the pH is kept below 4.

Table 4. Reagent concentrations that are tolerated by MALDI MS[a]

Substance	Tolerable conc.	Interference
NaCl, KCl	100 mM	[M + Na]+, [M + K]+ observed
NH₄HCO₃,	100 mM	Very slow evaporation,
NH₄CH₃CO₂,		heterogeneous, or no crystals
Tris, Hepes		
Phosphate buffer	10 mM	'Crusty' layer on top of crystals
Urea, GdmCl	0.5–1 M	Same as above
Glycerol, DMSO	1% (v/v)	'Greasy' film on crystals, no matrix, and peptide signals
Coomassie stain	0.01% (v/v)	Peaks at m/z 800–850, suppression of peptide signals
PEG	0.01% (v/v)	Distribution of signals spaced 44 Da apart
Pyridine	0.05% (v/v)	Glossy film of matrix is formed
SDS	None	Peaks spaced 288 Da apart at m/z 400–1500, no or weak analyte signals
Octylglucoside,	1% (v/v)	Low concentrations may improve spectral quality for large peptides
octylthioglycoside,		
heptanoyl-N-methylglucamide		

[a] Every effort should be made to reduce the reagent and buffer concentrations to as low a level as possible. The fast evaporation method (*Protocol 2*) is more robust towards sample components than the dried droplet method (*Protocol 1*). The final pH of the sample or sample/matrix solution should be less than 4. Note that certain non-ionic detergents actually may improve the mass spectral quality.

Note: an improved, robust matrix film can be prepared by mixing solutions of HCCA (40 g/litre, in acetone/isopropanol (1:1)) and nitrocellulose (10 g/litre, in acetone) in a 4:1 ratio (39). This matrix is deposited as described in *Protocol 2* for the regular fast evaporation method. The nitrocellulose binds peptides and protein very efficiently which allows thorough rinsing of the probe before mass analysis and may improve the mass spectrometric response for otherwise difficult or low abundance samples.

Protocol 3. Sample preparation for ES MS: flow injection analysis

Equipment
- 50–100 μm i.d. fused silica capillary (Polymicro Technologies)
- Syringe pump (Harvard or Isco brand)
- Injection valve (for example, Rheodyne HPLC model 8125 with a 5 μl sample loop)

Method
1. Fill the syringe pump with a suitable solvent mixture (*Table 3*), for example 1% formic acid/MeOH (1:1, v/v). Connect the syringe pump to the injection valve and to the ES ion source with the fused silica capillary.[a] Start the pump at a flow rate of 1–3 μl/min and check for air

bubbles and leaks. Turn on the ES mass spectrometer, and make certain that the spray (ion current) is stable.

2. Dissolve the sample to a final concentration of 1–10 µM protein in the solvent used as the mobile phase for the syringe pump. The concentration of involatile buffer constituents (salts) in the sample should be below 10 mM, preferably less than 1 mM. Use a desalting device to remove involatile low mol. wt contaminants (see main text).

3. Use a 5 µl or 10 µl Hamilton syringe (square-end needle) to inject the sample into an HPLC valve. Fill the injector loop completely to ensure reproducibility from injection to injection.

4. Begin data acquisition.

5. To avoid cross-contamination it may be necessary to rinse the injection valve and solvent lines to the ES MS by injection of 70% or 99% formic acid between analyte runs.

[a] The ES ion source operates at high voltages. Current may leak through the fused silica capillary lines. The injection valve and the syringe pump should be electrically connected to ground.

4.3 Interfacing chromatographic separation methods to an electrospray mass spectrometer

Gradient RP-HPLC can be interfaced to ES MS by using a pneumatically assisted ES source or a liquid sheath flow ES source to stabilize the spray as the solvent composition changes over time (26–28). Most ES sources can maintain a stable spray at flow rates up to 10–20 µl/min. Higher flow rate (100–1000 µl/min) ES sources have been developed recently. Narrowbore (2.1 mm i.d. column, 150–200 µl/min) and microbore (1 mm i.d. column, 30–40 µl/min) HPLC are easily interfaced to ES MS. Because the ES ion current is dependent on the analyte concentration, it is not necessary or desirable to introduce all the eluate to the ES source. Use a low dead volume splitting tee (Valco or Upchurch) to direct only 5–10% of the eluant to the ES MS—the remaining sample can be collected for further analysis. Connect the HPLC column to the ES source (via an UV detector and a splitting tee) by a 50–100 µm fused silica capillary.

Note: the ES ion source operates at high voltages. Current may leak through the FSC lines. The HPLC pumps, injection valve, and UV detector should be electrically connected to ground.

The choice of column depends on the sample to be analysed. Use short columns (1–2 cm) and a steep gradient (5-90% (v/v) organic modifier in 15 min) for rapid desalting/concentration and elution of protein samples. Short alkyl chain (C_4) reverse-phase resin is typically used. Such columns and cartridges are available from PerSeptive Biosystems, Pharmacia, The Separations group, Applied Biosystems, and others. When an actual separation is required, use longer C_8 or C_{18} reverse-phase columns (10–25 cm) and a shallow gradient (5-90% (v/v) organic modifier in 30–90 min) as in common HPLC practice.

A 1 mm i.d. column is adequate for LC-MS protein and peptide analysis down to the 10–20 picomol level. For specialized, high sensitivity applications capillary HPLC column (320 μm i.d.) should be used. They are relatively sturdy and simple to make (29) or can alternatively be purchased from LC-Packings.

It is possible to interface capillary zone electrophoresis to an ES mass spectrometer. However, the difficulties associated with the low sample utilization, and the presence of buffers have not yet made CZE–ES MS a routine technique for peptide characterization.

4.4 Sample preparation for nanoelectrospray MS

A miniaturized ES source, the NanoES, has recently been introduced (30,31). Due to its low flow rate of about 20 nl/min, it generates very small droplets that facilitate more efficient desolvation and ionization of analyte molecules. Advantages of this source include high spray stability, long measurement time, and a sensitivity of 0.1–0.01 pmol for peptide sequencing. The sample is injected directly into the spray needle—no pumps, valves or solvent lines are required. A new capillary spray needle is used for each experiment, so cross-contamination is avoided.

Sample preparation for NanoES may include a desalting step as is the case for regular ES. However, it is desirable to load the sample in only a 1–2 μl volume into the spraying needle, so correct sample handling is important. A desalting/concentration method based on a two-capillary system for efficient transfer of sample from a capillary column to the capillary spray needle has been developed and is routinely used in the authors' laboratory.

5. Interpretation of mass spectra

MALDI and ES are so-called soft ionization techniques because they mainly produce intact molecular ions. Little or no analyte fragmentation (decomposition of proteins in the mass spectrometer, that is) is evident under normal working conditions. In the positive ion mode, the charge usually originates by attachment of one or more protons (H^+) to the analyte molecule (M) creating molecular ions: $[M + nH]^{n+}$, $n = 1, 2, 3, ...$ In the negative ion mode, the molecular ion is denoted $[M - nH]^{n-}$ where n is the number of abstracted protons. In the following we will consider only the positive ion mode, unless otherwise stated.

The number of charges present on a molecular ion is mainly dependent on the ionization method and the size of the analyte molecule. MALDI MS analysis of small peptides (< 2 kDa) usually produces only singly charged molecular ions, $[M + H]^+$ i.e. ion species that give rise to one m/z signal per peptide. For larger peptides (2–5 kDa) both singly charged, $[M + H]^+$, and less abundant doubly charged, $[M + 2H]^{2+}$, molecular ions are detected, i.e. two m/z signals per analyte. Analysis of even larger peptides and proteins by MALDI produces multiple m/z values per analyte species, i.e. $[M + H]^+$, $[M + 2H]^{2+}$, $[M + 3H]^{3+}$, etc. Furthermore, ion signals corresponding to molecular clusters (dimers $[2M + H]^+$, trimers $[3M + H]^+$, etc.) may appear in MALDI mass spectra, especially at high analyte concentrations.

In ES MS, approximately one charging agent is attached per 1 kDa of the analyte mass. This is illustrated by the ES mass spectrum of T5 exonuclease shown in *Figure 1*. Such a spectrum contains redundant molecular weight information—each signal corresponds to one charge state of the intact protein. Algorithms have been developed that deconvolute the distribution of multiply charged molecular ions into a single signal per analyte (32) (*Figure 1*, insert). The excellent mass accuracy that can be obtained by electrospray mass spectrometry is due to the redundant m/z information in the multiple charge states of each analyte species in the ES spectrum.

The charging agent is not always a proton. For example, the presence of sodium (Na, 23 Da) and potassium (K, 39 Da) salts in reagent and buffer

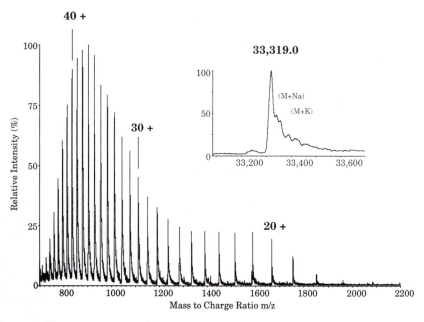

Figure 1. ES mass spectrum of bacteriophage T5 exonuclease (SwissProt accession no. P06229). The distribution of multiply charged ions, typical of an ES MS spectrum, can be deconvoluted into a single species (inset). The experimentally determined mass (33 319 Da) is in agreement with the mass calculated from the amino acid sequence without the N terminal Met residue (33 317 Da). The spectrum demonstrates the purity and correct processing of the purified protein.

solutions often results in abundant molecular ions corresponding to $[M + Na]^+$ and $[M + K]^+$. This may cause problems in MALDI and ES MS analysis of peptides and proteins because alkali metal ions interfere with ionization. Furthermore, the presence of charging agents other than protons can skew the analyte peak and cause a distribution of the ion intensity over several molecular ion species reducing the signal-to-background ratio and thereby the sensitivity of detection. Again, the highest grade reagents and solvents available should be used at any sample handling step prior to MS in order to minimize the amounts of alkali metal ions and other involatile non-protein substances. The tailing to the high mass side of the deconvoluted peak in *Figure 1* (insert) is due to Na^+ and K^+ adducts.

The presence of multiple *m/z* signals (adducts, multiple charge states, and stable isotopes) in a mass spectrum of a single analyte is often a source of confusion for the inexperienced user, leading to the belief that several different components are present in the sample. Refer to the *Appendix* at the end of this chapter for an explanation of isotopic distributions observed in mass spectrometry.

6. Mass analysis of intact proteins

As mentioned in the introduction the molecular weight of an intact protein is a key physicochemical characteristic because it reflects the amino acid composition and the modifications of a polypeptide. If the molecular weight calculated from the amino acid sequence derived from the cDNA agrees with the experimentally obtained value to within a few mass units then the researcher can be reasonably confident that the sequence is correct (note, however, that deamidation of Asn and Gln to Asp and Glu is difficult to detect at the protein level because the mass only increases by 1 Da). If the measured mass does not agree with expectation, or if more confidence is required, then a peptide mass map (described below) is an easy means of confirming a sequence. Measurement of the intact protein mass is commonly used for quality control of purified, recombinant proteins. It reveals the presence of contaminating or processed species (ragged termini, decomposition products, post-translational modifications, etc.). Thus intact molecular weight determination by mass spectrometry validates the identity, purity, and homogeneity of a protein. An example is shown in *Figure 1*. The mass of T5 exonuclease was determined by ES MS to agree with the calculated molecular weight to within 2 Da. Additionally, the mass spectrum does not contain signals from other protein species, indicating that the protein preparation is pure and that the N terminal Met residue has been quantitatively removed.

A list of amino acid masses and the procedure to calculate molecular weights are given in the *Appendix*.

7. Peptide mass mapping

Peptide mapping is an established procedure for the analysis of proteolytic digest mixtures derived from proteins (see *Chapter 5*). Compared to chromatographic or electrophoretic methods, mass spectrometric peptide mapping (peptide mass mapping) provides a wealth of information in a short time. The amino acid sequence of the analyte protein is often known, and a sequence-specific enzyme is usually employed for digestion so that the molecular weights of all expected peptides can be calculated ahead of time and compared to the experimentally obtained data. Mass spectrometry reveals molecular differences that give rise to a change of peptide mass. Peptide mass mapping applications range from sequence verification of recombinant proteins, to studies of conformational change upon ligand binding, and from localization of post-translational modifications, to identification of folding domains within proteins by determination of protease exposed regions.

Peptide mapping is often used to compare proteins that differ in only part of the sequence. Localized differences may be due to mutations or to chemical or post-translational modifications. Comparative peptide mass mapping of modified and unmodified protein species may therefore reveal the peptides that are modified (and those that are not modified) and may also suggest the type of modification based on the observed mass change. Methods for studying post-translational modifications are discussed in Section 9.

MALDI MS is especially useful for the analysis of crude peptide mixtures because it is fast, very sensitive, and tolerant to buffer constituents. Using the fast evaporation method for matrix preparation, 10–100 fmol of analyte loaded on to the probe is sufficient to obtain a good quality peptide mass map. For these reasons, MALDI MS is excellent for optimization of experimental conditions for enzymatic digests (best enzyme, enzyme/substrate ratio, incubation time and temperature, buffer, etc.). An array of different sample preparations can be analysed in a relatively short time. When the optimum conditions have been identified, a larger batch of protein can be processed. More detailed analysis of the resulting peptide mixture can be accomplished by RP-HPLC followed by MALDI MS, ES MS/MS, or by other techniques. *Figure 2* shows a MALDI peptide mass map obtained after tryptic digestion of a G_{ia} subunit isolated from bovine brain. The peptide map covers approximately 45% of the protein sequence. The peptides that are detected are from the N terminal and central region of the protein whereas the large hydrophobic peptides from the C terminal domain escape detection under the conditions used.

The matrices and solvents chosen for a sample preparation will affect the measured MALDI peptide mass map. Using the standard 'peptide matrix HCCA', as in the above mentioned example, peptides in the mass range from 600 to 2500 Da are mainly observed while larger peptides are frequently absent. By changing to the dried droplet method with HCCA in formic acid/

isopropanol/water (1:1:1, by vol.) or sinapinic acid in 0.1% TFA/acetonitrile (2:1, v/v), larger peptides may emerge in the mass spectrum. Several combinations of matrices, solvents, and detergents (*Tables 2* and *3*) can be tried for a given peptide mixture to improve the overall sequence coverage.

Traditionally, exhaustive peptide mapping is performed by RP-HPLC. ES MS allows on-line mass spectrometric detection of HPLC eluates as mentioned previously. The combination of liquid chromatography–mass spectrometry (LC-MS) provides an additional dimension of separation. Peptides are now characterized by their retention time and by their molecular weight. The latter identifies peptides derived from proteins of known sequences. LC-MS is the method of choice to produce near complete peptide maps. This is necessary for complete amino acid sequence verification or for characterizing all sites of modification. Higher amounts of protein (100–1000 picomoles) may be needed for such detailed analysis where several complementary methods (biochemical assay, Edman degradation, MALDI MS and LC-ES MS peptide mapping, tandem mass spectrometry) are to be used. Two independent cleavages by enzymes with different specificities improve the coverage of the protein sequence. Note that it may be difficult in LC-MS to determine the charge state of eluting peptides, introducing ambiguities in the interpretation of the result.

Mass spectrometry is also useful for off-line mass analysis of HPLC purified peptides. First, a mass determination may identify a peptide based on the above mentioned knowledge of the protein sequence and enzyme specificity. Secondly, the mass spectrum in many cases reveals the presence of several peptides in HPLC fractions that appear by UV trace to contain only one species. Purity is important if a peptide, known or unknown, is going to be used in bioassays or sequenced by Edman degradation (33).

Tandem mass spectrometry allows, in principle, peptides to be sequenced as they elute from the LC column. Coupling of microbore and capillary column HPLC to ES MS and -MS/MS make peptide separation and sequencing possible at picomole to subpicomole levels. LC-MS/MS is rarely used for sequencing of all peptides derived from a protein but rather is used for structural analysis of selected peptides: most peptides are identified by their mass alone, while others may require MS/MS analysis to be further characterized. For extensive sequencing projects, it is common practice to split the HPLC eluate 1:9 between the ES source and a fraction collector (see Section 4.3). Peptides are then sequenced individually by ES MS/MS or by Edman degradation.

7.1 Software that aids the interpretation of mass spectra

Computer programs can be employed to design or evaluate a peptide mass mapping experiment. Software packages are available for several computer

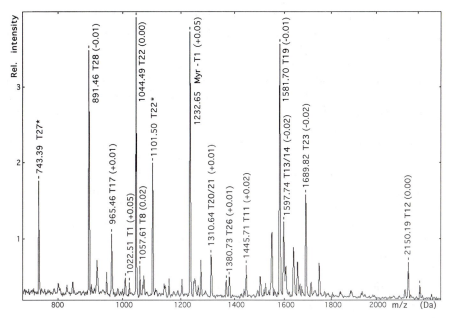

Figure 2. Tryptic peptide map of a G_{ia} subunit isolated from bovine brain (SwissProt accession no. P08239) obtained by MALDI reflector TOF MS. The fast evaporation method (*Protocol 2*) was used for sample preparation. The spectrum was mass calibrated by linear regression over seven matrix-related ion signals. Peptide identity and deviation from the calculated mass are indicated for the major ion signals. An ion signal corresponding to the myristoylated N terminal tryptic peptide is observed at *m/z* 1232 (see main text for details). Ion signals corresponding to over-alkylated peptides are indicated with an asterisk (*).

platforms. A list of freeware programs for biological mass spectrometry is maintained at our WWW site at `http://www.mann.embl-heidelberg.de`. Most instrument manufacturers include protein sequence analysis software with their MALDI or ES mass spectrometer systems. These programs accept a protein sequence as input and can generate specific peptide fragments, for example, tryptic peptides, and calculate their average and monoisotopic M_r, hydrophobicity indices, net charges, etc. They are also a valuable help for subsequent interpretation of peptide mass mapping data. Comparison of the experimental data with the predicted peptide masses allows a peptide to be tentatively assigned to *m/z* signals present in the mass spectrum. If no immediate peptide match is found for a given *m/z* value, the latter can be used to search the protein sequence for a matching peptide. Programs are available that allow protein sequence databases to be searched with mass spectrometry data for protein identification (see below).

Protocol 4. Denaturation, reduction, and S-alkylation of proteins[a]

Equipment and reagents

- Centrifuge cartridge
- 0.2 M NH$_4$HCO$_3$ pH ~ 8
- Fresh DTT solution, about 0.1 M
- Iodoacetamide solution, about 0.2 M
- Protease

Method

1. Dissolve the protein in a minimal volume of 0.2 M NH$_4$HCO$_3$ to a final concentration of 2–20 μM. If the protein is insoluble or tightly folded, include a denaturant such as 6 M GdmCl.[b]

2. Add freshly prepared solution of DTT[c] to a final concentration of 5 mM (or at least 50-fold molar excess over the number of cysteines present in the protein). Incubate at 50°C for 30 min. Cool to room temperature.

3. Add iodoacetamide to a final concentration of 25 mM (or 250-fold molar excess over cysteine residues). Incubate in the dark at room temperature for 30 min.

4. Optional: desalt and concentrate the sample using a centrifuge cartridge with an appropriate mol. wt cut-off; e.g. 3 kDa, 10 kDa, 30 kDa for Microcon (Amicon) ultrafiltration cartridges.

5. Dilute the solution so that the concentration of chaotropic agent is less than 1 M before adding the protease. Check the specification sheet for the protease to ensure that activity is maintained in the presence of chaotropic agents.

6. For tryptic digestion, go to *Protocol 5*, step 2.

[a] Reduction and *S*-alkylation prevent cysteine residues from forming mixed disulfides (see also Chapter 7). *S*-Alkylation generally improves the detectability of Cys-containing peptides in RP-HPLC and in MALDI and ES mass spectrometry.
[b] The use of chaotropic (unfolding/denaturing) agents in sample preparation for mass spectrometry is generally not advisable because these compounds often interfere with sample preparation and ionization. However, some proteins are resistant to proteolysis unless they are denatured by urea or GdmCl. A desalting procedure should be used after this step if the sample amounts allow it.
[c] Do not use β-mercaptoethanol—it can form mixed disulfides with proteins.

Protocol 5. Preparation of a tryptic protein digest for MS peptide mapping

Reagents

- 0.1 M NH$_4$HCO$_3$, 2 mM CaCl$_2$ pH 7.5
- Trypsin
- 5% (v/v) TFA

Method

1. Dissolve the protein[a] in 0.1 M NH_4HCO_3, 2 mM $CaCl_2$ pH 7.5, to a final concentration of 2–20 μM.

2. Add trypsin to obtain an enzyme to substrate ratio of 1 : 20 (w/w).

3. Incubate at 37 °C for 4 h.

4. Transfer an aliquot (0.5–1 μl) of the digest mixture to a 0.5 ml tube containing 2 μl of 5% TFA to quench the enzyme activity. Prepare 1 μl of this diluted peptide mixture for MALDI MS as described in *Protocol 1* or *2*.

5. If cleavage is incomplete, as judged by a MALDI spectrum, add another batch of enzyme and incubate overnight at 37 °C. Repeat step 4. If cleavage is still incomplete, add a denaturing agent to unfold the protein or use a different proteolytic enzyme.

6. When digestion is complete, dry the sample in a vacuum centrifuge and store at −20 °C, or below, until further analysis.

[a] We recommend reducing and alkylating the protein as outlined in *Protocol 4*.

7.2 Protein identification

A peptide map, obtained by RP-HPLC, or by mass spectrometry of a proteolytic digest mixture is a unique 'fingerprint' of a polypeptide, i.e. it can be used to identify individual proteins. Mass spectrometry, in particular, produces highly specific data in the form of a list of peptide molecular weights (the higher the accuracy, the more specific the data). This peptide mass data can be used to search protein sequence databases (34–38) and has proven extremely powerful for fast and error tolerant identification of proteins isolated by chromatographic methods and by one-dimensional or two-dimensional gel electrophoresis. It will be increasingly important as genome projects proceed. The method consists of the following steps:

(a) Tryptic digestion of the protein to be identified.

(b) Acquisition of a mass spectrometric peptide map (MALDI MS or ES MS) to generate a list of tryptic peptide masses.

(c) Searching an abstracted protein sequence database for all predicted tryptic peptides for all known proteins.

(d) Evaluation of retrieved sequences to identify the best match.

In the first step, proteins isolated by 1D or 2D gel electrophoresis are enzymatically digested either in-gel after excision of the protein spot of interest (*Protocol 6*) or on a membrane after electroblotting (for example on to nitrocellulose or PVDF membranes) (34) (see also Chapters 3 and 5).

The last step, the evaluation of a database search, is generally accomplished by taking into account the number of peptide matches and the mass accuracy of these matches (experimental versus calculated peptide mass). If necessary, the organism of origin of the protein, the intact protein molecular weight, and the pI (for proteins isolated by 2D PAGE) can also be considered.

In our experience with MALDI reflector TOF MS, at least five peptides should match within 0.1 Da for a positive protein identification. If a larger mass error is allowed, e.g. 1–2 Da, more than ten peptide matches are needed in addition to knowledge of origin, mass, and/or pI of the protein in order to retrieve an unambiguous match from a large protein database. High resolution peptide mass mapping by delayed extraction MALDI reflector TOF MS is now the method of choice for protein identification experiments (39). In cases where the stringent criteria (at least five peptide matches within a mass error of 50 p.p.m.) are not fulfilled, we turn to the use of ES MS/MS to obtain partial sequence information (peptide sequence tags, Section 8). Database searches with sequence tags are much more specific than the peptide map information and typically result in only one candidate being retrieved.

Protocol 6.   In-gel protein digestion

Reagents[a]

- 50% (v/v) acetonitrile
- 5% (v/v) formic acid
- 0.1 M NH$_4$HCO$_3$
- 10 mM DTT in 0.1 M NH$_4$HCO$_3$, freshly prepared

- Acetonitrile
- 55 mM iodoacetamide in 0.1 M NH$_4$HCO$_3$
- Digestion buffer: 12.5 mg trypsin/ml of 50 mM NH$_4$HCO$_3$, 5 mM CaCl$_2$

A. *Excision of protein bands from polyacrylamide gels[b]*

1. Wash the gel slab with water (two changes, 10 min each).[c]

2. Use a clean scalpel to excise the spot of interest from the gel. Cut as close to the protein band as possible to reduce the amount of 'background' gel.

3. Excise a gel piece of roughly the same size from a non-protein containing region of the gel for use as a control. Note: if the protein of interest appears to be contaminated with other proteins migrating nearby, excise the control blank below the band of interest to obtain a representative pattern of contaminations.

4. Cut the excised pieces into 1 × 1 mm cubes and transfer them to a 1.5 ml microcentrifuge tube.

B. *Washing of gel pieces*

1. Wash the gel particles with water and with 50% acetonitrile (one or two changes each, 15 min per change). Note: all the solvent volumes used in the washing steps should roughly equal two times the gel volume.

2. Remove all remaining liquid and add enough acetonitrile to cover the gel particles.

3. After the gel pieces have shrunk (they become white and stick together), remove the acetonitrile and rehydrate the gel pieces in 0.1 M NH_4HCO_3.

4. After 5 min, add an equal volume of acetonitrile.

5. Remove all liquid after 15 min of incubation.

6. Dry down gel particles in a vacuum centrifuge.

C. *Reduction and alkylation*

1. Swell the gel particles in 10 mM DTT/0.1 M NH_4HCO_3 and incubate for 45 min at 56°C to reduce the protein. Note: we recommend in-spot reduction even if the proteins were reduced (but not alkylated) prior to an electrophoresis run.

2. Chill the tubes to room temperature.

3. Remove excess liquid and replace it quickly with roughly the same volume of 55 mM iodoacetamide in 0.1 M NH_4HCO_3.

4. Incubate for 30 min at room temperature in the dark.

5. Remove iodoacetamide solution, and wash the gel particles with 0.1 M NH_4HCO_3 and acetonitrile as described in part B. Note: all the Coomassie blue should be removed at this time. The gel particles should appear completely transparent. If a large amount of protein is analysed (more than 10 pmol) residual staining may still be observed. In this case, an additional 0.1 M NH_4HCO_3/acetonitrile washing cycle should be performed.

D. *In-gel digestion*

1. Dry down the gel particles completely in a Speed Vac. If a large gel volume is analysed, shrink it by addition of acetonitrile before drying.

2. Rehydrate gel particles by a digestion buffer containing 50 mM NH_4HCO_3, 5 mM $CaCl_2$, and 12.5 μg/ml of trypsin[d] (e.g. Boehringer Mannheim, sequencing grade, or Promega, modified, sequencing grade) at 4°C (ice bucket). Add enough digestion buffer to cover the gel pieces. Add more buffer if all the initially added volume is absorbed by the gel pieces.

3. After 45 min remove the remaining supernatant and replace it with 5–20 μl of the same buffer—but without protease—to keep the gel pieces wet during enzymic cleavage (37°C, overnight).

4. Remove a 0.5 μl portion of the supernatant for direct peptide mixture analysis by MALDI MS. Inject the aliquot into a 0.5 μl droplet of 5% formic acid deposited on a matrix/nitrocellulose film as described on p. 38.

Protocol 6. *Continued*

E. *Extraction of peptides*

1. Extract the peptides from the gel by addition of a sufficient volume of 25 mM NH_4HCO_3 to cover the gel pieces.

2. Incubate for 15 min and add the same volume of acetonitrile.

3. Incubate for 15 min and recover the supernatant.

4. Repeat the extraction two times with 5% formic acid.

5. Pool all the extracts and dry the sample in a vacuum centrifuge.

F. *Mass spectrometric peptide mapping*

1. Redissolve the peptides in 10–30 μl of 5% formic acid.

2. Analyse by MALDI MS or by ES MS to obtain peptide mass information for database searches. See text for further details.

[a] Use the purest chemicals available at all stages of sample preparation, including gel casting.
[b] The protocol described is applicable without any modification to both one-dimensional and two-dimensional polyacrylamide gels of different thickness (routinely 0.5–1 mm), acrylamide concentration (7.5–18% of acrylamide), and protein band (spot) size. Gels are prepared by standard techniques using 0.1% SDS. Proteins are visualized by staining with Coomassie brilliant blue R-250 or G-250 (see Chapter 1), or alternatively by silver staining (39), or reverse staining (zinc–imidazole staining) (40). If the silver staining technique is used, then the extensive washing steps preceding sample reduction and alkylation can be omitted.
[c] Gloves should be worn to avoid contamination by human epidermal proteins (keratins). It is also important to rinse gloves with water to remove talcum powder and traces of dust.
[d] Trypsin is the enzyme of choice for subsequent mass spectrometric peptide sequencing, but it can be replaced by other enzymes.

8. Peptide sequencing by tandem mass spectrometry

As already mentioned, tandem mass spectrometry can be employed for amino acid sequencing because peptide ions preferentially fragment by cleavage of the backbone amide bonds. Tryptic peptides are especially suited for tandem mass spectrometric sequencing because they contain potential charge sites at the N terminus (amino group) and at the C terminus (Lys or Arg side chains), and therefore they produce abundant doubly charged ions. Collision-induced dissociation of such a peptide ion generates mainly singly charged ions corresponding to C terminal peptide fragments (y ions). In general, complementary y ion and b ion (from N terminal fragments) signals are used to assemble the peptide sequence. The nomenclature for peptide ion fragmentation can be found in (41,42). We note that much experience is needed to interpret tandem mass spectra and that, for high reliability sequencing, an additional derivatization step is usually required (43).

An example of a selected peptide is shown in *Figure 3*. The peptide map

(*Figure 3A*) was acquired by scanning the first mass analyser over the *m/z* range 400 to 1200. Note that the most prominent peaks correspond to doubly charged peptide ions. For the MS/MS sequencing experiment, the doubly charged peptide ion [T73]$^{2+}$ was selectively filtered through the first mass analyser and dissociated in the collision cell. The mass spectrum of the fragments (*Figure 3B*) acquired in the second mass spectrometer allows in this case determination of the complete amino acid sequence.

Often it is not possible to obtain a complete amino acid sequence of a peptide by tandem MS, and for most purposes this is not necessary anyway. A partial amino acid sequence (two to three consecutive residues) in combination with the mass of either terminus of the peptide constitutes a 'peptide sequence tag', (44) which is a highly specific probe for searching a protein sequence database. If a sequence match is retrieved, then the tandem mass spectrum can be used to confirm the assignment. Modified peptides can also be located by this method. This approach greatly improves sample throughput because the time-consuming task of complete sequence determination from a tandem mass spectrum is no longer necessary.

The NanoES ion source (described above) is excellent for MS/MS analysis of peptide mixtures because a 1 µl volume of sample can maintain ion production for > 30 min. Experimental conditions can be optimized for each peptide being sequenced by MS/MS. We routinely use this technique for sequencing of unseparated peptide mixtures generated by in-gel digestion of proteins isolated by 1D and 2D PAGE (45).

Table 5. Peptides identified in an in-gel tryptic digest of vacuolar sorting protein 1[a]

Peptide[a]	Monoisotopic mass (Da)		Sequence[b]
	Calculated	Measured	
T3	1072.6	1072.6	•SSVLENIVGR
T6	1220.7	1221.0	•RPLVLQLINR
T10,11	2171.1	2171.6	•VNQTANELIDLNINDDDKK
T20,21	2057.1	2057.4	ETDKVTGANSGISSVPINLR
T21	1583.9	1584.0	VTGANSGISSVPINLR
T23	1320.7	1320.8	•VPVGDQPPDIER
T32	1093.6	1093.6	•YGYIPVINR
T41,42	1342.8	1342.8	KLNSILLHHIR
T49	1269.4	1269.7	ISYVFHETFK
T60,61	2037.1	2036.7	YPALREAISNQFIQFLK
T63	1712.9	1713.2	•AEQTYINTAHPDLLK
T73	1233.7	1233.8	•LAALESPPPVLK
T78	869.5	869.6	•TIADIIPK
T84,85	1435.8	1436.1	•LYGKQDIEELTK

[a] See the legend to *Figure 3* for more details. Peaks correspond to those of *Figure 3A*.
[b] Peptides marked by bullets (•) were sequenced by NanoES MS/MS.

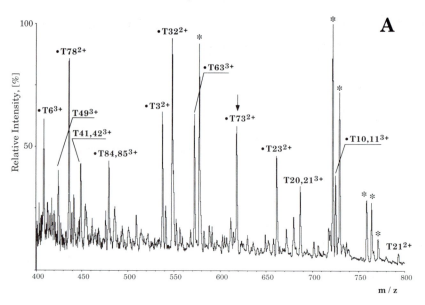

HO$_2$C- **K**   **L V P P P S E LAA L**-(NH$_2$)

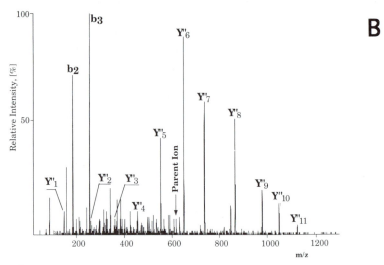

Figure 3. Peptide mapping and sequencing using MS. (A) Mass spectrum of the in-gel tryptic digest of a protein from budding yeast purified by one-dimensional polyacrylamide gel electrophoresis and visualized by silver staining (about 50 ng of the protein was loaded on the gel). On the basis of mass spectrometric data the protein was identified as vacuolar sorting protein 1 (SwissProt accession no. P21576). Sequences of corresponding tryptic peptides are listed in *Table 5*. Peaks marked by bullets (•) were sequenced by NanoES MS/MS.

9. Strategies for the analysis of modified proteins: phosphoproteins and peptides

A mass spectrometric strategy for the characterization of post-translationally modified proteins includes:

(a) Mass determination of the intact protein.

(b) Peptide mapping to identify modified peptides.

(c) Enzymatic or chemical removal of modifications, monitored by peptide mapping.

(d) Purification of modified peptides followed, if necessary, by tandem mass spectrometric analysis for structure elucidation.

The use of enzymes (for example, phosphatases, glycosylases, lipases) combined with mass analysis of the substrate before and after enzyme treatment is also a simple and effective approach to identify post-translational modifications (see also Chapter 4).

Phosphorylation is an important modulator of enzyme activity and signal transduction *in vivo*. Mass spectrometry is an excellent tool for localizing phosphorylation sites in proteins, which is a requirement to understand the mechanisms of regulation at the molecular level. ES MS can determine protein molecular weights with a mass accuracy and resolution that allows individual phosphorylation states (mass increases of 80 Da) to be revealed for intact proteins up to 50 kDa in size, provided that no buffer salts interfere with the analysis. Enzymatic digestion of a phosphoprotein by a sequence-specific protease produces a peptide mixture that also contains phosphopeptides. Several options for mass spectrometric analysis are now available. MALDI MS analysis of the peptide mixture before and after treatment with alkaline phosphatase will identify phosphopeptides by the mass decrease corresponding to the removal of phosphate groups by the phosphatase (46). Phosphopeptides may additionally be spotted in a MALDI MS peptide map because they form relatively high intensity $[M + Na]^+$ and $[M + K]^+$ ions as compared to non-phosphorylated peptides due to the presence of the negatively charged phosphate group. MALDI reflector TOF post-source decay analysis can be used to localize phosphorylation sites (47).

HPLC-ES MS is a powerful tool for phosphopeptide analysis. In the 'stepped orifice potential' method the quadrupole mass spectrometer is set to monitor the loss of phosphate (PO_3^-, 79 Da) from phosphopeptides by

---

Other peaks were assigned to tryptic peptides by their masses. Peaks designated by asterisks (*) originate from trypsin autolysis. (B) Tandem mass spectrum of the doubly charged ion of peptide T73 (*m/z* 617.9, marked by an *arrow* in A). The complete amino acid sequence of the peptide is confirmed by the fragment ion signals present in the mass spectrum. Peaks are designated in accordance with the nomenclature of Roepstorff (41).

increasing the ion internal energy in the entrance region (orifice or skimmer) of the mass spectrometer. At the same time a mass scan is performed to measure the peptide molecular weight (48). A strong signal at m/z 79 will be detected every time a phosphopeptide elutes from the HPLC column. A similar approach can detect glycosylated peptides (49).

Identification of phosphopeptides and glycopeptides in unseparated peptide mixtures can also be accomplished by NanoES tandem mass spectrometry (50). Further methods for characterizing phosphorylated proteins are described in Chapters 4 and 5.

Acknowledgements

We thank all members of the EMBL Protein & Peptide Group for providing data and advice for this manuscript.

Appendix

Calculating protein and peptide molecular weights

The molecular weight of a protein or peptide of known amino acid sequence can be calculated by adding the masses of each individual amino acid residue in the protein and adding the mass of the N and C terminal groups:

$$\mathbf{m}(\text{protein}) = \sum n_i \, \mathbf{m}(aa_i) + \mathbf{m}(\text{N term}) + \mathbf{m}(\text{C term})$$

where n_i is the total number of a given amino acid residue, aa_i, present in the protein sequence. $\mathbf{m}(x)$ is the average or the monoisotopic mol. wt of the species x. Note the mass difference, corresponding to one water molecule, between an *amino acid* and an *amino acid residue*. Amino acid residue masses are listed in *Table 6*.

The monoisotopic molecular weight of an organic compound is calculated from the elemental composition by using the masses of ^{12}C, ^{1}H, ^{14}N, ^{16}O, ^{31}P, and ^{32}S. However, peptides and proteins consist of so many atoms that the presence of naturally occurring heavy isotopes, mainly ^{13}C, adds significantly to the molecular weight (42). This results in a cluster of isotopic peaks spaced 1 Da apart for a singly charged peptide molecular ion. In protein and peptide mass analysis on quadrupole and time-of-flight instruments, the mass resolution is calculated as $(m/z)/\Delta(m/z)$, where m/z is the mass-to-charge ratio and $\Delta(m/z)$ is the full-width at half maximum of a peak in the mass spectrum. Depending on the mass analyser employed, mass spectrometry can be used to resolve the isotopic cluster for peptides with a mol. wt below 5000 (51). At sufficiently high resolution (e.g. $(m/z)/\Delta(m/z) \sim 1000\text{--}10\,000$) the isotopes can be separated, and the monoisotopic mass of peptides can be measured. If the isotopes are not resolved, the average mass of the unresolved signal is measured. The average mass is calculated from the average of the isotopic masses of the elements weighted for the abundances of each isotope (*Table 6*). This

Table 6. Amino acid residue masses[a]

| Amino | Three letter code | Single letter code | Residue mass | |
			Monoisotopic	Average
Glycine	Gly	G	57.02147	57.052
Alanine	Ala	A	71.03712	71.079
Serine	Ser	S	87.03203	87.078
Proline	Pro	P	97.05277	97.117
Valine	Val	V	99.06842	99.133
Threonine	Thr	T	101.04768	101.105
Cysteine	Cys	C	103.00919	103.144
Isoleucine	Ile	I	113.08407	113.160
Leucine	Leu	L	113.08407	113.160
Asparagine	Asn	N	114.04293	114.104
Aspartic acid	Asp	D	115.02695	115.089
Glutamine	Gln	Q	128.05858	128.131
Lysine	Lys	K	128.09497	128.174
Glutamic acid	Glu	E	129.04260	129.116
Methionine	Met	M	131.04049	131.198
Histidine	His	H	137.05891	137.142
Phenylalanine	Phe	F	147.06842	147.177
Arginine	Arg	R	156.10112	156.188
Tyrosine	Tyr	Y	163.06333	163.17
Tryptophan	Trp	W	186.07932	186.213
Homoserine lactone	Hsl		83.03712	83.090
Homoserine	Hse		101.04768	101.105
Pyroglutamic acid	Glp		111.03203	111.100
Carbamidomethylcysteine			160.03065	160.197
Carboxymethylcysteine			161.01466	161.181
Pyridylethylcysteine			208.06703	208.284

[a] Taken from ref. 42.

average mass, also called the chemical mass, is commonly used in peptide and protein mass spectrometry. Another feature in mass analysis above mol. wt ~ 2000 is that the peak corresponding to the monoisotopic elemental composition is no longer the most abundant peak in the isotopic cluster and is, thus, not easily identified. This is mainly due to the contribution of ^{13}C (natural abundance 1.10%) to the total mass. The difference between the monoisotopic mass and the average mass increases with the C + H content of the molecule, that is, it increases with the molecular weight of the protein.

Reporting molecular weight data

When reporting molecular weight data obtained by mass spectrometry it must be specified whether the mass value corresponds to the protonated molecular ion $[M + H]^+$ or to the neutral molecule (M), and whether the monoisotopic or average mass is meant. We recommend that the mass of the neutral molecule is reported, because this avoids ambiguities relating to positive or negative ion mass analysis and to multiple charging. If the mass analyser allows the monoisotopic mass of a molecule to be determined, this value should be reported; otherwise the average mass should be given.

The *mass* of a molecule is given in units of daltons (Da) or atomic mass units (amu or u). The *molecular weight* is a unitless quantity. For example, the average mass of horse heart cytochrome *c* is 12360.1 Da, while the average mol. wt of cytochrome *c* is 12360.1. The *relative molecular mass* (M_r) used in the biochemical literature to indicate the size of proteins, is identical to the *molecular weight* term. Similarly, the *molecular mass* is identical to the term *mass* as it is used above. These terms are used interchangeably. When referring to signals in a mass spectrum, it is convenient to use the *m/z* value. This unitless quantity can only be used to refer to peaks in a mass spectrum.

References

1. Carr, S. A., Hemling, M. E., Bean, M. F., and Roberts, G. D. (1991). *Anal. Chem.*, **63**, 2802.
2. Chait, B. T. and Kent, S. B. H. (1992). *Science*, **257**, 1885.
3. Andersen, J. S., Svensson, B., and Roepstorff, P. (1996). *Nature Biotech.*, **14**, 449.
4. Siuzdak, G. (1994). *Proc. Natl. Acad. Sci. USA*, **91**, 11290.
5. Wang, R. and Chait, B. T. (1994). *Curr. Opin. Biotech.*, **5**, 77.
6. Karas, M. and Hillenkamp, F. (1988). *Anal. Chem.*, **60**, 2299.
7. Hillenkamp, F., Karas, M., Beavis, R. C., and Chait, B. T. (1991). *Anal. Chem.*, **63**, 1193A.
8. Fenn, J. B., Mann, M., Meng, C. K., Wong, S. F., and Whitehouse, C. M. (1989). *Science*, **246**, 64.
9. Smith, R. D., Loo, J. A., Edmonds, C. G., Barinaga, C. J., and Udseth, H. R. (1990). *Anal. Chem.*, **62**, 882.
10. Mann, M. and Wilm, M. (1995). *Trends Biochem. Sci.*, **20**, 219.
11. Jennings, K. R. and Dolnikowski, G. G. (1991). In *Methods in enzymology* (ed. J. A. McCloskey), Vol. 193, p. 37. Academic Press, San Diego.
12. Yost, R. A. and Boyd, R. K. (1992). In *Methods in enzymology* (ed. J. A. McCloskey), Vol. 193, p. 154. Academic Press, San Diego.
13. Smith, R. D. and Light-Wahl, K. J. (1993). *Biol. Mass Spectrom.*, **22**, 493.
14. Miranker, A., Robinson, C. V., Radford, S. E., Aplin, R. T., and Dobson, C. M. (1993). *Science*, **262**, 896.
15. Robinson, C. V., Gross, M., Eyles, S. J., Ewbank, J. J., Mayhew, M., Hartl, F. U., *et al.* (1994). *Nature*, **372**, 646.
16. Loo, R. R. O., Dales, N., and Andrews, P. C. (1994). *Protein Sci.*, **3**, 1975.
17. Beavis, R. C., Chaudhary, T., and Chait, B. T. (1992). *Org. Mass Spectrom.*, **27**, 156.
18. Beavis, R. C. and Chait, B. T. (1989). *Rapid Commun. Mass Spectrom.*, **3**, 432.
19. Strupat, K., Karas, M., and Hillenkamp, F. (1991). *Int. J. Mass Spectrom. Ion Proc.*, **111**, 89.
20. Pieles, U., Zurchner, W., Schar, M., and Moser, H. E. (1993). *Nucleic Acids Res.*, **21**, 3191.

21. Wu, K. J., Steding, A., and Becker, C. H. (1993). *Rapid Commun. Mass Spectrom.*, **7**, 142.
22. Beavis, R. C. and Chait, B. T. (1989). *Rapid Commun. Mass Spectrom.*, **3**, 233.
23. Xiang, F. and Beavis, R. C. (1994). *Rapid Commun. Mass Spectrom.*, **8**, 199.
24. Vorm, O., Roepstorff, P., and Mann, M. (1994). *Anal. Chem.*, **66**, 3281.
25. Vorm, O. and Mann, M. (1994). *J. Am. Soc. Mass Spectrom.*, **5**, 955.
26. Ling, V., Guzzetta, A. W., Canova-Davis, E., Stults, J. T., Hancock, W. S., Covey, T. R., *et al.* (1991). *Anal. Chem.*, **63**, 2909.
27. Huang, E. C. and Henion, J. D. (1991). *Anal. Chem.*, **63**, 732.
28. Griffin, P. R., Coffman, J. A., Hood, L. E., and Yates III, J. R. (1991). *Int. J. Mass Spectrom. Ion Proc.*, **111**, 131.
29. Davis, M. T. and Lee, T. D. (1992). *Protein Sci.*, **1**, 935.
30. Wilm, M. S. and Mann, M. (1994). *Int. J. Mass Spectrom. Ion. Proc.*, **136**, 167.
31. Wilm, M. and Mann, M. (1996). *Anal. Chem.*, **68**, 1.
32. Mann, M., Meng, C. K., and Fenn, J. B. (1989). *Anal. Chem.*, **61**, 1702.
33. Elicone, C., Lui, M., Geromanos, S., Erdjument-Bromage, H., and Tempst, P. (1994). *J. Chromatogr. A*, **676**, 121.
34. Henzel, W. J., Billeci, T. M., Stults, J. T., Wong, S. C., Grimley, C., and Watanabe, C. (1993). *Proc. Natl. Acad. Sci. USA*, **90**, 5011.
35. Mann, M., Højrup, P., and Roepstorff, P. (1993). *Biol. Mass Spectrom.*, **22**, 338.
36. Pappin, D. J. C., Højrup, P., and Bleasby, A. J. (1993). *Curr. Biol.*, **3**, 327.
37. Yates, J. R., Speicher, S., Griffin, P. R., and Hunkapiller, T. (1993). *Anal. Biochem.*, **214**, 397.
38. James, P., Quadroni, M., Carafoli, E., and Gonnet, G. (1993). *Biochem. Biophys. Res. Commun.*, **195**, 58.
39. Shevchenko, A., Wilm, M., Vorm, O., and Mann, M. (1996). *Anal. Chem.*, **68**, 850.
39a. Jensen, O. N., Podtelejnikov, A., and Mann, M. (1996). *Rapid Commun. Mass Spectrom.*, **10**, 1371.
40. Fernandez-Patron, C., Calero, M., Collazo, P. R., Garcia, J. R., Madrazo, J., Musacchio, A., *et al.* (1995). *Anal. Biochem.*, **224**, 203.
41. Roepstorff, P. and Fohlmann, J. (1984). *Biomed. Mass Spectrom.*, **11**, 601.
42. Biemann, K. (1990). In *Methods in enzymology* (ed. J. A. McCloskey), Vol. 193, p. 886. Academic Press, San Diego.
43. Hunt, D. F., Yates, J. R., Shabanowitz, J., Winston, S., and Hauer, C. R. (1986). *Proc. Natl. Acad. Sci. USA*, **83**, 6233.
44. Mann, M. and Wilm, M. S. (1994). *Anal. Chem.*, **66**, 4390.
45. Wilm, M., Shevchenko, A., Houthaeve, T., Breit, S., Schweigerer, L., Fotsis, T., and Mann, M. (1996). *Nature*, **379**, 466.
46. Liao, P.-C., Leykam, J., Andrews, P. C., Gage, D. A., and Allison, J. (1994). *Anal. Biochem.*, **219**, 9.
47. Talbo, G. and Mann, M. (1994). *Techn. Protein Chem.*, **V**, 105.
48. Huddleston, M. J., Annan, R. S., Bean, M. F., and Carr, S. A. (1993). *J. Am. Soc. Mass Spectrom.*, **4**, 710.
49. Carr, S. A., Huddleston, M. J., and Bean, M. F. (1993). *Protein Sci.*, **2**, 183.
50. Wilm, M., Neubauer, G., and Mann, M. (1996). *Anal. Chem.*, **68**, 527.
51. Buchanan, M. V. and Hettich, R. L. (1993). *Anal. Chem.*, **65**, 245A.

3

Immunological detection of proteins of known sequence

KARL HEINZ SCHEIDTMANN, SILKE REINARTZ, and
HARALD SCHLEBUSCH

1. Introduction

Many proteins were first identified as antigens or neo-antigens, such as viral
or tumour-specific proteins in infected or transformed cells, which in the carrier organism induced an immune response. In those cases, sera from infected
animals or patients provided the tools for identification of the antigenic
proteins.

With the rapid progress in molecular biology, cellular genes are often identified, molecularly cloned, and sequenced without knowing their products.
The amino acid sequence of such gene products, however, can be deduced
with high probability by elucidating the correct reading frame and potential
processing signals. Proof for the existence of the predicted protein can then
be achieved by *in vitro* translation or by immunological means. In the latter
case, antibodies must be induced against an unkown antigen. This is generally
achieved by using as antigens either synthetic peptides, corresponding to a
certain segment of the protein, or the protein produced by expression of the
gene in bacteria, usually fused to a known bacterial protein. The initial identification of the gene product then allows further characterization of the protein with respect to its structure, function, mode of expression, subcellular
distribution, etc. Thus, there is a wide spectrum of possible applications and
investigations provided by immunological techniques.

This chapter concentrates on two aspects; first, the production of specific
antibodies against an unknown polypeptide by three different routes: induction of antibodies against synthetic peptides, against bacterially expressed
fusion proteins, or production of antibodies in bacteria using the phage
display system. The second part describes the potential use of these antibodies for the identification and characterization of the polypeptide searched
for.

2. Generation of specific antibodies

Once the coding region of a gene is cloned and sequenced, one can use two approaches to raise specific antibodies against the presumptive gene product, either to synthesize oligopeptides corresponding to certain regions of the protein, or to express (part of) the protein in bacteria and use this as antigen for immunization. The peptide approach offers some unique possibilities, which will be discussed below (for review, see ref. 1), but bears the risk of failure in that the antibodies obtained might recognize the peptide but not the protein. With bacterially expressed polypeptides as antigens this problem seems negligible, but the efficient expression itself may be a time-consuming problem.

3. Antibodies against synthetic peptides

Principle: a favourable segment of the protein consisting of at least eight amino acids is synthesized, coupled to a carrier protein, and injected into an animal to induce antibodies. These antibodies, it is hoped, will recognize the protein.

3.1 Criteria for selection of peptides

With this approach the most important question is which segment of the protein to choose. The first choice are peptides corresponding to the termini, particularly the carboxy terminus, which will usually yield 'good' antibodies, that is, ones which cross-react with the native protein (1,2). If antibodies against internal regions of a protein are to be generated, some considerations are necessary. Antigenic sites of proteins were thought to consist of short stretches of about six amino acids that may be arranged in the primary structure either continuously or discontinuously (3). Recent structural analyses, however, suggest that antigenic epitopes in native proteins are mostly of the discontinuous type, that is they are composed of several short stretches, mainly loops exposed on the surface (see ref. 4, for a discussion). Despite this, anti-peptide antibodies recognizing 'internal' regions of a protein have been successfully induced, perhaps due to the fact that even in composite epitopes one segment contributes most of the binding energy (4). The antigenic epitope may be determined by its sequence or by the conformation in which this sequence is embedded. Thus, it may be called a sequential (or linear) or a conformational epitope. When anti-peptide antibodies are to be induced, one needs to choose a site that will *per se* represent a sequential epitope within the protein (see refs 1, 4–6, and further references therein). These considerations require methods that allow the prediction of potential antigenic sites of a protein. The following criteria have been considered for the selection of peptides:

- exposure on the surface of the protein
- conformation
- immunogenicity

3.1.1 Surface exposure

Three parameters are correlated with a high probability for surface exposure: the hydrophilicity, the mobility and, as structural component, the potential to form turn or loop structures. In principal, more polar or hydrophilic regions should be on the surface, whereas apolar or hydrophobic regions should be in the interior of the protein. As for the mobility, data from nuclear magnetic resonance (NMR) studies and X-ray crystallography indicate that residues on the surface of a protein have higher mobility than those in the interior. Regarding the structural organization, more 'rigid' structures such as α helices and β strands are more likely to form the internal core of a protein, whereas turn and loop structures have a high probability for surface exposure (7).

For each of these parameters, algorithms have been developed in which the values for each individual amino acid are averaged over groups of six or seven neighbouring residues over the entire sequence (8–10). As a result, profiles for hydrophilicity or mobility are obtained. These usually show a high degree of coincidence. Similarly, segments with the potential for forming certain secondary structures can be predicted (11,12). In this case, regions of predicted α helix or β strand structures often coincide with negative peaks of hydrophilicity (13).

3.1.2 Conformation

The consideration of conformation is probably only a theoretical one. Although the sequence chosen as peptide may have a distinct conformation in the protein (which can be predicted with some probability), a short peptide can assume many conformations, only some of which coincide with that in the native protein (1). Thus, of the various antibodies that may be induced against different conformations of the petide, some should recognize the 'natural' conformation in the protein too. In general it seems favourable to use relatively long peptides of 15 or more residues, which may indeed mimic the 'natural' conformation (14,15).

3.1.3 Immunogenicity

Some amino acid residues seem to be more immunogenic than others, since they are found in antigenic epitopes with relatively high frequency. These are, in decreasing order, His, Lys, Ala, Leu, Asp, and Arg. Thus, antigenicity values can be derived for individual residues, and regions of high antigenic potential can be calculated (16).

An evaluation of the different prediction methods is given in ref. 17. Based on the above considerations, one determines first the peak regions from the calculated hydrophilicity and mobility profiles, and then selects for synthesis those segments that are enriched in antigenic residues. However, one has to keep in mind that these are theoretical and probabilistic considerations. In our experience, there is no way of predicting with certainty whether antibodies against a particular peptide will recognize this epitope in the native protein. Fortunately, these considerations are not so important for terminal regions.

3.1.4 Length of peptide

As pointed out above, peptides of 8–15 residues appear to be more 'successful' than short peptides (14,15). There might be two reasons for that: first, they might contain several overlapping epitopes, one of which coincides with a natural epitope on the protein; secondly, the longer peptides might mimic the natural conformation.

There are two considerations as to whether a residue at the border of a potential antigenic site should be included or not in the peptide to be synthesized. First, it might be favourable to extend a peptide in order to include immunogenic residues. Secondly, for immunization, peptides are usually coupled to a carrier protein (see Section 3.2). Coupling can use amino or acidic groups, or Tyr or Cys residues. The group used for coupling shoud be at one end of the peptide and should not also be present internally in the peptide. Thus, the peptide to be synthesized is extended until the next residue favourable for coupling is encountered, or such a residue (e.g. Tyr) is added to one end solely for coupling.

3.2 Coupling of peptide to carrier protein

Usually, peptides are coupled to a carrier protein in order to increase their immunogenicity and to prevent their rapid degradation in the animal. As carriers, bovine serum albumin (BSA) or keyhole limpet haemocyanin (KLH) are widely used. The latter might induce a higher immune response (perhaps by a factor of two) but tends to form insoluble aggregates with all coupling procedures, due to its high molecular weight. We routinely use BSA, which does not cause such problems. However, coupling to a carrier is not absolutely required. Polymerization of the peptide itself may render it immunogenic (18). Also, peptides can be synthesized in branched form that can be used directly for immunization (19).

The peptide should be coupled in its natural orientation: if it represents the carboxyl terminus of the protein, it should be coupled with its amino terminal residue, and vice versa; for 'internal' peptides either terminus might be used. The following bifunctional reagents serve as couplers:

(a) Glutaraldehyde couples via amino groups (α- and ε-amino groups).

(b) Water soluble carbodiimide couples via either amino or carboxyl groups, but with a preference for acidic groups.

(c) *Bis*-diaobenzidine (BDB) couples via Tyr residues (*Protocol 1*).

(d) *N*-succinimidyl-3-(2-pyridyl dithio)-propionate (SPDP) couples via amino groups of the carrier and Cys residues of the peptide (*Protocol 2*).

In a typical coupling reaction we use 10–25 mg of BSA (20 mg = 0.3 μmol) and a 20- to 40-fold molar excess of peptide. Of this, 20–60% may be coupled to the carrier depending on the coupling reagent and the peptide. Consequently, 5–20 mol of peptide may be linked to 1 mol of carrier. Uncoupled peptide and coupling reagents are removed by dialysis or gel filtration, depending on the volume of the reaction. If desired, the efficiency of coupling can be quantified by using radiolabelled peptide (for example iodinated at a Tyr residue) or by colorimetry as for the SPDP coupling. Protocols are given for BDB and SPDP coupling because they are most specific and efficient (*Protocols 1, 2*). (For the other methods, see refs 2 and 20.)

Protocol 1. Coupling via tyrosyl residues by bis-diazobenzidine (BDB)[a]

This is an easy and effective coupling method (coupling efficiency ~ 50–60%).[b]

Reagents
- Benzidine–HCl[a] (Sigma)
- NaNO$_2$
- PBS
- Borate buffer: 0.16 M Na borate pH 9, 0.14 M NaCl

Method

1. Dissolve 230 mg of benzidine–HCl in 45 ml of 0.2 M HCl.

2. Add 175 mg of NaNO$_2$ in 5 ml of water.

3. Stir at 4°C for 1 h.[c]

4. Dissolve 25 mg of BSA and a 30-fold molar excess of peptide in 1 ml of borate buffer; 4°C.

5. Add 1 ml of BDB dropwise, while stirring.[d]

6. Adjust to pH 9 with 0.5 M NaOH.

7. React at 4°C for 2 h, or until a precipitate starts to form.

8. Dialyse against PBS at 4°C for two days, with two or three changes.

[a] Caution, benzidine is carcinogenic!
[b] See ref. 2.
[c] The solution turns light yellow; it can be stored in 1 ml aliquots at –60°C (stable for > one year).
[d] The solution turns dark brown.

Protocol 2. Coupling with *N*-succinimidyl-3-(2-pyridyl dithio)propionate (SPDP)[a]

Equipment and reagents
- Sephadex G25 column
- *N*-succinimidyl-3-(2-pyridyl dithio)propionate (SPDP) (Sigma or Pierce)
- BSA
- PBS
- Phosphate buffer: 0.1 M NaPO$_4$ pH 7.5, 0.1 M NaCl[b]

Method

1. Prepare a 20 mM SPDP solution in ethanol.[c]

2. Dissolve 10 mg (0.15 μmol) of BSA in 1 ml of phosphate buffer.

3. Add a 20- to 30-fold molar excess of SPDP dropwise to the BSA solution, react at room temperature for 30 min.

4. Remove the free coupling reagent by gel filtration on Sephadex G25, equilibrated with phosphate buffer.

5. Measure the A$_{280}$ of the collected fractions to identify the activated BSA. After this step, the amount of SPDP linked to BSA may be determined for an aliquot as described in steps 6 and 7, otherwise proceed to step 8.

6. Add a threefold molar excess of DTT.[d]

7. Measure the A$_{343}$ of the released pyridine-2-thione.[e]

8. Dissolve 10 μmol of peptide (threefold molar excess over the activated groups) in 1 ml of phosphate buffer.

9. Mix activated BSA and peptide and react at room temperature for 2 h.[f]

10. Measure the A$_{343}$ to determine the coupling efficiency.[e,g]

11. Dialyse extensively against PBS.

---

[a] For the detailed reaction see ref. 21. SPDP is a heterobifunctional reagent and couples via amino and sulfhydryl residues. Amino groups (α and ε) of the carrier protein are first reacted with SPDP, which is then accessible to react with the Cys residue of a peptide in a thiol–disulfide exchange reaction, or vice versa. The method is efficient and yields a relatively homogeneous product, since there is no polymerization of the carrier protein itself. The coupling efficiency can be monitored by measuring A$_{343}$.

[b] Note: at higher pH or in Tris buffer, the compound is rapidly hydrolysed!

[c] Stable at room temperature for two weeks.

[d] This results in a disulfide exchange with concomitant release of pyridine-2-thione.

[e] Molar absorbance = 8.08 × 10^3/M cm); the amount of free pyridine-2-thione corresponds to the amount of thiol groups accessible for coupling.

[f] This results in the release of pyridine-2-thione as described in step 5 for the exchange reaction with DTT.

[g] The amount of free pyridine-2-thione corresponds to that replaced by the peptide.

4. Antibodies against bacterially expressed proteins

Newly identified and characterized genes, or parts of them, may be expressed in bacteria to obtain large quantities of protein for biochemical studies and/or for the induction of antisera. The present considerations will be restricted to those relevant for the generation of antibodies.

A suitable and widely used expression system is provided by the glutathione-*S*-transferase (GST) system (22) in which the polypeptide in question is expressed as a fusion protein in an inducible fashion under control of the *lac* promoter. Vectors are available that allow cloning of the foreign gene segment in the appropriate reading frames (Pharmacia). The GST portion usually allows stable expression in a soluble form, which can otherwise be a problem. Also, the GST part allows easy and efficient purification by affinity chromatography on glutathione–Sepharose (see *Protocol 3*). If favourable or necessary, the fusion partner can be proteolytically cleaved off with thrombin or factor X proteases (22), but this is usually not required for generation of antisera, since any antibodies directed against GST can be removed by adsorption to it.

If problems are encountered with the stability or solubility of a given fusion product, shorter, preferably carboxy terminal, fragments of the foreign polypeptide may be tried. 60 to 70 residues are sufficient to induce good antisera, but longer fragments are usually more favourable (23). Problems with solubility may be dealt with by trying denaturation/renaturation protocols. The methodology for the construction and handling of such expression systems is outside the scope of this article and the reader is referred to appropriate manuals (24–26). An instruction booklet on the GST expression system is also provided by Pharmacia, where the components of the system can be purchased.

4.1 Purification of fusion protein from bacteria

Bacteria can be extracted by sonication, and the fusion protein can be purified by affinity chromatography on glutathione–Sepharose. After elution with excess glutathione, the protein may be concentrated by ultrafiltration through desalting cartridges or further purified by preparative SDS–PAGE (see Chapter 1). If the protein is insoluble or poorly extractable it might be easier to prepare whole cell extracts by lysing bacteria in SDS sample buffer (see *Protocol 7*) and, after denaturation, to separate the protein by SDS–PAGE without prior purification. The protein is localized, eluted from the gel, and used for immunization as described in *Protocol 3*. One needs at least 0.1 mg of protein per rabbit.

Protocol 3. Purification of GST-fusion protein from bacteria[a]

Equipment and reagents

- Equipment to grow bacteria in suspension
- Sonifier
- LB medium containing 50 μg ampicillin
- IPTG (isopropyl-β-D-thiogalactoside)
- GSH
- Glutathione–Sepharose
- Triton X-100 or Nonidet P-40
- PBS

- Elution buffer: 50 mM Tris pH 7.5, 5 mM GSH
- SDS lysis buffer (same as sample buffer for SDS–PAGE): 0.125 Tris–HCl pH 6.8, 3.3% (w/v) SDS (Bio-Rad), 20 mM DTT (or 5% (v/v) 2-mercaptoethanol)
- Coomassie blue: 0.6% (w/v) Coomassie blue in 100 mM Tris–HCl pH 7.4, 20% (v/v) methanol

A. *Affinity chromatography of GST–fusion proteins[a]*

1. Start bacterial culture (200 ml) from overnight culture.

2. Measure A_{600} to follow the growth of the bacteria until an absorbance of 0.5 is reached.

3. Add 0.1 mM IPTG to induce the expression of the GST–fusion protein. Incubate for 2–3 h.

4. Harvest the cells by centrifugation at 3000 *g*, 4°C, for 10–20 min.

5. Resuspend in PBS at 0.1 of the original volume.[b]

6. Sonicate five to ten times for 30 sec on ice.[c]

7. Add Triton X-100 (or Nonidet P-40) to 1% (v/v) and incubate 5–10 min on ice to facilitate complete lysis and solubilization.

8. Centrifuge at 10 000 *g* at 4°C for 20 min.

9. Apply supernatant to glutathione–Sepharose (column or batch).[d]

10. Allow adsorption of the GST–fusion protein to the matrix at 4°C or room temperature for 30 min.

11. Wash twice with PBS containing 0.1% Triton X-100 then once without detergent.

12. Elute the bound proteins with elution buffer or SDS lysis buffer.

B. *Elution of proteins from polyacrylamide gels*

1. Apply the protein sample to a preparative SDS–PAGE gel (see Chapter 1).[e]

2. Localize the protein by staining in Coomassie blue.

3. Cut out the band of interest and chop into pieces.

4. Homogenize the gel in a small volume of 50 mM NH_4HCO_3, e.g. in a Dounce homogenizer or alternatively by pressing it through a needle.

5. Add 2-mercaptoethanol to 5% (v/v) and boil for 5 min.

6. Elute overnight at 37°C while stirring.

7. Spin down the gel homogenate at 13 000 *g* at room temperature for 20 min.

8. Lyophilize to remove mercaptoethanol.

9. Prepare emulsion with adjuvant (TiterMax) for immunization (see Section 5).

[a] Ref. 22.
[b] The suspension can be stored in aliquots at –70°C after this step.
[c] Between sonication allow the suspension to cool down for 30–60 sec.
[d] The amount of glutathione–Sepharose required has to be determined, but generally > 20 µl of Sepharose is required per 100 ml of extract.
[e] Up to 2 mg of protein in 1 ml of sample buffer can be loaded on a gel of 2 mm × 14 cm

5. Immunization

For immunization (see Chapter 13), the peptide conjugates, or purified antigen from other sources, are mixed with adjuvant, preferably TiterMax (Serva) which is less toxic than Freund's adjuvant to the animal but still yields high antibody titres. Usually, 0.5–1 mg of conjugate or 0.1–0.5 mg of purified antigen in about 2 ml of emulsion are applied per rabbit, or accordingly lesser amounts for mice or guinea-pigs. One should immunize at least two, and preferably four, animals per antigen, because occasionally some animals respond poorly or not at all. Before producing monoclonal antibodies it might be wise to test polyclonal sera for their specificitiy and ability to recognize the protein in question. The keeping and handling of animals usually requires authorized personnel and permission from the government. However, some companies offer the generation of poly- or monoclonal antibodies at reasonable prices.

6. Generation of recombinant antibody fragments in bacteria through recombinant phage

Recent developments in antibody engineering have been directed towards the expression of antibody fragments in bacterial (27,28) and phage display systems (29), leading to increasingly diverse applications in biology, clinical diagnosis, and therapy. The ability of the phage display system to express antibody fragments offers several advantages over hybridoma technology allowing rapid, economical production and manipulation of variable antibody genes to improve binding affinities, e.g. by site-directed mutagenesis. The technology relies on the expression of antibody fragments as fusion proteins displayed on the surface of filamentous phage, e.g. fd or M13. Two different fd/M13 phage proteins have been shown to be useful for phage display, namely the major coat protein g8p present in about 2800 copies per phage and the minor coat protein g3p present in three or five copies on the phage tip. Thus far, the g3p–fusion has been favoured in most phage display systems. Either single chain variable fragments (scFv) or Fab fragments have been expressed by this technology. ScFv fragments are composed of heavy and light chain variable domains connected covalently by a short flexible linker of about 15 amino acids that facilitates the functional folding of the antigen combining site. Because of their small size and high stability, scFv fragments represent interesting immunological agents of the future.

Hybridoma cell lines represent the most abundant and straightforward source from which antibody genes can be cloned, since they express the heavy and light chain genes for one single antibody. Splenocytes from immunized mice can alternatively be used for phage antibody expression, thus by-passing hybridoma technology (30). Since splenocytes express a whole repertoire of rearranged V genes, more extensive selection and screening are required to isolate a specific clone. Recently phage display libraries have been described based on a repertoire of rearranged V genes derived from peripheral blood lymphocytes of immunized and non-immunized human donors (31) and repertoires of V gene segments rearranged *in vitro* to build synthetic antibodies entirely outside the immune system (32).

The procedures described for rescue of V genes from mouse hybridomas (see *Protocol 4*) are essentially identical to those used for the amplification of libraries of V_H and V_L genes from splenocytes or peripheral blood lymphocytes. Gene rescue is performed by polymerase chain reaction (PCR) using V_H and V_L-specific primer mixtures. Amplified V_H and V_L genes are subsequently assembled into a scFv gene with a short linker DNA [$(Gly_4Ser)_3$] using another PCR procedure 'splicing by overlap extension'. This three-fragment assembly reaction appears to be a critical step in cloning a scFv gene, because, in some cases, only an exact equimolar ratio of V_H, V_L, and linker ensures a successful joining. The scFv gene is cloned in frame into *E. coli* expression vectors like the phagemid pCANTAB5E (Pharmacia) as fusion to the N terminus of the mature g3p. By interposing an amber stop codon between the scFv gene and gIII gene the pCANTAB5E vector allows the expression of both phage displayed and soluble scFv fragments, depending on the host strain (33).

E. coli supE strains like TG1 are infected with M13KO7 helper phage to rescue recombinant phage, which can be selected in a panning reaction with the antigen bound to a solid phase (see *Protocol 5*). By elimination of the non-binding phages enrichment factors of 20 up to 1000 can be reached after one round of panning (31), and subsequent rounds enable the isolation of even rare phage clones from libraries.

Once an antigen positive phage clone has been identified in ELISA and isolated, soluble scFv can be produced by infection of non-supressor *E. coli* strains like HB2151 (see *Protocol 6*). For detection of soluble scFv, it is necessary to have additional sequences, i.e. Etag- or other tag sequences, fused to the scFv, which serve as targets for monoclonal antibodies. Functional soluble scFv may be detected either in the bacterial periplasm or in the extracellular fraction, whereas non-binding soluble scFv is found in the cytoplasm due to the reducing environment which minimizes formation of disulfide bonds. Yields of soluble scFv expression vary greatly depending upon the antibody clone.

Protocol 4. Cloning and expression of a phage-displayed scFv

Equipment and reagents

- Thermocycler
- Enzymes, primers,[a] vectors,[a] etc. for PCR and molecular cloning
- SOB[b]
- SOB-AG[b]: SOB plus 100 μg/ml ampicillin, 2% (w/v) glucose

- YT[b]
- YT-AG[b]: YT plus 100 μg/ml ampicillin, 2% (w/v) glucose
- YT-AK[b]: YT plus 100 μg/ml ampicillin, 50 μg/ml kanamycin

Method

1. Isolate mRNA from hybridomas (use 6×10^5 cells).

2. Transcribe mRNA into cDNA with murine reverse transcriptase using random hexadeoxyribonucleotides.

3. Separately amplify the V_H and V_L genes with specific primer mixtures[c] by PCR through 30 cycles of polymerization (94°C 1 min, 55°C 2 min, 72°C 2 min).

4. Check the PCR products by agarose gel electrophoresis.[d]

5. Gel purify the DNA of appropriate size and determine the exact concentration of purified V_L, V_H, and linker.[e]

6. Use V_H, V_L, and linker primers[c] at an equimolar ratio and cycle 20 times (94°C 1 min, 63°C 4 min).[f]

7. PCR amplify 2 μl of the assembled product using a mixture of specific V_H primers containing a *Sfi*I site at the 5' end and of V_L specific primers with a *Not*I site at the 3' end.

8. Gel purify the assembled scFv gene and subsequently digest first with *Sfi*I for 1 h at 50°C followed by *Not*I for 1 h at 37°C. Gel purify the digested product again.

9. Ligate the digested scFv gene in the phagemid vector pCANTAB5E (Pharmacia) and transform into competent *E. coli* TG1 cells. Plate transformed cells on SOB-AG plates and grow overnight at 37°C.

10. Scrape the cells from the plate after flooding the plate with 5 ml 2 × YT-AG medium. Dilute to an A_{600} of 0.5. Infect bacteria with M13KO7 helper phage (3×10^{10} p.f.u./15 ml culture) for 1 h at 37°C with shaking at 250 r.p.m.[g]

Protocol 4. *Continued*

11. Pellet the cells by centrifugation at 1000 *g* for 10 min and remove the supernatant.

12. Resuspend the bacteria in 2 × YT-AK medium and grow overnight at 37 °C, 250 r.p.m.

13. Harvest the phage supernatants by centrifugation at 1000 *g* for 20 min and filter through a 0.45 μm sterile filter. The phage supernatants can be stored at 4 °C.

[a] Primers, plasmids, phages, bacteria, including Instruction booklet can be purchased from Pharmacia.
[b] See refs 24–26.
[c] See refs 30, 34, 35.
[d] The expected size is approximately 340 bp for V_H and 325 bp for V_L.
[e] Note: a visual estimation of the amounts based on the band intensities has proved to be insufficient in some cases.
[f] Alternatively a 'touch down PCR' gives even better results: 94 °C 1 min, 65 °C 4 min, 94 °C 1min, 60 °C 2 min, 72 °C 2 min, 94 °C 1 min, 55 °C 2 min, 72 °C 2 min, (94 °C 1 min, 50 °C 2 min, 72 °C 2 min) 20 times.
[g] Note: polypropylene tubes are recommended since phage may adsorb non-specifically to other plastics.

Protocol 5. Affinity selection of phage-displayed scFv (panning)[a]

Equipment and reagents

- Polypropylene plates[b]
- Polyethylene glycol (PEG 8000)
- PBS
- Tween 20
- BSA
- YT-AG and YT-AK media (see *Protocol 4*)

Method

1. Coat a 25 cm² culture flask with 5 ml of antigen (diluted in PBS to 10 μg/ml) for 1 h at 37 °C, 250 r.p.m.

2. Wash the flask three times with PBS.

3. Block remaining sites of the flask with PBS containing 2% (w/v) BSA for 1 h at room temperature.

4. Wash the flask three times with PBS.

5. Dilute the phage supernatant (from *Protocol* 4) 1:2 with PBS/2% BSA, incubate for 15 min at room temperature, and add to the coated flask. Incubate for 2 h at 37 °C.

6. Wash 20 times with PBS and 20 times with PBS containing 0.1% (v/v) Tween 20 to eliminate non-specifically bound phage.

7. Prepare *E. coli* TG1 log phase cells in 2 × YT medium and incubate with bound phage for reinfection for 1 h at 37 °C.

8. Plate 100 μl of bacteria on SOB-AG plates and grow overnight at 30°C.

9. Pick single recombinant clones and incubate in 200 μl 2 × YT-AG medium in a microtitre plate overnight at 30°C, 100–150 r.p.m.

10. Infect 20 μl of saturated cultures of single clones with 200 μl M13K07 helper phage (5 × 10[8] p.f.u./ml) for 2 h at 37°C, 100–150 r.p.m.[b]

11. Centrifuge the bacteria at 800 g for 10 min and resuspend the pellets in 2 × YT-AK medium. Grow overnight at 37°C, 100–150 r.p.m.

12. Repeat centrifugation at 800 g for 10 min. Remove phage supernatants of single clones and screen for binding to the antigen by ELISA. Detect antigen binding phage antibodies with POD-conjugated antibodies directed against the g8p major coat protein of M13.

[a] A phage titre of 10[9] p.f.u./ml is required for optimal results. The phage supernatants can be concentrated by PEG precipitation if the titre is too low.
[b] Note: use polypropylene plates or treat polystyrene plates with PBS containing 2% (w/v) BSA.

Protocol 6. Expression of soluble scFv

Equipment and reagents

- As in *Protocols 4* and *5*
- SOB-AG-N plates: SOB-AG plus 100 μg/ml nalidixic acid
- SB-AG medium: SB medium plus 100 μg/ml ampicillin plus 2% (w/v) glucose
- SB-AI medium: SB medium plus 100 μg/ml ampicillin plus 1 mM IPTG

Method

1. Prepare log phase *E. coli* HB2151 cells in 2 × YT medium (approx. 3 h at 37°C, 250 r.p.m. until A_{600} = 0.3–0.5), and infect with recombinant phage supernatant previously screened for binding to the antigen by ELISA.

2. Streak infected *E. coli* HB2151 cells on SOB-AG-N plates and grow overnight at 37°C.

3. Pick one single colony, resuspend it in 5 ml SB-AG medium, and grow overnight at 30°C, 250 r.p.m.

4. Add 50 ml freshly prepared SB-AG medium and incubate for 1 h at 30°C, 250 r.p.m.

5. Pellet the cells by centrifugation at 1500 g for 15 min, and resuspend the pellet in SB-AI medium. Induce expression for 3 h at 30°C with shaking.

6. Centrifuge at 1500 g for 15 min. Store supernatants at −20°C until use.

Protocol 6. *Continued*

7. Resuspend the cells in PBS plus 1 mM EDTA and incubate for 10 min on ice. Centrifuge at 15 000 *g* for 20 min and store the supernatant at −20 °C. This fraction represents the periplasmic fraction.

8. To prepare whole cell extracts, resuspend the cells in PBS, boil for 5 min, and centrifuge at 15 000 *g* for 20 min. Remove the supernatant representing the whole cell extract and store at −20 °C.

9. Check extracellular, periplasmic, and whole cell extracts for the presence of soluble scFv by Western blot and ELISA.[a]

[a] Use the anti-Etag antibody (Pharmacia) for detection which recognizes the Etag peptide fused to the scFv protein. (Only the periplasmic and extracellular fraction may contain functional soluble scFv of about 30 kDa.)

7. Testing antisera

Antisera are first tested for antibodies against the peptide by enzyme-linked immunosorbent assay (ELISA), and then for recognition of the protein by one of the techniques described below. Of course, these techniques are applicable to any antiserum.

7.1 ELISA

The ELISA is described in detail in Chapter 13. For the present pupose it is important to consider that the antiserum raised against a peptide conjugate contains antibodies against the peptide, the carrier, and in some cases, against the coupling reagent. (This latter case must be considered when monoclonal antibodies are to be selected.) Therefore, when screening by ELISA, the peptide is coupled to a 'second carrier' that is different from that used for immunization, for example, ovalbumin instead of BSA (if direct coupling of the peptide to the ELISA plate is not possible). In this way, antibodies against the initial carrier will not react. Additional controls include:

(a) The second carrier protein, either native or treated with the coupling reagent.

(b) If available, a different peptide.

(c) Pre-immune serum (collected from the animal prior to immunization).

(d) Immune serum in the presence of peptide (competitive ELISA) which should inhibit or reduce binding of the first antibody.

This latter control is unique to anti-peptide antibodies and is applicable to other detection methods too (2,36).

7.2 Testing for antibodies recognizing the protein

As mentioned above, antibodies against the peptide will not necessarily react with the protein. Testing for antibodies which do so is not simple if the protein in question is not known, especially if it is a rare species. It is therefore recommended to employ more than one of the available techniques to ascertain the specificity of the antibodies and, at the same time, the existence of the protein searched for. These techniques are described in Sections 9 and 10.

There are two principal ways to detect an unknown protein by immunological means: in the direct way, the antigen is detected and/or purified, i.e. by immunoprecipitation or immune affinity chromatography in conjunction with SDS–PAGE. In the indirect way, the existence of the antigen is deduced from its reaction with specific antibodies, as in ELISA, immunofluorescence or immunohistochemistry, or 'Western analysis'. The choice of a particular technique depends of course on the specific problem and the specificity and sensitivity required.

8. Detection of proteins in cell extracts

A particular protein may be first detected in or isolated from whole cell extracts, because soluble and insoluble, i.e. structure-associated proteins, can be analysed simultaneously. Subsequently, it may be desirable to employ cell fractionation procedures to achieve a higher degree of purification or to obtain further information about the subcellular localization, distribution, or turnover between different compartments. Some proteins might be tightly associated with insoluble structures such as cytoskeleton, chromatin, or nuclear matrix. In those cases special extraction conditions are required. In any case, the method of extraction depends on the type of detection and analysis to be used.

8.1 Preparation of whole cell extracts

Cells and tissues can be lysed directly in SDS sample buffer (see *Protocol 7*), which will result in complete denaturation and solubilization of all proteins. The protein mixture is boiled and applied directly to SDS poyacrylamide gels (see Chapter 1). Subsequent detection may be possible by Western analysis (see Section 9.1, *Protocol 12*) or, after partial renaturation, by immunoprecipitation (see Section 9.2, *Protocol 13*).

Protocol 7. Preparation of whole cell extract

Reagents

- SDS lysis buffer (same as sample buffer for SDS–PAGE): 0.125 Tris pH 6.8, 3.3% (w/v) SDS (Bio-Rad), 20 mM DTT (or 5% (v/v) 2-mercaptoethanol)
- PBS or TBS

Protocol 7. *Continued*

Method

1. Wash the cells two or three times with PBS or TBS at room temperature.

2. Remove the buffer completely.

3. Add 0.5 ml of SDS sample buffer per 9 cm dish.[a]

4. Distribute over the dish and incubate at room temperature for 5 min.[b]

5. Transfer the lysate to a glass tube either with an inverted pipette or by decanting.

6. Sonicate for 15 sec.

7. Boil for 5 min.

8. Add glycerol to 10% (v/v) and bromphenol blue to 0.1% (w/v).

9. Centrifuge at 13 000 g for 3 min.

10. Apply to a polyacrylamide gel.

[a] Cells growing attached to a Petri dish can be lysed directly on the plate as described above. For tissues or suspension cultures that are collected by centrifugation, it is advisable to resuspend the cell pellet first in Tris buffer without SDS and DTT and then to add both afterwards; otherwise a sticky insoluble pellet will form and lysis will be incomplete. Instead of sonicating the extract, the DNA can also be removed by ultracentrifugation at 100 000 g for 30 min.
[b] During this time, a viscous lysate will form due to denaturation of the chromatin.

8.2 Preparation of 'soluble' cell extracts

The most convenient way to prepare soluble cell extracts is to use an isotonic buffer containing non-ionic detergents such as Nonidet P-40 or Triton X-100. The detergent solubilizes extra- and intracellular membranes, and all soluble proteins, including nucleoplasmic proteins, are released. Usually, the lysis buffer is supplemented with reducing agents, such as dithiothreitol (DTT), and protease inhibitors to prevent degradation of proteins during extraction. The compositions and preparation of some standard extraction buffers are in the respective protocols. To further extract insoluble structures, either high salt or denaturing detergents are employed (see *Protocols 8–9*).

More rigorous extraction buffers containing deoxycholate and SDS (such as RIPA buffer, *Protocol 13*) solubilize the cytoskeleton and nuclear structures. In this way, extraction is more complete but the release of the chromatin results in a high viscosity that might reduce the efficiency of the subsequent immunoprecipitation (see Section 9.2).

Protocol 8.   Preparation of soluble cell extracts

Reagents

- Isotonic lysis buffer: 10 mM Tris–HCl (or NaPO$_4$) pH 8,[a] 140 mM NaCl, 3 mM MgCl$_2$, 1 mM DTT, 0.5% Nonidet P-40, one or more of the following protease inhibitors: 50 μM leupeptin, 1 mM phenylmethylsulfonyl fluoride (PMSF), 10 μg BPTI/ml[b]
- PBS or TBS

A. *Monolayer cultures*

1. Place the cells on ice.

2. Remove the culture medium and wash three times with ice-cold TBS or PBS; 3 ml per wash per 9 cm dish is sufficient.

3. Remove the buffer completely and add 0.5–1 ml of lysis buffer per 9 cm dish.[c]

4. Incubate for 15 min on ice.

5. During this time, scrape off the lysed cells with a rubber policeman and transfer to a centrifuge tube.

6. Centrifuge at approx. 800 *g*, 4 °C for 3 min to sediment the nuclei.

7. Remove the supernatant, which represents the 'soluble cell extract' and contains the soluble cytoplasmic as well as nucleoplasmic proteins; the nuclei may be discarded or further extracted (see *Protocol 9*).

B. *Suspension cultures[d]*

1. Sediment the cells by low speed centrifugation (about 800 *g*) at room temperature for 3–10 min, depending on the total volume.

2. Place on ice and wash twice with 0.1 of the original volume of ice-cold PBS.

3. Remove PBS completely and resuspend the cells in lysis buffer, about 1–2 × 10^7 cells/ml.[e]

4. Continue as in part A, steps 4–7.

[a] The pH may be as high as 9, which may facilitate extraction of nuclear proteins; moreover, some proteases may be inactive at higher pH.
[b] When phosphoproteins are to be analysed, sodium phosphate buffer should be used instead of or in addition to Tris, to inhibit phosphatases; also, 10 mM NaF, NaVO$_4$, or NaMoO$_4$ may be included.
[c] This is just sufficient to cover the dish.
[d] Note: this procedure can also be used to prepare concentrated extracts from monolayer cultures.
[e] This is the maximum to obtain complete lysis; for higher concentrations add more detergent.

8.3 Extraction of nuclei

If a protein is not detected in a soluble extract, it might represent a rare species that is not detectable in the bulk of total protein, or it might be tightly associated with nuclear structures, such as chromatin or nuclear matrix, and resist normal extraction. Thus, one should either prepare a whole cell extract and perform a Western analysis, or the nuclei may be re-extracted by nuclease treatment and/or by using buffers with high salt or different detergents as described in *Protocol 9*. This sequential extraction has the advantage that proteins can be enriched within a certain subcellular fraction.

Protocol 9.   Extraction of nuclei[a]

Reagents

- DNase I
- RNase A
- Isotonic lysis buffer (see *Protocol 8*)
- Matrix buffer: 10 mM Tris, 140 mM NaCl, 0.5% SDS, 10 mM *N*-ethylmaleimide, 1 mM DTT, and 10 μg/ml BPTI
- Buffer B: lysis buffer without NaCl
- Buffer C: 10 mM Tris pH 7.4, 10 mM NaCl, 3 mM $MgCl_2$, containing 1% (v/v) Tween 20 and 0.5% (v/v) sodium deoxycholate
- Empigen BB (Calbiochem-Novabiochem)

A. *Method 1*

1. Resuspend the nuclei in 0.2 vol. of isotonic lysis buffer.
2. Add NaCl to a final concentration of 0.5–1 M and mix immediately by vortexing.[b]
3. Dilute three- to fivefold with buffer B and centrifuge immediately at 13 000–100 000 g, 4°C, for 5–10 min to sediment the chromatin.
4. Remove the supernatant carefully and proceed as with the normal extract.[c]

B. *Method 2*[d]

1. Resuspend the nuclei in buffer C.
2. Vortex and centrifuge at 2500 g for 5 min; the supernatant contains the cytoskeleton.
3. Add 50 μg/ml each of DNase I and RNase A and incubate at 37°C for 15 min.
4. Centrifuge at 3000 g for 5 min and remove the supernatant containing the 'chromatin fraction'.
5. Resuspend and extract the residual matrix fraction in matrix buffer.

[a] There are many different ways to extract nuclei—two versions are given. The first leads to a general nuclear extract that contains cytoskeletal proteins in addition, but might still lack some nuclear proteins that are very tightly associated with chromatin. The second procedure leads to fractionation of the nuclei.
[b] This results in immediate dissociation of nuclear structures, including chromatin.
[c] Note: instead of increasing the salt concentration, one can also add additional detergents (e.g. 1% sodium deoxycholate, 0.1% SDS, or 2%. Empigen BB (37).
[d] See ref. 23.

8.4 Opening cells by 'Douncing'

In cases where cytoplasmic or membrane fractions are to be isolated or where enzymatic assays are to be performed, the use of detergents should be avoided. Cells are first broken by Douncing in hypotonic buffers in which cells swell and become very fragile against shearing forces. Then further fractionation might be employed. The procedure is described in *Protocol 10*.

Protocol 10. Extraction by Dounce homogenization

Equipment and reagents

- Hypotonic buffer: 20 mM Hepes–KOH pH 7.5, 5 mM KCl, 1.5 mM MgCl$_2$, 1 mM DTT, 1 mM PMSF[a]
- Dounce homogenizer of appropriate size, with tight-fitting pestle

Method

1. Wash the cells twice with ice-cold PBS and scrape them off the plate with a rubber policeman.

2. Centrifuge at 1500 r.p.m. at 4°C for 3 min.

3. Resuspend in hypotonic buffer and centrifuge again.

4. Resuspend in hypotonic buffer at 2–5 × 10^7 cells/ml.

5. Incubate on ice for 10 min; during this time transfer the suspension to a Dounce homogenizer.

6. Homogenize the cells with a tight-fitting pestle by 5–20 strokes.[b] Alternatively, the cells can be disrupted by passing them 10–20 times through a 28 gauge needle.[c]

7. Remove the nuclei and cell debris by centrifugation at 800 *g*, 4°C.

8. The supernatant (cytoplasmic fraction) and the pellet (nuclear and membrane fractions) may be further fractionated or processed as desired.

[a] Other protease inhibitors might be added.
[b] This is the most critical step, which should be exercised with care for reproducible results. The disruption of cells should be monitored by inspection in a phase-contrast microscope; homogenization is stopped when about 90% of the cells are disrupted, otherwise, the nuclei will also break.
[c] Ref. 38.

8.5 Cell fractionation

Initial lysis of cells with mild detergents such as Nonidet P-40 or Triton X-100 allows separation of a soluble 'cytoplasmic' and an insoluble nucleus-associated fraction. When standard procedures are employed (as in *Protocol 8*), the soluble 'cytoplasmic' fraction also contains nucleoplasmic proteins that leak out of the residual nucleus during extraction. The extent of leakage depends on the type and concentration of detergent and the incubation time. Opening of cells by Douncing minimizes nuclear leakage. However, the procedure is time-consuming and requires some experience. In particular, when radio-labelled cells have to be extracted, Douncing is not practical. Therefore, fractionation procedures have been developed that are based on sequential detergent extractions. For such special purposes, see ref. 39 (separation of cytoplasmic and nuclear fractions) or ref. 37 (separation of nuclear sub-fractions; *Protocol 9*).

8.6 Radiolabelling of cells

Radiolabel still provides the highest sensitivity, although the introduction of chemiluminescence or the use of various sandwich techniques has greatly improved the sensitivities of other techniques. Moreover, the investigation of certain questions (e.g. to trace the fate of a protein or to determine its half-life) depends on the ability to label a protein radioactively. Radiolabelling of proteins is most efficient with [^{35}S]methionine, which can be obtained in high specific activity, has a relatively long half-life, and has an energy sufficient for easy detection. In those rare cases where a protein does not contain methionine or cysteine, another essential amino acid that is readily taken up by the cell (e.g. leucine or arginine) may be used. Phosphoproteins may be labelled with ^{32}P$_i$ or ^{33}P$_i$, (see *Protocol 11*).

Protocol 11. Radiolabelling of cells[a]

Equipment and reagents
- Facilities and permission for work with radioactivity
- Dialysed FCS
- Deficient media, such as DMEM without methionine, etc.

Method

1. Remove the medium and wash twice with pre-warmed deficient medium (medium lacking the compound supplemented in radioactive form).[b]

2. Add a small volume (1.5 ml per 9 cm dish) of deficient medium containing dialysed FCS.[c]

3. For very short pulses (e.g. 5 min), pre-incubate the cells in deficient medium for 30 min (or longer) to use up or reduce the endogenous pools.[d]

4. Add radioactive compound (we use about 9 MBq of [^{35}S]methionine, or up to 37 MBq of either ^{3}H-labelled amino acids or ^{32}P$_i$ per 9 cm plate.[e]

---

5. Incubate for the desired time, while gently rocking.

6. If pulse labelling is followed by a chase, remove the radioactive medium (dispose of properly!), wash twice with complete medium (containing excess of the unlabelled compound used for labelling), and incubate further for the desired time.

7. Harvest as described in *Protocol 8.*

---

[a] Radiolabelling of cells can be performed for a short time (pulse labelling) in deficient medium or for long time periods (under steady state conditions) in complete medium. In deficient media, the specific activity of the labelled protein will be higher, but after extended incubation, the overall rate of protein synthesis decreases, which may result in distortions of the protein metabolism. This protocol is for pulse labelling of monolayer cultures.

[b] Note: some cells are very sensitive against deficient media and easily detach from the plate! In this case, add dialysed FCS.

[c] Dialysed against TBS to remove endogenous amino acids or phosphate, etc.

[d] This is also recommended for ^{32}P-labelling shorter than 2 h.

[e] Note: some amino acids are in ethanol and might need to be lyophilized prior to use.

---

9. Evaluation of detection methods

The detection of proteins within cell extracts can be achieved by Western blotting or immunoprecipitation.

9.1 Western blotting

Western analysis is certainly the first choice when searching for an unknown protein. In principle, a protein mixture from a cell extract is applied to denaturing SDS–PAGE (see Chapter 1) and electrophoretically transferred to a membrane (nitrocellulose or nytran, etc.). Subsequently, the membrane is reacted with the first antibody, that directed against the protein (or a peptide derived from it), and then with a second antibody that is directed against the first one and to which an enzyme, such as peroxidase or alkaline phosphatase, is covalently linked. A colour reaction carried out by the enzyme will indicate the position of the protein (40). The method is also applicable to two-dimensional gel electrophoresis.

The method has several advantages:

(a) Proteins can be extracted from tissue culture cells as well as from tissues.

(b) They need not be labelled.

(c) No special precautions are required to maintain the native conformation of a protein, so rigorous extraction procedures can be employed.

(d) Regions of a protein that are buried in the native conformation might be exposed upon denaturing electrophoresis; this is a special concern if anti-peptide antibodies are used.

Prerequisites for Western analysis are that the protein is present in sufficient amounts, that there is not too much cross-reaction by the antibody, and that interaction with the antibody is not destroyed by denaturation. Thus, the greatest sensitivity should be used initially. The sensitivity can be increased in several ways:

(a) With the streptavidin–biotin system, which leads to an amplification of the signal (41).

(b) With enhanced chemiluminescence (ECL), in which chemical energy is converted to light, which in turn can be monitored by exposure to X-ray films for various lengths of time (42,43).

(c) Instead of using an enzymatic reaction for detection, radiolabelled second antibody, biotin, or protein A can be employed.

To work out the best conditions and to include the proper controls, it is favourable to run cell extract proteins on a whole slab gel and transfer them to the membrane. A strip of the gel and the membrane are stained with Coomassie blue or Ponceau, respectively, to monitor the efficiency of the transfer and eventually to determine the positions of marker proteins. The rest of the membrane is cut into narrow strips that can be individually reacted under various conditions. Details are given in *Protocol 12*); another procedure is given in Chapter 5, *Protocol 8*.

Protocol 12. Electroblotting and Western analysis[a]

Equipment and reagents

- Power supply with high current output (> 500 mA, 100 V)
- Transfer chamber
- Holders (plastic grids) to mount gel and filters
- Nitrocellulose membranes, e.g. BA85 (Schleicher & Schuell) or other membranes (Immobilon: Millipore); the latter have higher binding capacities than conventional nitrocellulose membranes
- Filter paper: 3MM (Whatman) or GB002 (Schleicher & Schuell)
- X-ray film (X-Omat AR; Kodak)
- Transfer buffer: 25 mM Tris–HCl pH 8.3, 150 mM glycine
- Blocking buffer: 0.5% (v/v) Tween 20 in PBS
- Washing buffer: 0.05% (v/v) Tween 20 in PBS

- Peroxidase buffer: 50 mM sodium acetate pH 5
- Coomassie brilliant blue: 0.1% (w/v) in 10% acetic acid, 25% methanol
- Ponceau stain: 0.2% (w/v) Ponceau in 3% TCA
- Anti-Ig–peroxidase conjugate (alternatively anti-Ig–alkaline phosphatase) (Nordic Immunological Laboratories)
- 3-Amino-9-ethylcarbazole (AEC): 0.4% (w/v) stock solution in dimethylformamide
- H_2O_2
- Substrate solution: 0.5 ml 0.4% AEC in dimethylformamide, 9.5 ml sodium acetate, 10 μl of 30% H_2O_2
- Luminol, 4-iodophenol, H_2O_2 (for enhanced chemiluminescence) (ECL)

A. *Electrotransfer from gel to membrane*

  1. Cut out the membrane and the filter paper in the appropriate size.[b]

  2. Soak the nitrocellulose briefly in distilled water and then in transfer buffer.[c]

3. Soak the polyacrylamide gel in transfer buffer for 5 min.

4. Cut off one lane from the gel and stain with Coomassie blue.

5. Mount filter paper, gel, and membrane in the following order: plastic grid, scotch brite, filter paper, gel, membrane, filter paper, scotch brite, plastic grid.[d,e]

6. Assemble in a holder and mount into the transfer chamber with the anode on the side of the membrane.

7. Apply a current of 0.3 A and a voltage of 5–10 V/cm in the cold room for at least 4 h or overnight.

B. *Immunostaining*

1. Cut off a strip of the whole membrane and stain with Ponceau to ensure that the proteins have been properly transferred.

2. Wash the membrane briefly in transfer buffer.

3. Incubate in blocking buffer for at least 2 h.[f]

4. Wash briefly with distilled water.[g]

5. For further processing, cut the membrane into strips.

6. If the membrane has been dried, re-wet it in blocking buffer.

7. Incubate the individual strips with antiserum diluted in washing buffer[h] at room temperature for at least 2 h, while continuously agitating.

8. Wash three times with washing buffer for 15 min.

9. Incubate with anti-Ig–peroxidase diluted in washing buffer at room temperature for 30 min.[i]

10. Wash as in step 8.

11. Incubate in substrate buffer (e.g. 50 mM sodium acetate) for 10 min.

12. Prepare the substrate solution.

13. Incubate the membrane with substrate solution for up to 15 min.[j]

14. Stop the reaction with distilled water and dry the membranes.

---

[a] See ref. 40.

[b] Note: wear gloves and avoid contact and scratching of the membrane; handle with blunt-ended forceps.

[c] Immobilon membrane is hydrophobic and must be soaked in methanol for a few seconds, then for 5 min in water, and transfer buffer.

[d] Note: even a brief contact between gel and membrane will result in transfer of substantial amounts of protein; therefore, the membrane must be placed on the gel only once and in its final position.

[e] Remove any air bubbles between the different layers! They can be removed by rolling a sterile glass or plastic pipette over the membrane.

[f] This is to block the surface of the membrane against further non-specific binding; some investigators use gelatin, BSA, dry milk, or casein for blocking; we found PBS/0.5% Tween sufficient.

[g] At this stage, the membrane can be dried between two leaves of filter paper and stored, but the results are usually better if staining is continued.

Protocol 12. *Continued*

[h] The dilution of antiserum has to be determined, usually 1 : 20 to 1 : 200.

[i] The dilution of anti-Ig–conjugate has to be determined; is usually in the range between 1:500 and 1:5000. At this step, biotinylated anti-Ig, streptavidin, and biotin labelled peroxidase or radiolabelled biotin may be employed to increase the sensitivity.

[j] If chemiluminescence is employed, the membrane is immediately sealed in a plastic bag and exposed to X-ray film for various times (seconds to 1 h) (see manufactorer's instructions). The following controls may be included: if available, extract from cells that do not express the protein (e.g. uninfected versus virus-infected cells); pre-immune serum or serum with different specificity; if anti-peptide sera are employed, immune serum in the presence of excess peptide; as positive control, a known protein with its specific antiserum may be included or the antiserum is simply reacted with the second antibody–enzyme conjugate.

9.2 Immunoprecipitation

Usually, it is desirable to isolate a protein in its native conformation in order to perform additional biochemical analyses. This can be achieved by immunoprecipitation or by immunoaffinity chromatography. Both require that the cell extract be prepared so that the proteins remain in their native state. Subsequently, the protein is reacted with antiserum and immobilized by adsorption to an insoluble matrix, which can be washed and from which it can be eluted in the desired form, and either applied to various biochemical analyses or analysed by SDS–PAGE. The method is rapid and mild, the purity of the isolated protein may vary from 5–90%, depending on the abundance of the protein, and on the specificity and affinity of the antibodies.

The matrix may be fixed *Staphylococcus aureus* bacteria carrying a surface protein (protein A or G) to which certain antibody classes bind with their Fc domain (44). Cleaner results are usually obtained with purified protein A (or G) covalently coupled to Sepharose or agarose. Elution from the immune complex can be achieved with SDS sample buffer (for subsequent SDS–PAGE) or with buffers of low or high pH (2.5 or 10.8 respectively) or at high ionic strength, or with detergents like Empigen, which would dissociate the immune complex but not denature the purified protein, depending on the kind of subsequent analyses. The controls are similar to those described for Western analysis (see *Protocol 12*).

9.2.1 Buffers

Usually, detergent-containing buffers such as RIPA buffer (*Protocol 13*) are employed, first to keep the proteins soluble and secondly to minimize non-specific interactions. For low affinity antibodies, buffers with less detergent (0.05% Nonidet P-40) may be used. Buffers should contain reducing agents, protease inhibitors, chelators, or phosphatase inhibitors.

9.2.2 Antiserum

Antiserum is stored in aliquots at $-20\,°C$ or for an extended time at $-70\,°C$. Prior to use, it may be incubated once at $56\,°C$ for 30 minutes to inactivate proteases. This is usually not necessary for hybridoma supernatants containing monoclonal antibodies, but it should be checked. The amount of serum, or antibodies, required for complete precipitation of a particular protein has to be determined for each individual batch of serum. Using too much serum will result in high backgrounds and even reduced precipitation.

9.2.3 Immunosorbents: *S. aureus* cells and protein A–Sepharose

Fixed *S. aureus* bacteria can be purchased as a 10% suspension (Pansorbin, containing protein A, or Omnisorb, containing protein G; Calbiochem). Prior to use they are washed and then resuspended in the original volume. The amount required to bind the antibodies quantitatively (usually two- to three-fold excess over the serum) has to be determined by titration for each individual batch of both serum and bacteria. If the antibodies do not bind to protein A, or do so only poorly, protein G may be tried instead.

Alternatively, protein A–Sepharose (or protein G–agarose) may be used, which is more expensive but yields cleaner results. Protein A–Sepharose suspension is prepared in RIPA or lysis buffer (about 1:10, w/v) and allowed to swell for at least 30 minutes. The suspension is washed twice and kept as a 1:1 suspension (settled Sepharose : buffer) to facilitate pipetting. For storage at 4°C, add 3 mM NaN_3. For pipetting, cut off the tip of the micropipettes. Some classes of immunoglobulins bind poorly to protein A, depending on the Ig class and the species the antibodies are derived from. In this case, protein G–agarose may be used. Alternatively, one can covalently couple 'second antibodies' to protein A–Sepharose and use this as the matrix.

Protocol 13. Immunoprecipitation[a]

Equipment and reagents

- Pansorbin, or Omnisorb (Calbiochem), or protein A–Sepharose, or protein G–agarose (Pharmacia or Sigma)
- Correction mix: 2.5% (v/v) Nonidet P-40, 5% (w/v) sodium deoxycholate, 0.5% (w/v) SDS
- Washing buffer 2: 10 mM Tris–HCl pH 7.5

- Washing buffer 1 (RIPA buffer): 10 mM Tris–HCl pH 7.5, 140 mM NaCl, 1% (v/v) Nonidet P-40, 1% (w/v) sodium deoxycholate, 0.1% (w/v) SDS, 1 μg/ml BPTI[b]
- Sample buffer (see *Protocol 3*)

Method

1. Centrifuge the cell extract and the antiserum at 13000 *g*, at 4°C, for 10–30 min.[c]

2. Remove the supernatant carefully with a Pasteur pipette and transfer it to a siliconized glass tube or an Eppendorf tube.

3. Add 0.2 vol. of correction mix (to increase the concentration of detergents).

4. Add the antiserum in appropriate amounts, mix well, and incubate on ice for at least 60 min.

Protocol 13. *Continued*

5. During this time, wash Pansorbin (or prepare protein A–Sepharose) as described below.[d]

6. Add immunosorbent and incubate at 4°C for 60 min.[e]

7. Wash three times with washing buffer 1.[f]

8. Wash once with washing buffer 2 (this is to remove the detergents, deoxycholate in particular, which would reduce the quality of SDS–PAGE).

9. Add a small volume (30–50 μl) of sample buffer, resuspend, and heat to 60°C for 5 min to dissociate the immune complex.

10. Sediment the immunosorbent and remove the supernatant containing antigen and antibodies in dissociated form.

11. Boil for 3 min for complete denaturation and reduction.[g]

12. Add 10% (v/v) glycerol containing 0.1% (w/v) bromphenol blue.

13. Apply to an SDS–polyacrylamide gel (see Chapter 1), or store at −20°C.

14. Before loading on a gel, centrifuge the sample at 13 000 g for 3 min.[h]

---

[a] Ref. 45.

[b] For low affinity antibodies, buffers with reduced detergent concentrations (e.g. 0.05% Nonidet P-40) are recommended.

[c] This is to remove all insoluble and particulate material, which would otherwise be co-precipitated and cause high backgrounds!

[d] Washing can be done in siliconized glass tubes, centrifugation at 4°C, 1600 g for 5–20 min, depending on the volume; alternatively in Eppendorf tubes in a microcentrifuge at 10 000 g for 30–60 sec.

[e] Note: if Sepharose or cellulose is used, the reaction mixture has to be agitated continuously.

[f] See comment in footnote [d]; take care that the pellet is resuspended completely after each washing step, otherwise high backgrounds are obtained.

[g] Note: in some cases the protein should be alkylated to prevent re-oxidation of the sulfydryl residues; then 1/3 vol. of a 2.5% (w/v) N-ethylmaleimide solution is added after boiling.

[h] After SDS electrophoresis, radiolabelled proteins are visualized by fluorography or autoradiography; unlabelled proteins are stained with Coomassie blue or silver stain; the latter is about ten times more sensitive (see Chapter 1).

10. Detection of proteins *in situ*

The detection of proteins *in situ*, that is in the cells where they are expressed, offers the opportunity to investigate in what fraction of a tissue or a cell culture and in which cellular compartment a particular protein is located. In principle, the cells or tissue sections are fixed and reacted with the first antibody specific for the protein in question; then like in Western or ELISA analyses, a second antibody directed against the first one is used to which is linked either a fluorescent dye such as rhodamine-β-isothiocyanate or fluorescein isothiocyanate (FITC) for detection by a fluorescence micro-scope, or an enzyme such as alkaline phosphatase or peroxidase, which catal-yse coloured reactions.

10.1 Immunofluorescence

This method is not as sensitive as Western analysis or immunoprecipitation of radiolabelled protein, because a particular protein may be distributed over the entire cell and may require a certain threshold for detection, whereas in the latter cases a relatively large amount of the protein is concentrated in a small area of a membrane filter or a gel. On the other hand, immunofluorescence offers some unique possibilities such as investigation of co-localization of two putatively interacting proteins within the cell, by using anti-Ig antibodies carrying different fluorescent labels (e.g. rhodamine, FITC), or to determine the orientation of a given protein within the membrane by using anti-peptide antibodies generated against different domains of the protein. For immunofluorescence, cells may be grown on plastic dishes for immediate inspection or on glass coverslips, which can be conserved. For cells growing in suspension, special adhesion slides, i.e. polylysine coated, are available (Sigma). Cells can be fixed either with glutaraldehyde for surface staining or with methanol/acetone, which leads to permeabilization of the plasma membrane and staining of intracellular components. Since surface staining of cells requires some experience, only internal staining will be described (see *Protocol 14*). The controls are similar to the other detection methods described above. The sensitivity can be increased with the biotin–streptavidin system and by using phycoerythrin as fluorescent dye.

Protocol 14. Immunofluorescence[a]

Reagents

- PBS
- Methanol/acetone, 1:1 (v/v), stored at −20°C
- Anti-Ig–FITC (antibody against primary antiserum, fluorescently labelled with FITC)[b]

Method

1. Wash the cells twice with ice-cold PBS.

2. Remove the buffer completely.

3. Add the methanol/acetone mixture (−20°C) and incubate at −20°C for 5 min.

4. Remove the fixation mixture and leave to air dry.[c]

5. For subsequent staining, mark a small area on the plate.

6. Put a circular wet filter paper on the top of the Petri dish to give a humid atmosphere.

Protocol 14. *Continued*

7. Centrifuge the antiserum and the anti-Ig–FITC conjugate at 4°C, 13 000 *g* for 10 min.

8. Add the antiserum at the appropriate dilution on to the marked area.[d]

9. Incubate in a humidified chamber at 37°C for 1 h.

10. Wash with PBS and dry under an air current.

11. Incubate with the anti-Ig–FITC conjugate[e] at 37°C for at least 15 min.

12. Add PBS and observe directly in a fluorescence microscope.

13. If cover slips are used, wash them briefly with PBS, dry, and embed them inverted in Citifluor on a glass slide;[f] observe in oil.

As pointed out for the other indirect methods, the sensitivity can be increased by using biotinylated anti-Ig and FITC with streptavidin as bridge.

[a] Ref. 46.
[b] Or other fluorescent dye such as rhodamine.
[c] If the cells are grown on coverslips, these steps are carried out in a small tray; the plates or coverslips may be kept at this stage in a desiccator over $CaCl_2$ at −20°C.
[d] Dilution may be 20- to 500-fold in 0.5% Tween/PBS.
[e] Diluted as was the antiserum.
[f] Citifluor prevents bleaching of the fluorescence.

10.2 Other staining procedures

A number of staining procedures have been developed for detection of antigens in biological specimens. They are particularly useful for surface staining of cells or tissue sections. They are all based on the same principle as described for Western analysis and generally use the biotin–streptavidin system for amplification of the signal. The biotin may be radiolabelled, coupled with an enzyme such as peroxidase, or with colloidal gold. The sensitivity of peroxidase or gold staining may be further increased by additional silver staining (47) (see Chapter 1). Most of these detection systems are available as kits including detailed instruction manuals. Therefore, detailed protocols are not included here.

11. Further applications of antibodies

A number of further applications and investigations are provided by immunological tools, particularly with anti-peptide or monoclonal antibodies that are directed against a single, definable site on a protein. These methods are only briefly described to point out the possibilities. For detailed instructions, the reader is referred to the original references or more appropriate laboratory manuals in the *Practical approach* series or similar books.

11.1 Immunoaffinity chromatography

This technique provides a potent means for purification of proteins in a native and active form. The method can be used with anti-peptide or mono-clonal antibodies. In both cases, the antibodies are covalently coupled to a matrix; cell extract is passed over it in buffer of appropriate ionic strength and pH, and the desired protein will bind to it. Either batch or column pro-cedures are possible. Elution of the bound protein can be achieved in differ-ent ways. From an anti-peptide antibody matrix, the bound protein can be eluted with excess peptide under otherwise physiological conditions (48,49). From a monoclonal antibody matrix elution can be achieved by altering the pH (pH 2.5 or 10.8) or the salt concentration (50); this requires immediate neutralization to avoid irreversible denaturation.

11.2 Detection of nucleic acids

Several companies offer non-radioactive labelling kits for *in vitro* labelling of nucleic acids to circumvent the problems with radioactivity. Biotin or digoxi-genin labelled nucleoside triphosphates are incorporated into DNA or RNA *in vitro*, and the labelled nucleic acids can be used as probes for Southern or Northern analyses using antibodies against biotin or digoxigenin and second antibody–enzyme conjugates for detection. Antibodies against BrdU are also available, thus allowing incorporation of BrdU to be monitored *in vivo* and to detect actively proliferating cells within a population.

11.3 Assays for enzymatic or DNA-binding activities

From sequence data of protein families with common functions, such as DNA-binding proteins or protein kinases, some common structural features have been found that allow predictions as to whether the newly discovered protein belongs to this category. The prediction can be experimentally tested. The protein can be isolated from cell extracts by immunoprecipitation and assayed, for example, for protein kinase activity, either in an autophosphory-lation reaction or with heterologous substrates (51,52). Similarly, the immune complex can be incubated with DNA fragments that might contain binding sites for the precipitated protein. Those fragments bound by the protein can be eluted and analysed (53,54). A further possibilty is to employ antibodies in 'electrophoretic mobility shift assays'. In this assay, labelled oligonucleotides containing binding motifs are allowed to bind to their respective DNA-bind-ing protein and the free and protein-bound oligonucleotide are separated by native polyacrylamide gel electrophoresis; the DNA–protein complex exhibits a retardation of the electrophoretic mobility with respect to the free oligonucleotide. Antibodies against the binding protein may either interfere with its binding or cause a supershift of the mobility of the complex (55).

Once the structural and functional organization of a protein is elucidated,

Karl Heinz Scheidtmann et al.

one can induce antibodies against defined regions and perform functional studies with those antibodies, *in vitro*, as outlined above, or *in vivo,* by microinjection, etc. Moreover, such antibodies can provide information about the conformational alterations, in that some epitopes are accessible only in certain forms of the protein. For instance, a wide panel of antibodies has been generated and selected against the tumour suppressor protein p53, which can be used to distinguish wild-type and mutant forms of the protein in tumours or in biochemical studies (56,57).

References

1. Walter, G. and Doolittle, R. F. (1983). In *Genetic engineering: principles and methods* (ed. J. K. Setlow and A. Hollaender), Vol. 5, p. 61. Plenum, New York.
2. Walter, G., Scheidtmann, K. H., Carbone, A., Laudano, A. P., and Doolittle, R. F. (1980). *Proc. Natl. Acad. Sci. USA*, **77**, 5197.
3. Atassi, M. Z. (1975). *Immunochemistry*, **12**, 423.
4. Laver, W. G., Air, G. M., Webster, R. G., and Smith-Gill, S. J. (1990). *Cell*, **61**, 553.
5. Benjamin, D. C., Berzofsky, J. A., East, I. J., Gurd, F. R. N., Hannun, C., Leach, S. J., *et al.* (1984). *Annu. Rev. Immunol.*, **2**, 67.
6. Atassi, M. Z. (1980). *Mol. Cell Biochem.*, **32**, 21.
7. Hopp, T. P. (1986). *J. Immunol. Methods*, **88**, 1.
8. Kyte, J. and Doolittle, R. F. (1982). *J. Mol. Biol.*, **157**, 105.
9. Hopp, T. P. and Woods, K. R. (1981). *Proc. Natl. Acad. Sci. USA*, **88**, 3824.
10. Karplus, P. A. and Schulz, G. E. (1985). *Naturwissenschaften*, **72**, 212.
11. Chou, P. J. and Fasman, G. D. (1978). *Adv. Enzymol.*, **47**, 97.
12. Garnier, J., Osguthorpe, D. J., and Robson, B. (1978). *J. Mol. Biol.*, **120**, 97.
13. Rose, G. D., Gierasch, L. M., and Smith, J. A. (1985). *Adv. Protein Chem.*, **37**, 1.
14. Welling, G. W. and Fries, H. (1985). *FEBS Lett.*, **182**, 81.
15. Friedrich, U., Scheidtmann, K. H., and Walter, G. (1986). *Immunol. Lett.*, **12**, 207.
16. Welling, G. W., Weijer, W. J., van der Zee, R., and Welling-Wester, S. (1985). *FEBS Lett.*, **188**, 215.
17. Van Regenmortel, M. H. V. and de Marcillac, G. D. (1988). *Immunol. Lett.*, **17**, 95.
18. Borras-Cuesta, F., Fedon, Y., and Petit-Camurdan, A. (1988). *Eur. J. Immunol.*, **18**, 199.
19. Tam, J. P. (1988). *Proc. Natl. Acad. Sci. USA*, **85**, 5409.
20. Goodfriend, T. L., Levine, L., and Fasman, G. (1964). *Science*, **143**, 1344.
21. Carlson, J., Drevin, H., and Axen, R. (1978). *Biochemistry*, **173**, 723.
22. Smith, D. B. and Johnson, K. S. (1988). *Gene*, **67**, 31.
23. Sauter, M. and Müller-Lantzsch, N. (1987). *Virus Res.*, **8**, 141.
24. Maniatis, T., Fritsch, E. F., and Sambrook, J. (ed.) (1989). *Molecular cloning: a laboratory manual* (2nd edn). Cold Spring Harbor Laboratory Press, NY.
25. Glover, D. M. (ed.) (1985). *DNA cloning: a practical approach*. IRL Press, Oxford.
26. *Current protocols in molecular biology.* John Wiley, New York.

27. Better, M., Chang, C. P., Robinson, R. R., and Horwitz, A. H. (1988). *Science*, **240**, 1041.
28. Plückthun, A. and Skerra, A. (1989). In *Methods in enzymology* (ed. J. J. Langone), Vol. 178, p. 497. Academic Press, New York.
29. McCafferty, J., Griffiths, A. D., Winter, G., and Chiswell, D. J. (1990). *Nature*, **348**, 552.
30. Clackson, T., Hoogenboom, H. R., Griffiths, A. D., and Winter, G. (1991). *Nature*, **352**, 64.
31. Marks, J. D., Hoogenboom, H. R., Bonnert, T. P., McCafferty, J., Griffiths, A. D., and Winter, G. (1991). *J. Mol. Biol.*, **222**, 581.
32. Griffiths, A. D., Malmqvist, M., Marks, J. D., Bye, J. M., Embleton, M. J., McCafferty, J., *et al.* (1993). *EMBO J.*, **12**, 725.
33. Hoogenboom, H. R., Griffiths, A. D., Johnson, K. S., Chiswell, D. J., Hudson, P., and Winter, G. (1991). *Nucleic Acids Res.*, **19**, 4133.
34. Orlandi, R., Gussow, D. H., Jones, P. T., and Winter, G. (1989). *Proc. Natl. Acad. Sci. USA*, **86**, 3833.
35. Ward, E. S., Gussow, D., Griffiths, A. D., Jones, P. T., and Winter, G. (1989). *Nature*, **341**, 544.
36. MacArthur, H. and Walter, G. (1984). *J. Virol.*, **52**, 483.
37. Staufenbiel, M. and Deppert, W. (1984). *J. Cell Biol.*, **98**, 1886.
38. Ludlow, J. W. (1992). *Oncogene*, **7**, 1011.
39. Schickedanz, J., Scheidtmann, K. H., and Walter, G. (1986). *Virology*, **148**, 47.
40. Towbin, H., Staehelin, T., and Gordon, J. (1979). *Proc. Natl. Acad. Sci. USA*, **76**, 4350.
41. Bayer, E. A. and Wilchek, M. (1980). *Methods Biochem. Anal.*, **26**, 1.
42. Bronstein, I. and McGrath, P. (1989). *Nature*, **338**, 599.
43. Schaap, A. P., Akhaven, H., and Romano, L. J. (1989). *Clin. Chem.*, **35**, 1863.
44. Goding, W. E. (1978). *J. Immunol. Methods*, **20**, 241.
45. Mann, K., Hunter, T., Walter, G., and Linke, H. K. (1977). *J. Virol.*, **24,** 151.
46. Deppert, W. and Walter, G. (1982). *Virology*, **122**, 56.
47. Gallyas, F., Görcs, T., and Merchenthaler, I. (1982). *J. Histochem. Cytochem.*, **30**, 183.
48. Scheidtmann, K. H., Hardung, M., Echle, B., and Walter, G. (1984). *J. Virol.*, **50**, 1.
49. Grussenmeyer, T., Scheidtmann, K. H., Hutchinson, M. A., Eckhart, W., and Walter, G. (1985). *Proc. Natl. Acad. Sci. USA*, **82**, 7952.
50. Simanis, V. and Lane, D. P. (1985). *Virology*, **144**, 925.
51. Eckhart, W., Hutchinson, M. A., and Hunter, T. (1979). *Cell*, **18**, 925.
52. Müller, E., Boldyreff, B., and Scheidtmann, K. H. (1993). *Oncogene*, **8**, 2193.
53. McKay, R. D. G. (1981). *J. Mol. Biol.*, **145**, 471.
54. Hinzpeter, M., Fanning, E., and Deppert, W. (1986). *Virology*, **148**, 159.
55. Hupp, T. R., Meek, D. W., Midgley, C. A., and Lane, D. P. (1992). *Cell*, **71**, 875.
56. Hall, A. R. and Milner, J. (1990). *Oncogene*, **5**, 1683.
57. Zhang, W., Hu, G., Estey, E., Hester, J., and Deisseroth, A. (1992). *Oncogene*, **7**, 1645.

4

Identification of common post-translational modifications

RADHA G. KRISHNA and FINN WOLD

1. Introduction

Considering that some of the 20 primary, encoded amino (imino) acids can be converted to something close to 200 secondary amino acids in post-translational reactions (1), it should not be surprising to find that the number and complexity of the methods involved in recognizing and characterizing these secondary amino acids are quite impressive. The chemical properties of the secondary amino acid derivatives vary broadly and complicate any attempts to extract some general rules and approaches.

The first logical step must be to establish that an unusual amino acid indeed is present in the protein. The incentive to look for the modification may come from predictions based on specific structural features observed in the cDNA sequence, from unusual structural features observed in the protein, or directly from serendipitous observations by alert investigators. The next two steps, which can be taken in any order, are to determine the derivative's chemical structure and to establish its location in the polypeptide chain. It would be nice if there were a single procedure for each of these steps; these three procedures could be presented, and this chapter would be done. Obviously, it is not that simple, and the main purpose of this chapter is to attempt to seek general methods or common principles that may apply to different methods. Right from the start, it needs to be stated that to accomplish the complete characterization according to the above goals, some fairly elaborate equipment will be required. As a minimum, amino acid analysis and amino acid sequencing are obligatory procedural components. These are most obviously carried out with the instruments dedicated to those purposes, but newer methods and applications, such as mass spectroscopy (see Chapter 2) and NMR, also requiring sophisticated instrumentation are supplementing and even replacing the original methods. It is probably not realistic to tackle the characterization of most post-translational modifications without having access to this type of instrumentation.

2. General considerations

What kind of general methods do we consider in exploring the presence and the identification of post-translationally modified amino acid residues in a protein?

2.1 Hydrolysis and amino acid analysis

The polypeptide chain can be hydrolysed and the modified amino acids identified by any one of several procedures. The standard procedure for peptide bond cleavage is hydrolysis with 6 M HCl at 110 °C for 20–24 h. Asn and Gln residues are completely converted to Asp and Glu, and, unless special precautions are taken by adding phenol and/or reducing adents, Trp is also destroyed, probably through chlorination. For special cases, hydrolysis in 5 M NaOH (or Ba(OH)$_2$) at 110 °C for 5–20 h can also be used, but it is not a very common procedure since Ser, Thr, Cys, and Arg are destroyed during hydrolysis, in addition to Asn and Gln. Standard acid hydrolysis will destroy a large number of derivatized amino acids, precluding the use of this procedure in the isolation of many post-translational products. The digestion of proteins with proteases can also be used and will leave the derivatives intact, so it is the method of choice, even if it may be difficult to achieve total hydrolysis of some proteins.

With the free amino acids and derivatives in hand, the next problem is to recognize the derivative. If the modification represents a single residue in a 50 kDa protein, for example, the main problem will be to establish analytical procedures that will pick up the single residue among the 400–500 other amino acid residues. The analysis these days is likely to involve automated chromatographic procedures developed for optimal recovery and resolution of the normal, primary amino acids, and if the properties of the derivative are similar to those of the other amino acids, the derivative can be observed; if it is very different, it may not appear within the limits of the chromatogram. So there are two problems to be aware of in studying acid hydrolysates: the stability of the derivative and its behaviour in the standard analysis procedures. Once some information regarding the chemical properties of the amino acid derivative becomes available, it may be possible to modify the hydrolysis conditions and the analytical procedures to permit the direct identification of the derivative in the hydrolysate.

2.2 Protein sequencing

Standard Edman peptide sequencing is another obvious approach toward recognizing the presence of a derivatized amino acid and, at the same time, establishing its location in the polypeptide chain. Again there are problems with the general applicability of the method. Since a free α-amino group is required for the reaction with phenylisothiocyanate in the first step of the sequencing reaction, most of the derivatives of the N terminus that do not have such a free amino group (N^α-monoalkyl derivatives do react with isothiocyanate) cannot be analysed by this procedure. In the sequencing of a protein or peptide, the identification of the amino acid derivative obtained in each cycle requires that the anilinothiazolinone (ATZ) derivative can be extracted into the organic phase, converted to the phenylthiohydantoin (PTH) derivative, and then observed on the standard HPLC chromatogram. Highly hydrophilic derivatives, such as phosphorylated or glycosylated amino acids, will fail on the first criterion, some derivatives are destroyed during certain standard conversion conditions (PTH-Ser, PTH-Thr, PTH-Cys are generally identified as dehydrated decomposition products) and an unknown derivative could also be destroyed and be missed because of this step, and a large number of the normally extracted and converted derivatives will be missed because they elute outside the monitored region of the standard chomatogram. Again, it is possible to modify both the sequencing conditions and the analytical procedures to broaden the spectrum of derivatives that can be detected and, if proper standards are available, identified by the sequencing procedures.

2.3 Mass spectrometry of modified proteins, peptides, and liberated amino acids

With the improved accessibility and ease of operation of mass spectrometry (MS) facilities, MS analysis has become the most powerful and universally applicable approach to the characterization of modified amino acids in proteins (see Chapter 2). If we know the exact amino acid composition of a protein (from cDNA analysis) or of a peptide (from direct amino acid analysis), the theoretical mass can be compared with the one observed by MS, and any difference may then suggest the presence of a post-translational modification. Time-of-flight (TOF) instruments based on electrospray (ES) or matrix-assisted laser desorption (MALDI) techniques permit the mass determination of proteins of up to 100 kDa and the detection of derivatized amino acids in terms of unexpected mass discrepancies at the level of the intact protein. With a mass resolution of about 0.1%, small discrepancies would not be readily detected in the larger proteins. However, by analysing small, proteolytically produced peptides with known amino acid composition, either by the same TOF methods or by fast atom bombardment (FAB) MS, better than unit mass resolution can be achieved, and the mass discrepancies can be determined very precisely and used along with other chemical (stability, chromatographic behaviour) and physical (absorbance and fluorescence) properties for definitive identification of the derivative. MS is the topic of Chapter 2 and will not be discussed further here.

Concentrating on these three general approaches, it may be useful to review briefly how a number of post-translationally modified amino acids would fare when subjected to the common hydrolysis/analysis, sequencing, and mass determination. A list of known amino acid derivatives, selected to illustrate the variety of derivatives that can be found in proteins, is given in *Table 1* along with some of their relevant properties. The references in *Table 1* have been selected to include either recent reviews or methodological descriptions and may not properly acknowledge the discoverers of the various derivatives. The delta mass data are taken from ref. 2. A much more complete tabulation, compiled and maintained by Dr Ken Mitchelhill, University of Melbourne, can be found on the following home page: `http://www.medstv.unimelb.edu.au/WWWDOCS/SVIMRDocs/MassSpec/Delta-MassV2.html`

3. Specific cases

In the following, a few fairly common post-translational modifications will be considered in some detail. D-Amino acids need to be mentioned as unique products. The recognition and identification of glycans in glycoproteins is a quite common problem, and recent developments in both protein and glycan analysis have provided the necessary methodology to tackle the problem in a fairly straightforward manner. The blocked N terminus is another rather vexing problem, for which there is no simple, direct solution at this time. One procedure will be described as a possible method and as an illustration of the problems involved. Protein phosphorylation is among the oldest and best studied modification reactions and will be considered briefly as a mostly chemical characterization. One of the most common post-translational modifications, the formation of disulfide bonds is discussed in Chapter 7. The need for major instrumentation is quite obvious in the characterization of most of the derivatives; in many instances the experimental procedures are quite complex and, as in the case of blocked N termini, may not always yield clean answers.

3.1 D-Amino acids

Some derivatives are not readily recognized by any of the standard methods; the most obvious examples are the D-amino acids. It is becoming clear that, in addition to the slow, non-enzymatic production of D-amino acids that has been observed in long-lived proteins, especially for Asp and Ser, several other amino acids, such as Ala, Ile, Leu, Met, and Phe are converted relatively rapidly from the L- to the D-enantiomer in some proteins (95). These proteins have been shown to be derived from high molecular weight precursors synthesized on polysomes; analysis of the cDNA of the precursor shows that the D-amino acids are encoded as the normal L-enantiomers. In some cases it appears that the conversion is complete, in others a racemic mixture is observed. The D-amino acids typically appear in the second position of shorter peptides derived from large precursor proteins, and, although there is

Table 1. Some properties of post-translational derivatives

Derivative (ref.)	Derived from[a]	Delta mass[b]	Stability[c]	
THE N-TERMINUS: H_2N-				
N-Pyrrolidone carboxyl- (3)	Gln	-17	A$-$, B$-$	
N-α-Ketoacyl- (4)	Ser	-16	A$-$, B$-$	
N-Methyl- (5)		14	A$+$	P$+$
N-Formyl- (6)		28	A$-$, B$-$	
N-Acetyl- (7)		42	A$-$, B$-$	
N-Glycosyl- (8)		162	A$+$ (after BH_4^-)	
N-Lauroyl- (9)		182	A$-$, B$-$	
N-Tetradeca (mono and di)enoyl- (9)		208, 206	A$-$, B$-$	
N-Myristoyl-[d] (10)		210	A$-$, B$-$	
THE C-TERMINUS: $-\overset{\overset{\text{O}}{\|}}{\text{C}}-\text{OH}$				
-Amide (11)		-1	A$-$, B$-$	
O-Methyl- (12)		14	A$-$, B$-$	
-(N^α-Tyr) (13)		163	A$+$	P$+$
O-(ADP-ribosyl)- (14)		541	A$-$, B$-$	
O-(*N*-ethanolamine-glycan-phosphoinositides)[d] (15)		$> 1907^d$	A$-$, B$-$	
ARGININE: $H_2N-\overset{\overset{\text{NH}}{\|}}{\underset{\omega}{\text{C}}}-NH-CH_2-CH_2-CH_2-$				
Ornithine (16)		-42	A$+$, B$+$	P$+$
Citrulline (17)		1	A$+$	
N^ω-Methyl- (18)		14	A$+$, B$-$	
N^ω-Dimethyl-; N^ω, $N^{\omega'}$, -dimethyl- (18)		28	A$+$, B$-$	
Pentosidine (19)	Arg, Lys, (Ribose)	59	A$\pm$	
N^ω-(ADP-ribosyl)- (20)		541	A$-$, B$-$	
ASPARAGINE: $H_2N-\overset{\overset{\text{O}}{\|}}{\text{C}}-\underset{\beta}{CH_2}-$				
N^ε-(β-Aspartyl)-lysine (21)	Asn, Lys	-17	A$-$, B$-$	
Aspartate (22)		1	(A$+$, B$+$)	P$+$
N-Methyl- (23)		14	A$-$, B$-$	
N-(ADP-ribosyl)- (24)		541	A$-$, B$-$	
N-Glycosyl-[d] (25)		$(892$–$2770)^d$	A$-$, B$-$	
ASPARTATE: $HO-\overset{\overset{\text{O}}{\|}}{\text{C}}-\underset{\beta}{CH_2}-$				
O-Methyl- (26)		14	A$-$, B$-$	P$+$
Erythro-β-hydroxy- (27)		16	A$+$	
β-Carboxy- (28)		44	A$-$, B$+$	
O-Phosphoryl- (29)		80	A$-$, B$-$	
CYSTEINE: $HS-CH_2-$				
Dehydroalanine (30)		-34	A$-$, B$-$	

Table 1. Continued

Lysinoalanine (30)	Cys, Lys	−34	A+, B+
Lanthionine (30)	2 Cys	−34	A+, B+
S-γ-Glutamyl- (31)	Cys, Glu	−18	A−, B−
Cystine (32)	2 Cys	−2	A−, B−
S-(2-Histidyl)- (33)	Cys, His	−2	A−
S-(3-Tyr) (34)	Cys, Tyr	−2	A−
S-(sn-1-Glyceryl)- (35)		74	A±
S-Farnesyl- (36)		206	A±, B+
S-Palmitoyl- (37)		238	A−, B−
S-p-Coumaroyl- (38)		249	A−, B−
S-Geranylgeranyl- (36)		276	A±, B+
S-(sn-1-Dipalmitoyl-glyceryl)-^{d} (39)		524	A±, B−
S-(ADP)-ribosyl)- (40)		541	A−
S-Phycocyanobilin^d (41)		587	A±
S-Haem$^{d.}$ (42)		617	A±
S-(sn-1-Di-O-[3′,7′,11′, 15′-tetramethyl-hexadecyl]-glyceryl)- (43)		623	A±
S-(8α-Flavin [FAD]) (44)		784	A−, B−

GLUTAMATE: $\underset{\gamma}{HO-\overset{\displaystyle O}{\overset{\|}{C}}-CH_2-CH_2-}$

O-Methyl- (6,45)		14	A−, B−
γ-Carboxy- (46)		44	A−, B±
O-(ADP-ribosyl)- (20)		541	A−, B−
N^α-(γ-Glutamyl)-Glu$_{1-5}$ (47)		129–645	A−, B−
N^α-(γ-Glutamyl)-Gly$_{3-34}$ (48)		171–1936	A−, B−

GLUTAMINE: $\underset{\gamma}{H_2N-\overset{\displaystyle O}{\overset{\|}{C}}-CH_2-CH_2-}$

N^ε-(γ-Glutamyl)-lysine (49)	Gln, Lys	−17	A−, B−	
Glutamate (22)		1	(A+, B+)	P+
N^5-methyl- (50)		14	A−, B−	

HISTIDINE:

N^π-Methyl- (51)		14	A+
N-Phosphoryl- (52)		80	A−, B−
Diphthamide (53)		107	A±
4-Iodo- and diido- (54)		128 and 256	A−
N^τ-(ADP-ribosyl) diphthamide (55)		648	A−, B−
N^τ- and N^π-(8α-Flavin [FAD]) (44)		784	A−

LYSINE: $\underset{\varepsilon \quad\; \delta}{H_2N-CH_2-CH_2-CH_2-CH_2-}$

Desmosine (56)	4 Lys	−58	A±
Lysylpyridinoline (57)	3 Lys	−39	A±

Table 1. Continued

Merodesmosine (56)	3 Lys	−36	A±	
Aldol-histidine (57)	2 Lys, His	−20	A±, B+	
Lysinonorleucine (57)	2 Lys	−14	A+, B+	
Aldol (57)	2 Lys	−2	A+, B+	
Allysine (58)		−1	A−, B−	
Syndesine (59)	2 Lys	14	A+, B+	
δ-Hydroxyallysine (60)		15	A−, B−	
δ-Hydroxy- (60)		16	A+, B+	P+
N^ε-Acetyl- (61)		42	A−, B−	
N^ε-Lipoyl- (62)		188	A−, B−	
N^ε-Biotinyl- (63)		226	A−, B−	
N^ε-Ubiquitinyl-^{d} (64)		$(8566)_n$	A−, B−	
N^ε-Glycosyl- (65)		162	A±	
N^ε-Mono, di, trimethyl- (66)		14, 28, 42	A+	
Hypusine: N^ε-(4-amino, 2-hydroxybutyl)- (67)		87	A+	
δ-Hexosyloxy-^{d} (68)		177	A−, B±	

METHIONINE: $H_3C-S-CH_2-CH_2-$

Sulfoxide (69,70)		16	A−

PHENYLALANINE: (benzene ring)$-CH_2-$
β

β-Glycosyloxy-^{d} (71)		177	A−

PROLINE: (proline ring structure with positions labeled 4 H_2C, 3 CH_2, 5 H_2C, 2 $CH-$, N_1^+ H_2)

3-Hydroxy- (72)	16	A+, B+	P+
4-Hydroxy- (73)	16	A+, B+	P+
3,4-Dihydroxy- (74)	32	A+, B+	
4-Arabinosyloxy-^{d} (75)	157	A−, B±	
O^4-Hexosyloxy-^{d} (75)	177	A−, B±	

SERINE: $HO-CH_2-$

Dehydroalanine (76)		−18	A−
Lanthionine (77)	Ser, Cys	−18	A+
Alanino (τ- or π-histidine) (76)	Ser, His	−18	A+
O-Methyl- (78)		14	A+
O-Acetyl- (79)		42	A−, B−
Selenocysteine (80)		64	A−
O-Phosphoryl- (81)		80	A±, B−
O-(GlcNAc-1-phosphoryl)- (82)		268	A−, B−
O-Pantetheinephosphoryl- (83)		324	A±
O-Glycosyl-^{d} (84)		≥ 162	A−, B−

THREONINE: (H_3C and HO attached to $CH-$)

O-Methyl- (78)		14	A+

Table 1. Continued

β-Methyl-lanthionine (77)	Thr, Cys	18	A+
O-Phosphoryl- (81)		80	A±, B−
O-Glycosyl-[d] (84)		≥ 162	A−, B−

TRYPTOPHAN:

2,4'-BisTrp-6',7'-dione (85)	2 Trp	28	A−
C2-Aldohexopyranosyl- (86)		162	A−

TYROSINE: HO— —CH₂—

3,3'-Bityr (87)		−2	A+
IsodiTyr (88)		−2	A+
3,4-Dihydroxy-Phe (DOPA) (89)		16	A±
3,4,6-Trihydroxy-Phe (TOPA) (85)		32	A−
Halogenated derivatives[e] (90)			
3-chloro-		34, 36	A+
3,5-dichloro-		68, 70, 72	A+
3-bromo-		78, 80	A+
O-Phosphate (91)		80	A±, B±
O-Sulfate (92)		80	A−, B±
3-iodo-		126	A+
3,5-dibromo-[e]		156, 158, 160	A+
3,5-diido-		252	A+
O-Uridylyl- (93)		306	A−
O-Adenylyl- (93)		329	A−
triiodo-thyronine (94)		470	A±, B±
tetraiodo-thyronine (94)		596	A±, B±
O-(8α-Flavin [FAD]) (44)		783	A−

[a] In the case of cross-linking derivatives, all the amino acid precursors have been identified; for the others the precursor is identified as the primary amino acid under which the derivative is listed. It is essential to know the identity of the parent amino acid in defining the mass changes; dehydroalanine has a delta mass of −34 if it is derived from Cys, and −18 if derived from Ser.

[b] The delta mass values represent the changes in mass that would be observed if the primary amino acid has been converted to the specific derivative identified (e.g. Asn (radical mass 114) → Man₃Glc-NAc₂Asn (radical mass 1006), delta mass 892; Cys (103) + Ser (87) → lanthionine (172), delta mass −18).

[c] The chemical stability is recorded very broadly in terms of the standard hydrolysis conditions (A, 6 M HCl, 110°C, 20–24 h; B, 5 M NaOH, 110°C, 5–20 h). + means that a significant quantity of the derivative will survive these conditions; − that the derivative will be destroyed; and ± that it is possible to modify the acid or base hydrolysis conditions to allow some of the derivative to remain after all or most of the peptide bonds have been cleaved. P+ indicates that the PTH derivative can be observed and analysed directly by the standard sequencing conditions.

[d] Several different derivatives are possible in this group. In the case of the glycosylated derivatives, the main saccharides and their radical masses are pentose, 132; deoxyhexose, 146; hexose, 162; NAc-hexosamine, 203; sialic acid, 291. For derivatives involving fatty acids, different acids than those given will give different masses. Some derivatives such as haem and phycocyanobilins have closely related isomers; ubiquitin, a polypeptide with radical mass 8566, could add different masses according to the number of ubiquitin chains that are involved in a given derivative.

[e] For Cl and Br, both of the most abundant masses (^{35}Cl, ^{37}Cl, ^{79}Br, and ^{81}Br) must be considered in estimating the mass of the halogenated derivatives.

no direct evidence on this point, it is thought that the reactions may be enzyme catalysed.

The analytical procedures for the determination of D-amino acids in a protein are quite unique in that all the chemical and physical properties, except for optical rotation, are identical for the D- and L-enantiomers. Several solutions to this problem have been explored and can be used with relatively simple instrument and chemical requirements. One is to produce diastereo-isomers by reacting with a chiral reagent (96,97) and to separate and analyse the resulting two diastereoisomers formed from a mixture of D- and L-amino acids (L-leucyl-L-serine and L-leucyl-D-serine are readily separated by HPLC). Another method is to separate the amino acid enantiomers directly by 'chiral chromatography' on columns with a chiral matrix (98). A third method is the classical one in which DL-amino acids are converted to the acyl or the amide derivatives and the products are treated with either acylase or carboxypeptidase (amidase) to liberate all the L-amino acids as free acids, leaving the D-amino acid derivatives intact (99). The latter can now be readily separated and analysed. The constant problem with all these methods is that it is usually necessary to detect and quantify a small amount of the D-enantiomer in the presence of large amounts of the L-enantiomer.

3.2 Glycoprotein characterization

The following glycopeptide characterization was designed to elucidate the position of *N*-glycosylation sites in a glycoprotein and to determine the basic glycan structures associated with each site (100). The general procedure is to digest the fully reduced and Cys blocked protein (see Chapter 7) with chymotrypsin and to isolate the glycopeptides by affinity chromatography on concanavalin A–Sepharose. The glycopeptide mixture is next fractionated by reverse-phase HPLC, taking advantage of the interesting fact that the retention time for a glycopeptide reflects that of the peptide and is essentially independent of the glycan components. The purified glycopeptides are subjected to sequencing and to MS (FAB-MS was used in the experiments discussed). If the entire sequence can be determined, the complete amino acid composition becomes established, and the theoretical mass can be calculated. Of course, the sequence will contain a 'hole' in the position of the glycosyl–Asn residue, and the theoretical mass must be calculated using a hypothetical Asn for that position. If the glycopeptide contains more than one glycan structure, there will be more than one mass observed, and the difference between each observed mass and the theoretical one should represent the glycan mass. To confirm the data, the glycopeptide is treated with glycoamidase, an enzyme that cleaves the amide bond of glycosyl–Asn to yield Asp and glycosyl–amine, which rapidly hydrolyses to the reducing glycan. The reaction mixture is subjected to HPLC; the liberated glycans are recovered in the non-retarded flow-through fraction, and the free peptides at their respective retention times. After desalting the glycan fraction, it can be subjected to MS. The free peptides can be sequenced showing Asp in the position of the hole in the intact glycopeptide sequence, and also subjected to MS to confirm the peptide mass.

The data obtained from a glycoprotein isolated as the major component of the secretion from the Rathke's glands of sea turtles are given in *Table 2*. The four chymotryptic glycopeptides obtained by affinity chromatography and preparative HPLC were sufficiently pure to permit detailed analysis by sequencing and FAB-MS. In all cases, it was possible to determine the entire sequence of the peptides; even with the expected preponderance of hydrophobic residues at the C terminus of the chymotryptic peptides, the background was sufficiently low to permit unequivocal identification of all the residues. Two of the peptides, GP1 and GP3, contained a chemically modified Cys residue, and sequence analysis showed that the two peptides came from the same sequence; they differed only in the presence of an N terminal Leu in GP3, which was absent in GP1. In all cases the mass of the glycopeptide (GP) is the sum of the glycan (G) and the peptide (P) (G + P = GP–18); thus if one determines the mass of GP and P and subtracts the latter from the former, the mass of the glycan radical (M–18) is obtained. The results in *Table 2* are reported in this way, and the deduced glycan masses are all radical masses. Keep in mind that GP contains Asn–glycan, while the free peptide, after the glycoamidase treatment, contains Asp. In the manipulation of the data when G is subtracted from GP, the resulting mass of P contains Asn (radical mass 114.1), while the mass determined for the peptide is for P containing Asp (radical mass 115.1). Thus, before the observed value of P is subtracted from that of GP, it must be reduced by one. For example, for peptide P1 the value subtracted is 1148.2 rather than 1149.2, for P2 1483.6 rather than 1484.6, etc. With these considerations, the elucidation of the glycan structures worked well. The calculated mass for the established peptide sequences agreed extremely well with the observed values, and when these values were subtracted from the observed multiple peak glycopeptides, the resulting major glycan radical mass values (italics in *Table 2*) were internally consistent from the different glycosylation sites, and mostly corresponded to reasonable structures. In converting the mass data to actual structures, it must be emphasized that the mass measurements cannot distinguish between individual hexoses (all have the same radical mass of 162) or *N*-acetylhexosamines (all have the same radical mass of 203). In this particular case, the carbohydrate analyses obtained from the hydrolysates of intact glycoprotein showed the presence of only two sugars, glucosamine and mannose. Quantitatively, the ratio of the two sugars varied somewhat for the different glycoprotein preparations and, as expected, was sensitive to the hydrolysis conditions. All samples showed a glucosamine:mannose ratio in the range from 2:1 to 3:1. The chromatograms were remarkably free of other sugars; only two minor peaks corresponding to less than 5% of the glucosamine peak

Table 2. Amino acid sequence and molecular mass of glycopeptides (GP), peptides (P), and glycans from the *N*-glycosylation sites in turtle glycoprotein[a]

Peptide/glycan	Observed sequence[b]	Calculated (M + H⁺)	Observed (M + H⁺)		
GP 1[c]	G(ABD-)CSDGXNTAL		2042.3	2448.6	2650.4
P 1[d]	G(ABD-)CSDGDNTAL	1149.4	1149.2		
Glycan (deduced radical mass)[d]			*894.1*	*1300.4*	*1502.2*
GP 2[c]	ASVVGSXHTEAEVAL		2376.1	2783.0	2985.5
P 2[d]	ASVVGSDHTEAEVAL	1483.9	1484.6		
Glycan (deduced radical mass)[d]			*892.5*	*1299.4*	*1501.9*
GP 3[c]	LG(ABD-)CSDGXNTAL		2559.6	2763.5	
P 3[d]	LG(ABD-)CSDGDNTAL	1262.5	1262.4		
Glycan (deduced radical mass)[d]			*1298.2*	*1502.1*	
GP 4[c]	NLEXASSIGW	2225.9	2389.4	2592.6	2796.3
P 4[d]	NLEDASSIGW	1091.5	1091.5		
Glycan (deduced radical mass)[d]	*1135.4*	*1298.9*	*1502.1*	*1705.8*	

[a] From ref. 100.

[b] The amino acid sequences are given in the usual one-letter code (see Chapter 2, *Table 6*). The identified peaks correspond to multiple glycoforms. Some additional peaks were observed, but could not be identified. Known glycans with their calculated radical masses are $Man_3GlcNAc_2$-, 892.9; $GlcNAcMan_3GlcNAc_2$-, 1096.1; $GlcNAc_2Man_2GlcNAc_2$-, 1137.3; $GlcNAc_2Man_3GlcNAc_2$-, 1299.3; $GlcNAc_3Man_3GlcNAc_2$-, 1502.5; $GlcNAc_4Man_3GlcNAc_2$-, 1705.7.

[c] The sequence of the glycopeptide shows a 'hole' (X) for the glycosylated Asn residue. After treatment with glycoamidase, the glycosyl–Asn in the glycopeptide is converted to and identified as Asp. The following glycan masses were observed for the mixed glycans from all sites, 1094.5, 1299.5, 1502.6, and 1705.7

[d] The glycan radical mass is deduced by subtracting the mass of the peptide corrected (by subtracting one) for the presence of Asn instead of Asp in the glycopeptide: Glycan radical mass = obs. glycopeptide mass–(obs. peptide mass–1).

were discernible in elution positions corresponding to galactosamine and xylose. This observation, along with the consistent co-chromatography of the two sugar peaks from the glycoprotein with authentic glucosamine and mannose, clearly established that these two sugars are the primary building blocks of the glycans and is the basis for the assignment of only these two sugars to all the observed mass peaks.

After treating the glycopeptides with glycoamidase, the non-retarded fractions from HPLC were pooled and desalted on BioGel P-2, and the concentrated, desalted fraction was subjected to FAB-MS analysis. Two major and two minor fractions were observed. The two major fractions (1299.5 and 1502.6) correspond to the two observed in all the individual glycopeptides (*Table 2*); one minor peak with a deduced radical mass of 1094.5, corresponding to a likely processing intermediate structure, $GlcNAcMan_3GlcNAc_2$-, was not observed in any of the glycopeptides, and one with a deduced radical mass of 1705.5 ($GlcNAc_4Man_3GlcNAc_2$-) was observed only in GP4.

The primary conclusion from the results is that the glycans in the turtle glycoprotein, like those in other secreted glycoproteins, are heterogeneous, due to different *in vivo* processing of different glycosylation sites.

Protocol 1. Identification of glycosylation sites and glycan composition in a glycoprotein[a]

Equipment and reagents

- LKB Alpha Plus amino acid analyser (Pharmacia Biosystems); numerous alternative instruments are available
- Applied Biosystems 477A gas phase sequencer
- Dionex BioLC carbohydrate analyser with pulsed amperometric detection, CarboPak PA 1 column
- Kratos MS50RF mass spectrometer; a variety of other high resolution mass spectrometers are available
- Gilson 203 'Smart' fraction collector permitting time- or absorption-directed collection of HPLC fractions
- HPLC: standard reverse-phase system with C-18 column and low wavelength (215–230 nm) detection; all analyses employ increasing acetonitrile gradients in 0.1% (v/v) TFA
- Concanavalin A–Sepharose affinity chromatography column (Pharmacia Biotech)
- α-Methylmannoside
- BioGel P-2 (Bio-Rad Laboratories) gel filtration column for desalting

- GdmCl buffer: 0.5 M Tris–HCl pH 8.4, containing 6 M guanidinium chloride and 2 mM EDTA
- 4[Aminosulfonyl]-7-fluoro-2,1,3-benzoxadiazole (ABD-F) (Wako Chemicals) in the GdmCl buffer in amounts sufficient to give a 5–50-fold molar excess over the Cys residues present
- Tributylphosphine (Bu₃P), 5% (v/v) in methanol or 1-propanol

  Caution: Bu₃P is toxic and pyrophoric, and it should be handled with care in a fumehood. It is readily air oxidized and is supplied under nitrogen in sealed bottles. Alcohol solutions are quite stable when stored under nitrogen.
- Acids for hydrolysis: 6 M HCl, 2 M HCl, 2 M TFA for the analysis of amino acids, glucosamine, and monosaccharides, respectively
- Glycoamidase (*N*-glycosidase F, EC 3.2.2.18) (Boehringer Mannheim)

A. *Demonstrating the presence of glycans*

1. Hydrolyse 1–5 nmol of protein in 6 M HCl for 21 h at 110 °C, and in 2 M HCl for 4 h at 110 °C (to optimize yield of glucosamine); subject the samples to amino acid analysis.

2. Hydrolyse 1–5 nmol of protein with 2 M TFA for 3 h at 100 °C and subject the sample to sugar analysis on the Dionex carbohydrate analyser, eluting isocratically with 4 mM NaOH.

B. *Production and purification of glycopeptides*

1. Treat 1–5 nmol of glycoprotein in the GdmCl buffer containing 30–150 nmol of ABD-F with 1–5 µl of the 5% Bu₃P solution; incubate at 50 °C for about 1 h.

2. Dialyse the sample exhaustively against water and finally against 0.2 M Tris buffer pH 8.0.

3. Add chymotrypsin (1:50, w/w) to the turbid solution and incubate at 37 °C overnight (12–15 h). The progress of the digestion can conveniently be monitored by HPLC, using a 1 h 10–70% (v/v) acetonitrile gradient and a 1 ml/min flow rate with a C-18 column.

4. Inactivate the chymotrypsin by brief heating at 100°C.

5. Apply the reaction mixture to the concanavalin A affinity column, wash with Tris buffer until no more 230 nm absorbing material elutes, and then elute with a solution of 0.2 M α-methylmannoside.

6. Lyophilize the mannoside fractions, and subject a concentrated solution of the product to gel filtration on BioGel P-2.

7. Analyse both the fractions from the wash and from the mannoside elution for carbohydrate content. If the former contains significant amounts of carbohydrate it could mean that the capacity of the affinity column was exceeded or that the glycans present in the glycoprotein are not bound by concanavalin A; in this case, repeat the chromatography on the same column or on a different lectin column.

8. Lyophilize the mannoside eluted fraction, dissolve it in a small volume of water, place it on the HPLC C-18 column, and elute with a suitable (e.g. 15–60%) acetonitrile gradient (generally narrower than the one given above for monitoring the proteolysis).

9. Collect the 230 nm absorbing peaks separately as the glycopeptide products. Multiple runs may be required to accumulate enough material for the subsequent characterization.

C. *Characterization of glycopeptides*

1. Divide each of the collected glycopeptide samples into two fractions, A and B.

2. Subject the A fractions to sequencing and to MS (see Chapter 2). For FAB-MS, a salt-free concentrated aqueous solution is needed; a 1 μl sample, containing 1–3 nmol of glycan, is mixed with 1 μl of either a 9:1 (v/v) mixture of glycerol and a saturated oxalic acid solution or a 1:1 mixture of glycerol and thioglycerol. The entire sample is then applied to the probe of the mass spectrometer. The spectra are obtained using Xe as the bombarding gas, and are calibrated with caesium iodide.

3. Subject the B fractions to digestion with 1 U of glycoamidase in 0.1 M sodium phosphate buffer pH 7.2 at 37°C overnight.

4. Fractionate the reaction mixture using the standard HPLC procedure. The liberated glycans appear in the flow-through fraction and can be analysed by FAB-MS, but the peak intensities of the free glycans are considerably less than those of corresponding glycopeptides, so more material is required for significant results.

5. Subject the peptide fractions to sequencing.

[a] The procedures are mainly from ref. 95.

3.3 Unblocking the N terminus in acetylated proteins

One of the most discouraging events in protein characterization is to purify a protein for the purpose of obtaining sequence information, as a basis for the design of appropriate RNA probes for example, and then to discover that the N terminus is blocked. A good deal of effort has been expended in trying to solve this problem, and a variety of methods are available that can be used for a given protein, but which may not work for another. It is not possible to remove the acetyl (Ac) or formyl group directly from an intact protein by acid hydrolysis; the conditions required, especially for acetyl removal, are stronger than those designed to cleave only acid labile (e.g. Asp) peptide bonds in proteins, and the hydrolysis conditions would consequently yield a large number of peptides in addition to the unblocked N terminus. If a peptide containing the acetylated N terminus and no Asp or other acid labile bonds is obtained, mild acid treatment (1 M HCl, 100–110°C, 10–20 min) may give a reasonable yield of unblocked peptide free of contaminating peptides (101). In the case of N terminal AcSer and AcThr, treatment of the protein adsorbed on to filter disks with anhydrous TFA has been used successfully (102).

The ideal solution to this problem is obviously to have a specific enzyme that will remove the N terminal acetyl group of the protein without touching any of the peptide bonds. Unfortunately, such an enzyme is not available, in spite of a good deal of work trying to find it. What is available is an acylaminoacyl-peptide hydrolase, an enzyme that removes the N terminal acylamino acid and leaves residue two in the original sequence as the new, free N terminus. This enzyme will be referred to as just 'hydrolase' in the following. The enzyme does not act on intact proteins, and a relatively short N terminal peptide must be used as substrate if a reasonable amount of unblocked product is to be obtained. For an acetylated protein available only in minute quantities, a procedure is needed that will permit the production of a suitable peptide substrate for the hydrolase and to sequence it without further purification. This can be accomplished as follows. The acetylated protein is digested with a protease (or treated with CNBr) to yield a mixture of peptides (see Chapter 5), one of which is acetylated and the rest of which all contain unblocked N termini. These amino groups are then blocked by a reagent whose product is resistant to the hydrolase. Phenylisothiocyanate has been used, in which case the resulting thiocarbamates must be oxidized to the corresponding carbamates by treatment with performic acid. The carbamates are inert in the subsequent sequencing steps which are based on the chemistry of the thiocarbamates (103). Another procedure, the one that will be described in *Protocol 2*, involves the blocking of the new peptides with succinic anhydride (104). After acylation of the peptide mixture with a large excess of succinic anhydride, the reaction mixture is acidified and extracted exhaustively with ether to remove the excess succinic acid. The ether insoluble fraction is then treated with the hydrolase to remove the acetylamino acid from the original N terminus, and that reaction mixture is in turn subjected to sequencing. Since the hydrolase itself is an acetylated protein, it does not interfere with the sequencing reaction. It is essential that the removal of succinic acid is complete and that the phosphate buffer used in the hydrolase reaction is at the lowest possible concentration. The buffering capacity of both of these anions is such that they can easily interfere with the sequencing cycles if their concentrations are too high.

One aspect of this procedure is quite empirical. In the absence of any information about the protein to be studied, the selection of the method for protein fragmentation may have to be determined by trial and error, comparing the results with several different proteolytic enzymes and CNBr cleavage. The ideal product of the fragmentation is the longest possible N terminal peptide that will be an acceptable substrate for the hydrolase and will yield optimal sequence information after unblocking. The hydrolase is sensitive to peptide length and also to the presence of certain amino acids, such as Pro, Arg, and Lys, near the N terminus, so the 'ideal product' may turn out to be quite elusive indeed. During succinylation Lys is converted to N^ε-succinyllysine, which no longer interferes with the hydrolase activity. N^ε-Succinyllysine can be identified directly in the sequencing as a PTH derivative eluting at 9.23 min, relative to PTH-Glu at 8.38 min and PTH-Ala at 10.93 min.

It is possible to identify the acetylated N terminal amino acid in a separate experiment. The peptide mixture after hydrolase treatment contains the succinylated peptides, the unblocked N terminal peptide, and the acetylamino acid released by the hydrolase. If this mixture is treated with the enzyme acylase, an enzyme that catalyses the hydrolysis of all acetylated amino acids except AcAsp, AcPro, and AcGly, and subjected to amino acid analysis, the liberated N terminal amino acid can be observed directly as the only free amino acid in the reaction mixture.

The method outlined in *Protocol 2* has been applied to a number of known proteins and one unknown one, the rabbit muscle hydrolase itself. Some results are given in *Table 3*, and illustrate how variable the results may be. Other reports on the use of the hydrolase suggest that longer peptides are satisfactory substrates for it. In general, acetylated peptides with three to eight amino acids are essentially completely hydrolysed under standard conditions unless they contain positively charged amino acids or Pro; longer peptides such as the acetylated 16 residue fibrinopeptide A (unpublished results) and the natural 13 residue Ac-α-melanocyte stimulating hormone were hydrolysed about 50% in 12 h, while a 14 residue Ac-renin substrate was only hydrolysed about 2% in 20 h with 4 μg of hydrolase (105). One report has described a microsequencing variation for the isothiocyanate blocking method (103).

Table 3. Sequencing data from several *N*-acetylated proteins subjected to protein fragmentation, blocking with succinic anhydride, unblocking the N terminal peptide with acylaminoacyl-peptide hydrolase, and sequencing of the entire reaction mixture[a]

Protein	Known sequence[b]	Found sequences (means of cleavage)[c]
Parvalbumin	Ac-AMTELLNAEDIKKAI—	MTEL (Ch); MTELLNAEDIK (Tr)
Enolase (rabbit)	Ac-AMQKIFARE—	MQKIF (Ch)
Myelin basic protein	Ac-AAQKRPSQRSKYLAS—	AQ (Ch, Tr)
Superoxide dismutase	Ac-ATKAVCVLKGDGPVQ—	TKAV (V8)
α-Crystallin	Ac-MDIAIQHP**W**FK—	DIAIQHP (Ch, Tr)
Cytochrome *c*	Ac-GDVEKDKKIF—	DVEKGKK (Ch)
Ovalbumin	Ac-GSIGAASMEFCFDVFK—	SIGAAS (CNBr), SIGAA (Tr), SIGAA(SME) (Ch)
Acylaminoacyl-peptide hydrolase	Ac-MERQVLLSE—	ERQVL (Ch)

[a] From ref. 104.
[b] The amino acid sequences are given in the usual one-letter code (see Chapter 2, *Table 6*).
[c] The various cleavage methods used were chymotrypsin (Ch), trypsin (Tr), *Staphylococcus aureus* strain V8 endoproteinase (V8), and cyanogen bromide (CNBr). Based on the known specificity of these reagents, the most likely cleavage points have been indicated by showing the generated C terminal amino acids in bold letters. The N terminal acetylamino acid could be determined for all the proteins by direct amino acid analysis of an aliquot of the hydrolase reaction mixture after treatment with acylase I.

Protocol 2.   Removal of N terminal acetylamino acids with acylaminoacyl-peptide hydrolase[a]

Equipment and reagents

- LKB Alpha Plus amino acid analyser (Pharmacia Biosystems); numerous alternative instruments are available
- Applied Biosystems 477A gas phase sequencer
- 0.9 × 7.5 cm glass test-tubes
- Acylaminoacyl-peptide hydrolase (EC 3.4.19.1): (Pierce, Takara Biochemical, Boehringer Mannheim, or Sigma)
- Acylase I (3.5.1.14) (Sigma and other sources)
- Hydrolase buffer: 50 mM Na phosphate buffer pH 7.2, containing 1 mM EDTA and 2 mM $MgCl_2$
- Acylase buffer: 100 mM Na phosphate buffer pH 7.0
- Succinic anhydride
- Triethylamine, 12% (v/v) aqueous solution
- TFA, 20% (v/v) aqueous solution
- 1–5 nmol of the protein or peptide to be analysed at a concentration of approx. 1 nmol/10 μl in a suitable volatile solvent

A. *Fragmenting the protein*

1. Transfer 1–2 nmol of blocked protein to a 0.9 × 7.5 cm glass test-tube and lyophilize.

2. Digest the protein with an endoprotease of known specificity at an enzyme:substrate ratio of 1:20 (w/w) in 50 μl of the appropriate buffer, generally for a relatively short time at 37 °C (see Chapter 5).

3. Inactivate the protease by heating at 100°C for 5 min or by acidifying with 20% acetic acid when carbonate buffers are used.

4. Lyophilize the reaction mixture.

B. *Blocking the α- and ε-NH$_2$ groups by succinylation*

1. Dissolve the lyophilized products from part A in 50 μl of H$_2$O or carbonate buffer.

2. Add solid succinic anhydride in a 200-fold molar excess over the estimated total amount of α- and ε-NH$_2$ groups. Add the anhydride in small portions over a span of 1 h with vigorous mixing on a vortex stirrer and with the pH maintained at 10.0 by the addition of 12% triethylamine (see also Chapter 6, *Protocol 5*).

3. Leave the reaction mixture at pH 10 at room temperature for several hours (overnight), acidify with 20% TFA to pH 2–2.5, and lyophilize thoroughly.

4. Extract the residue with ether. Multiple extractions (six to ten times) with 2 ml portions of ether are required to remove all succinic acid and salts of triethylamine.

5. Dry the now invisible residue. If desired, the combined ether extracts can be air dried and screened for any possible presence of peptides by reverse-phase HPLC on a C18–100/300 Å column and/or by amino acid analysis after hydrolysis.

C. *Unblocking the N terminal peptide and sequencing*

1. Add 50 μl of hydrolase buffer to the tube containing the blocked peptides.

2. Adjust to pH 7.2 with 1–3 μl of 0.5 M NaOH.

3. Incubate the total peptide mixture with 2–5 μg of acylaminoacyl-peptide hydrolase for 6 h at 37°C. More enzyme can be used as the enzyme itself is an acetylated protein and hence does not give any background contribution in sequencing. When pure enzyme is used, incubation times of up to 12 h do not give any side reactions. Although the hydrolase is quite active on short peptides, the larger quantities of enzyme and the longer incubation times may well be needed for longer peptides.

4. Heat inactivate the reaction mixture at 100°C for 5 min and lyophilize.

D. *Sequencing*

1. Dissolve the products in 50 μl of 70% (v/v) formic acid.

2. Add a portion equivalent to 0.5 nmol of the original protein to the sequencer.

Protocol 2. *Continued*

3. Determine the N terminal sequence from the second residue onwards. As indicated in the text, Lys is detected as the PTH derivative of N^{ε}-succinyllysine, which elutes about 1 min in front of PTH-Ala (about at the same time as N^{ε}-acetyllysine) in the standard analysis.

E. *Identification of the blocked N terminal amino acid*

1. Lyophilize the remaining material (typically corresponding to 0.5–1.5 nmol of the starting protein after removing the required portion for sequencing).

2. Dissolve it in 50 µl of acylase buffer.

3. Digest with 5 U of acylase I at pH 7.0 and 37 °C for 6 h.

4. Lyophilize the reaction mixture, dissolve the residue in the standard LKB sample buffer (citrate pH 2.2), and apply it *in toto* to the amino acid analyser. The amino acid analyses in the work from which this protocol is derived were carried out with a ninhydrin-based analyser requiring at least 0.5–1 nmol of amino acid for a significant analysis. With more sensitive amino acid analysis methods, the amount can obviously be reduced.

[a] The procedures are mainly from ref. 104.

3.4 Characterization of phosphorylated proteins

Phosphoproteins present some unique problems in that so many amino acids may be phosphorylated. As shown in *Table 1*, Ser, Thr, Tyr, His, Arg, Lys, and Asp residues may all have a phosphate attached, and knowing that a protein contains phosphate is not the same as knowing what the derivative is. The derivatives vary from simple monoesters in Ser, Thr, and Tyr (the Tyr derivative involving the fairly acidic phenolic group actually has some anhydride character too), through phosphoamidates in Arg, His, and Lys, to a pure anhydride in the Asp derivative. Special cases are observed with diester derivatives such as $SerOP(O,O^-)OX$, where X may be another Ser or Thr residue or a non-protein compound containing an alcohol function to esterify. The general chemical properties of these phosphorylated amino acids can be used to establish some fairly simple diagnostic tests for what kind of phosphate derivative is present, and with this information specific methods can be assigned to study each one. Other methods for characterizing phosphorylated proteins are described in Chapters 2 and 5.

3.4.1 Establishing the presence of phosphate

There are two basic methods. One uses the ^{32}P radioisotope as the experimental marker by which the presence of phosphate can be detected in a protein (e.g. after incubation with kinases and phosphate donors such as ^{32}P-labelled ATP or after purification of proteins from cells grown on ^{32}P phosphate), in a peptide or, after hydrolysis, as free phosphate or free phosphorylated amino acid derivatives. One of the advantages of the ^{32}P techniques is that the almost ubiquitous inorganic phosphate (from buffers and glassware) will not interfere; the sensitivity is also high, but good specific activity data need to be established for quantitative determinations of total phosphate in a protein.

The second method (*Protocol 3*) uses a colorimetric procedure, in which the phosphate is reacted with ammonium molybdate in 1.2 M H_2SO_4 in the presence of ascorbic acid to yield a coloured product that can be quantified by its absorbance at 820 nm. The method gives good colour yields with 2–10 nmol of phosphate. If the unknown sample contains substances that interfere with the molybdate reaction, advantage can be taken of the high solubility of the product phosphomolybdic acid in organic solvents. Thus, the reaction mixture can be extracted into isobutanol/benzene (1:1, v/v) and the colour of the organic phase can be used for the quantification. In combination with the use of ^{32}P, using radioactivity rather than colour production for the quantification, the sensitivity is even better. In the case of the straight colorimetric assay, it is obviously essential that the sample to be assayed is completely free of contaminating phosphate.

3.4.2 Preliminary and direct tests for different derivatives (52,81)

i. Acid and base lability

Treatment of a phosphorylated protein, peptide, or amino acid with mild acid (e.g. 0.1 M HCl, 60°C, 10 min) will essentially completely hydrolyse all the phosphoamidates and the phosphate anhydrates, while the phosphate esters will remain intact. Strong acid (e.g. 6 M HCl, 110°C, 20 h) will give a low yield of Ser-P, a somewhat higher yield of Thr-P, and essentially no Tyr-P. Treatment with base (e.g. 3 M NaOH, 120°C, 3 h) will destroy P-anhydrides, Arg-P (as well as Arg), and Ser-P and Thr-P (by β-elimination rather than hydrolysis), but leave the other derivatives pretty much intact. It needs to be stated that the stability of a given phosphorylated amino acid is not a constant. Each Ser-P, Thr-P, etc. residue exists in a different environment in the protein, and, even if the protein is completely denatured, the neighbouring amino acids in the primary chain will affect the stability of each individual phosphorylated derivative.

ii. ^{31}P NMR studies

If sufficient quantities of phosphorylated protein or peptide are available (0.1–1.0 μmol) the characteristic chemical shifts and their dependence on pH and metal ion binding can provide direct identification of the phosphorylated derivatives.

iii. MALDI-TOF-MS

To obtain optimal sensitivity, the phosphoprotein is digested with proteases and the MS analysis can then be carried out with the total peptide mixture. The procedure is to compare the mass of the peptides before and after the treatment with a non-specific phosphatase. Any phosphopeptide will lose 80 mass units for each phosphate removed by the phosphatases, and can be recognized on that basis (106).

iv. Partial hydrolysis of the protein and direct detection of specific phosphate esters

Moderate acid hydrolysis conditions have been explored (e.g. 1 M HCl, 100°C for different lengths of time) and found to give about 25% recovery of a given phosphate ester under conditions where 50% of the total phosphate in the phosphoprotein has been released as free phosphate and 25% remains in unhydrolysed peptides. Alkaline hydrolysis of proteins at high temperature (e.g. 5 M KOH at 155°C) will give complete hydrolysis of the peptide bonds in about 30 min, but leave Tyr-P intact. All three *O*-phosphate derivatives elute together near the flow-through volume in the standard cation exchange amino acid analyser procedure. They can be separated and quantified, however, using modified elution protocols (81).

3.4.3 Establishing the location of the phosphorylated derivative in the protein

i. Peptide mapping

The obvious approach is to subject the phosphoprotein to proteolytic cleavage and to separate the phosphate-containing peptides using the same types of experiments as those used for the glycopeptide separation above (Section 3.2 and Chapter 5). The peptides are separated by HPLC and any labelled with ^{32}P are readily identified by radioactive counting. Phosphoamidates and anhydrides may require neutral or mildly alkaline chromatography because of their acid lability. The resulting phosphopeptides are then subjected to sequencing. However, the standard Edman sequencing procedures, either with the spinning cup or with the gas phase instruments, give essentially zero yield of the phosphorylated amino acids. They are simply too hydrophilic to be extracted into the organic phase, and special procedures are required for the detection of the phosphorylated derivatives. Since Ser and Thr residues are themselves often modified in the standard procedure, the method of choice is to use radiochemical sequencing, attaching the phosphorylated peptide to derivatized polyvinylidene difluoride (PVDF) membranes that can retain the peptide/protein quantitatively (an arylamine PVDF membrane was used in ref. 107) and introducing into the sequencing protocol an acetonitrile/0.1% (v/v) aqueous TFA (25:75, v/v) solvent to extract the hydrophilic phosphoamino acid derivatives for analysis. The radioactive count at each cycle identifies the location of the phosphorylated derivatives (107).

ii. Direct sequencing after β-elimination and derivatization of the resulting dehydroamino acid

This method applies to Ser-P and Thr-P residues. One procedure simply uses 2-mercaptoethanol to convert the dehydro amino acids (dehydroalanine from Ser-P and β-methyl-dehydroalanine from Thr-P) to the corresponding thioethers. These products are stable, and their PTH derivatives can be observed and quantified directly by the standard sequencing protocol (108).

Protocol 3. Analysis of base labile phosphate esters (Ser-P and Thr-P) by colorimetric assay and by ^{32}P determination[a]

Equipment and reagents

- LKB Alpha Plus amino acid analyser; numerous alternative instruments are available
- 12 ml conical centrifuge tubes
- Ames phosphate reagent: 0.5% (v/v) ammonium molybdate in 0.6 M H_2SO_4—add ascorbic acid to a concentration of 2% (w/v) in the cold to prepare the fresh Ames reagent (the reagent should be prepared daily)
- Phosphomolybdate reagents (all reagents are stable at room temperature for several months)
- 5% (w/v) silicotungstic acid·H_2O in 1 M H_2SO_4, containing 50 nmol/ml of phosphate
- 5% (w/v) ammonium molybdate·4H_2O in 1 M H_2SO_4
- Isobutanol/benzene (1:1, v/v)

A. *Liberation of alkali labile phosphate*

1. Place the phosphoprotein (2–10 nmol of phosphate) in a volume of 1 ml or less in an Eppendorf tube.

2. Add enough cold 100% (w/v) TCA to give a 20% (w/v) solution; keep the mixture for 10 min in the cold.

3. Centrifuge at about 10 000 *g* for 2 min; remove the supernatant.

4. Add 0.5 ml of ice-cold 0.1 M NaOH to the pellet, mix on a vortex mixer, and immediately reprecipitate with TCA as described in step 2.

5. Wash the pellet twice by resuspending in 0.5 ml of 20% (w/v) TCA, and centrifuging.

6. Dissolve the final pellet in 0.3 ml of 1 M NaOH, seal the tubes, and incubate at the appropriate time and temperature (e.g. 37 °C for 1–10 h).

7. Stop the reaction by the addition of 0.075 ml of 100% TCA and leave on ice for 10 min before centrifuging. For ^{32}P determination, the same general procedure applies, except since the total protein is likely to be much reduced, it is recommended to add 0.2–0.5 mg of a carrier protein (such as serum albumin) to ensure optimal separation of protein from solubilized phosphate in the trichloroacetic acid precipitations.

Radha G. Krishna and Finn Wold

Protocol 3. *Continued*

B. *Colorimetric phosphate assay*

1. Mix 0.3 ml of sample, blank, and standard (0.5 mM KH_2PO_4 in 5 mM HCl) with 0.1 ml of fresh, ice-cold Ames reagent.

2. Seal the tubes, incubate for 20 min at 45°C, and then for 1 h at room temperature, for colour development.

3. Measure the absorbance at 820 nm.

C. ^{32}P *assay*

1. Place 0.5 ml sample (blank, standards) in a conical centrifuge tube.

2. Add 0.25 ml of the silicotungstic reagent, then 1 ml of the isobutanol/benzene reagent, and finally 0.25 ml of the ammonium molybdate reagent.

3. Mix the content of the tubes well on a vortex mixer, and centrifuge at about 1000 g to separate the phases.

4. Remove 0.5–0.7 ml of the upper phase for determination of total radioactivity by scintillation counting.

[a] The procedures are mainly from ref. 81.

4. Concluding remarks

The number and variety of methods that are required and that have been developed to characterize post-translational modifications in proteins are quite impressive. The examples discussed here represent only a very small sample of the total. If we consider only a single, common reaction, disulfide bond formation in proteins (Chapter 7), and take stock of the the amount of work that has been directed toward the understanding of the reaction itself, developing analytical procedures, and using the topographical information from disulfide bond position to the study of protein structure, it is clear that this has been and continues to be a very active field of activity in biochemistry, and has produced a truly remarkable volume of outstanding research contributions over the years. Similar, but often considerably smaller, volumes of research contributions exist for all the other post-translational reactions; the most important contribution in each case is undoubtedly the one that documented the discovery of the new derivative. Rarely did the researchers set out with the idea of looking for new amino acid derivatives in proteins. They were generally asking quite different questions, but through ingenuity, careful observation, clever experimentation, and sometimes with some help from the princes of Serendip, they were able to convert unexpected observations into new discoveries. It is regrettable that this feature of the post-translational

112

reactions had to be omitted in this chapter, but hopefully the case histories discussed give some impression of the variety of thinking and experimentation that has been, and will continue to be, applied to this field.

References

1. Krishna, R. G. and Wold, F. (1993). *Adv. Enzymol.*, **67**, 265.
2. Krishna, R. G. and Wold, F. (1992). In *Methods in protein sequence analysis* (ed. K. Imahori and F. Sakiyama), p. 167. Plenum Press, New York.
3. Orlowski, M. and Meister, A. (1971). In *The enzymes* (3rd edn) (ed. P. D. Boyer), Vol. IV, p. 123. Academic Press, New York.
4. van Poelje, P. D. and Snell, E. E. (1990). *Annu. Rev. Biochem.*, **59**, 29.
5. Siegel, F. L. (1988). In *Advances in post-translational modification of proteins and aging* (ed. V. Zappia, P. Galletti, R. Porta, and F. Wold), p. 341. Plenum Press, New York.
6. Sugino, Y., Tsunasawa, S., Yutani, K., Ogasahara, K., and Suzuki, M. (1980). *J. Biochem.*, **87**, 351.
7. Arfin, S. M. and Bradshaw, R. A. (1988). *Biochemistry*, **27**, 7979.
8. Kennedy, L. and Baynes, J. W. (1984). *Diabetologia*, **26**, 93.
9. Neubert, T. A., Johnson, R. S., Hurley, J. B., and Walsh, K. A. (1992). *J. Biol. Chem.*, **267**, 18274.
10. Johnson, D. R., Bhatnagar, R. S., Knoll, L. J., and Gordon, J. I. (1994). *Annu. Rev. Biochem.*, **63**, 869.
11. Kreil, G. (1984). In *Methods in enzymology* (ed. F. Wold and K. Moldave), Vol. 106, p. 218. Academic Press, New York.
12. Clarke, S. (1992). *Annu. Rev. Biochem.*, **61**, 355.
13. Flavin, M. and Murofushi, H. (1984). In *Methods in enzymology* (ed. F. Wold and K. Moldave), Vol. 106, p. 223. Academic Press, New York.
14. Hayaishi, O. and Ueda, K. (1984). In *Methods in enzymology* (ed. F. Wold and K. Moldave), Vol. 106, p. 450. Academic Press, New York.
15. Udenfriend, S. and Kodukula, K. (1995). *Annu. Rev. Biochem.*, **64**, 563.
16. Sletten, K. and Aakesson, I. (1971). *Nature New Biol.*, **231**, 118.
17. Rothnagel, J. A. and Rogers, G. E. (1984). In *Methods in enzymology* (ed. F. Wold and K. Moldave), Vol. 107, p. 624. Academic Press, New York.
18. Kim, S., Chanderkar, L. P., Ghosh, S. K., Park, J. O., and Paik, W. K. (1988). In *Advances in post-translational modification of proteins and aging* (ed. V. Zappia, P. Galletti, R. Porta, and F. Wold), p. 327. Plenum Press, New York.
19. Grandhee, S. K. and Monnier, V. M. (1991). *J. Biol. Chem.*, **266**, 11649.
20. Ueda, K. and Hayaishi, O. (1985). *Annu. Rev. Biochem.*, **54**, 73.
21. Klostermeyer, H. (1984). In *Methods in enzymology* (ed. F. Wold and K. Moldave), Vol. 107, p. 258. Academic Press, New York.
22. Robinson, A. B., Scotchler, J. W., and McKerrow, J. H. (1973). *J. Am. Chem. Soc.*, **95**, 8156.
23. Klotz, A. V. and Glazer, A. N. (1985). *J. Biol. Chem.*, **262**, 17350.
24. Sekine, A., Fujiwara, M., and Narumiya, S. (1989). *J. Biol. Chem.*, **264**, 8602.
25. Kornfeld, R. and Kornfeld, S. (1985). *Annu. Rev. Biochem.*, **54**, 631.
26. Clarke, S. (1988). In *Advances in post-translational modification of proteins and*

aging (ed. V. Zappia, P. Galletti, R. Porta, and F. Wold), p. 213. Plenum Press, New York.

27. Stenflo, J., Holme, E., Lindstedt, S., Chandramouli, N., Huang, L. H. T., Tam, J. P., *et al.* (1989). *Proc. Natl. Acad. Sci. USA*, **86**, 444.

28. Koch,T. H., Christy, M. R., Barkley, R. M., Sluski, R., Bohemier, D., VanBuskirk, J. A., *et al.* (1984). In *Methods in enzymology* (ed. F. Wold and K. Moldave), Vol. 107, p. 563. Academic Press, New York.

29. Degani, C. and Boyer, P. D. (1973). *J. Biol. Chem.*, **248**, 8222.

30. Steinert, P. M. and Idler, W. W. (1979). *Biochemistry*, **18**, 5664.

31. Tack, B. F., Harrison, R. A., Janatova, J., Thomas, M. L., and Prahl, J. W. (1980). *Proc. Natl. Acad. Sci. USA*, **77**, 5764.

32. Creighton, T. E. (1984). In *Methods in enzymology* (ed. F. Wold and K. Moldave), Vol. 107, p. 305. Academic Press, New York.

33. Lerch, K. (1984). In *Methods in enzymology* (ed. F. Wold and K. Moldave), Vol. 106, p. 355. Academic Press, New York.

34. Ito, N., Phillips, S. E. V., Stevens, C., Ogel, Z. B., McPherson, M. J., Keen, J. N., *et al.* (1991). *Nature*, **350**, 87.

35. Lampen, J. O. and Nielsen, J. B. K. (1984). In *Methods in enzymology* (ed. F. Wold and K. Moldave), Vol. 106, p. 365. Academic Press, New York.

36. Glomset, J. A., Gelb, M. H., and Farnsworth, C. C. (1990). *Trends Biochem. Sci.*, **15**, 139.

37. Hancock, J. F., Magee, A. I., Childs, J. E., and Marshall, C. J. (1989). *Cell*, **57**, 1167.

38. Hoff, W. D., Dux, P., Hard, K., Devreese, B., Nugteren-Roodzant, I. M., Crielaard, W., *et al.* (1994). *Biochemistry*, **33**, 13960.

39. Hantke, K. and Braun, V. (1973). *Eur. J. Biochem.*, **34**, 284.

40. West, R. A. Jr., Moss, J., Vaughan, M., Liu, T., and Liu, T.-Y. (1985). *J. Biol. Chem.*, **260**, 14428.

41. Glazer, A. N. (1984). In *Methods in enzymology* (ed. F. Wold and K. Moldave), Vol. 106, p. 359. Academic Press, New York.

42. Margoliash, E. and Schejter, A. (1966). *Adv. Protein Chem.*, **21**, 113.

43. Sagami, H., Kikuchi, A., and Ogura, K. (1995). *J. Biol. Chem.*, **270**, 14851.

44. Singer, T. P. and McIntire, W. S. (1984). In *Methods in enzymology* (ed. F. Wold and K. Moldave), Vol. 106, p. 369. Academic Press, New York.

45. Clarke, S. (1985). *Annu. Rev. Biochem.*, **54**, 479.

46. Suttie, J. W. (1985). *Annu. Rev. Biochem.*, **54**, 459.

47. Edde, B., Rossier, J., Le Caer, J.-P., Desbruyeres, E., Gros, F., and Denoulet, P. (1992). *Science*, **247**, 83.

48. Redeker, V., Levilliers, N., Schmitter, J.-M., Le Caer, J.-P., Rossier, J., Adoutte, A., *et al.* (1994). *Science*, **266**, 1688.

49. Greenberg, C. S., Birrckbichler, P. J., and Rice, R. H. (1991). *FASEB J.*, **5**, 3071.

50. Lhoest, J. and Colson, C. (1977). *Mol. Gen. Genet.*, **154**, 175.

51. Huszar, G. (1984). In *Methods in enzymology* (ed. F. Wold and K. Moldave), Vol. 106, p. 287. Academic Press, New York.

52. Fujitaki, J. M. and Smith, R. A. (1984). In *Methods in enzymology* (ed. F. Wold and K. Moldave), Vol. 107, p. 23. Academic Press, New York.

53. Van Ness, B. G., Howard, J. B., and Bodley, J. W. (1980). *J. Biol. Chem.*, **255**, 10710.

54. Wolff, J. and Covelli, I. (1969). *Eur. J. Biochem.*, **9**, 371.
55. Van Ness, B. G., Howard, J. B., and Bodley, J. W. (1980). *J. Biol. Chem.*, **255**, 10717.
56. Portridge, S. M. (1989). In *Elastin and elastases* (ed. L. Robert and W. Hornebeck), Vol. 1, p. 127. CRC Press, Boca Raton.
57. Eyre, D. (1987). In *Methods in enzymology* (ed. L. W. Cunningham), Vol. 144, p. 115. Academic Press, New York.
58. Mirelman, D. and Siegel, R. C. (1979). *J. Biol. Chem.*, **254**, 571.
59. Quaglino, D., Fornieri, C., Nanney, L. B., and Davidson, J. M. (1993). *Matrix*, **13**, 481.
60. Rucker, R. B. and Murray, J. (1978). *Am. J. Clin. Nutr.*, **31**, 1221.
61. Allfrey, V. G., DiPaola, E. A., and Sterner, R. (1984). In *Methods in enzymology* (ed. F. Wold and K. Moldave), Vol. 107, p. 224. Academic Press, New York.
62. Hale, G. and Perham, R. N. (1980). *Biochem. J.*, **187**, 905.
63. Goss, N. H. and Wood, H. G. (1984). In *Methods in enzymology* (ed. F. Wold and K. Moldave), Vol. 107, p. 261. Academic Press, New York.
64. Hershko, A. (1991). *Trends Biochem. Sci.*, **16**, 265.
65. Baynes, J. W., Thorpe, S. R., and Murtiashaw, M. H. (1984). In *Methods in enzymology* (ed. F. Wold and K. Moldave), Vol. 106, p. 88. Academic Press, New York.
66. Paik, W. K. and Dimaria, P. (1984). In *Methods in enzymology* (ed. F. Wold and K. Moldave), Vol. 106, p. 274. Academic Press, New York.
67. Park, M. H., Wolff, C. E., and Folk, J. J. (1993). *Biofactors*, **4**, 95.
68. Levine, M. J. and Spiro, R. G. (1979). *J. Biol. Chem.*, **254**, 8121.
69. Brot, N., Fliss, H., Coleman, T., and Weissbach, H. (1984). In *Methods in enzymology* (ed. F. Wold and K. Moldave), Vol. 107, p. 352. Academic Press, New York.
70. Kikuchi, Y. and Tamiya, N. (1992). In *Frontiers and new horizons in amino acid research* (ed. K. Takai), p. 309. Elsevier, Tokyo.
71. Lin, T. S. and Kolattukudy, P. E. (1980). *Eur. J. Biochem.*, **106**, 341.
72. Bornstein, P. (1974). *Annu. Rev. Biochem.*, **43**, 567.
73. Kivirikko, K., Myllyla, R., and Philajaniemi, T. (1989). *FASEB J.*, **3**, 1609.
74. Nordwig, A. and Pfab, F. K. (1969). *Biochim. Biophys. Acta*, **181**, 52.
75. Lamport, D. T. A. (1984). In *Methods in enzymology* (ed. F. Wold and K. Moldave), Vol. 106, p. 523. Academic Press, New York.
76. Sass, R. L. and Marsh, M. E. (1984). In *Methods in enzymology* (ed. F. Wold and K. Moldave), Vol. 106, p. 351. Academic Press, New York.
77. Kaletta, C., Entian, K.-D., and Jung, G. (1991). *Eur. J. Biochem.*, **199**, 411.
78. Swanson, R. J. and Applebury, M. (1983). *J. Biol. Chem.*, **258**, 10599.
79. Rudman, D., Chawla, R. K., and Hollins, B. M. (1979). *J. Biol. Chem.*, **254**, 10102.
80. Stadtman, T. C. (1991). *J. Biol. Chem.*, **266**, 16257.
81. Martensen, T. D. (1984). In *Methods in enzymology* (ed. F. Wold and K. Moldave), Vol. 107, p. 3. Academic Press, New York.
82. Gustafson, G. L. and Gander, J. E. (1984). In *Methods in enzymology* (ed. F. Wold and K. Moldave), Vol. 107, p. 172. Academic Press, New York.
83. Vagelos, P. R. (1973). In *The enzymes* (3rd edn) (ed. P. D. Boyer), Vol. VIII, p. 155. Academic Press, New York.
84. Paulsen, H. (1990). *Angew. Chem. Int. Ed. Engl.*, **29**, 823.

85. Klinman, J. P. and Mu, D. (1994). *Annu. Rev. Biochem.*, **63**, 299.
86. Hofsteenge, J., Muller, D. R., Beer, T., Loffler, A., Richter, W. J., and Vliegenthart, J. F. G. (1994). *Biochemistry*, **33**, 13524.
87. Amado, R., Aeschbach, R., and Neukom, H. (1984). In *Methods in enzymology* (ed. F. Wold and K. Moldave), Vol. 107, p. 377. Academic Press, New York.
88. Fry, S. C. (1984). In *Methods in enzymology* (ed. F. Wold and K. Moldave), Vol. 107, p. 388. Academic Press, New York.
89. Waite, J. H. and Benedict, C. V. (1984). In *Methods in enzymology* (ed. F. Wold and K. Moldave), Vol. 107, p. 397. Academic Press, New York.
90. Hunt, S. (1984). In *Methods in enzymology* (ed. F. Wold and K. Moldave), Vol. 107, p. 413. Academic Press, New York.
91. Hunter, T. and Cooper, J. A. (1985). *Annu. Rev. Biochem.*, **54**, 897.
92. Huttner, W. B. (1984). In *Methods in enzymology* (ed. F. Wold and K. Moldave), Vol. 107, p. 200. Academic Press, New York.
93. Rhee, S. G. (1984). In *Methods in enzymology* (ed. F. Wold and K. Moldave), Vol. 107, p. 183. Academic Press, New York.
94. Nunez, J. (1984). In *Methods in enzymology* (ed. F. Wold and K. Moldave), Vol. 107, p. 476. Academic Press, New York.
95. Kreil, G. (1994). *J. Biol. Chem.*, **269**, 10967.
96. Bada, J. L. (1984). In *Methods in enzymology* (ed. F. Wold and K. Moldave), Vol. 106, p. 98. Academic Press, New York.
97. Scaloni, A., Simmaco, M., and Bossa, F. (1991). *Anal. Biochem.*, **197**, 305.
98. Davanko, V., Bochkov, A., Kurganov, A., Roumeliotis, P., and Unger, K. (1980). *Chromatographia*, **13**, 677.
99. Greenstein, J. P. (1957). In *Methods in enzymology* (ed. S. P. Colowick and N. O. Kaplan), Vol III, p. 554. Academic Press, New York.
100. Chin, C. C. Q., Krishna, R. G., Weldon, P. J., and Wold, F. (1995). *Anal. Biochem.*,
101. Chin, C. C. Q. and Wold, F. (1985). *Biosci. Rep.*, **5**, 847.
102. Wellner, D., Panneerselvam, C., and Horecker, B. L. (1990). *Proc. Natl. Acad. Sci. USA*, **87**, 1947.
103. Hirano, H., Komatsu, S., Kajiwara, H., Takagi, Y., and Tsunasawa, S. (1993). *Electrophoresis*, **14**, 839.
104. Krishna, R. G., Chin, C. C. Q., and Wold, F. (1991). *Anal. Biochem.*, **199**, 45.
105. Kobayashi, K. and Smith, J. (1987). *J. Biol. Chem.*, **262**, 11435.
106. Yip, T.-T. and Hutchens, T. W. (1993). In *Techniques in protein chemistry* (ed. R. Hogue-Angeletti), Vol. IV, p. 201. Academic Press, New York.
107. Parten, B. F., McDowell, J. H., Nawrocki, J. P., and Hargrave, P. A. (1994). In *Techniques in protein chemistry* (ed. J. W. Crabb), Vol. V, p. 159. Academic Press, New York.
108. Scaloni, A., Barra, D., and Bossa, F. (1994). *Anal. Biochem.*, **218**, 226.

Peptide mapping

NICK A. MORRICE and ELIZABETH A. CARREY

1. Introduction: strategies

This chapter will cover the processes necessary to generate a peptide map, including the fragmentation of the protein of interest, the separation of the resulting peptides (oligopeptides of various lengths), and their identification and analysis. Strategies that generate a literally two-dimensional 'map' using electrophoretic methods will be described in Section 3.5. Every biochemistry textbook describes the use of such maps, for example the early use of paper chromatography to demonstrate that there is a difference in sequence between normal haemoglobin and the mutant form that causes sickle-cell anaemia. Improvements in the separation and analytical methods now available to us mean that peptide 'maps' may be used to obtain information, such as the actual sequence, from the individual peptides. A series of separation steps, based on standard protein chemistry, can each contribute a 'dimension' to the map and inform us about the properties of the peptide. The strategy to be adopted therefore has implications for the results that can be obtained, since the second or third dimension uses material that has been generated in an earlier separation step.

Many of the examples used in this chapter refer to work on CAD, a multi-enzyme polypeptide of 240 kDa that is obtained from the cytoplasm of an overproducing mammalian cell line (1). This protein is composed of a series of discretely folded domains that can be isolated after controlled proteolysis, cutting at the exposed regions between the folded domains (2). There are two phosphorylation sites, one in an exposed region and one in a folded part; these sites were mapped using many of the approaches described in the first edition of this chapter (3,4). In this revised version, we take account of advances in many of the individual steps that can be applied to peptide mapping and to the detection of phosphorylated peptides. Other books in the *Practical approach* series cover protein sequencing and the protein kinases in more detail.

2. Cleavage of peptide bonds

Hydrolysis of a protein may produce peptides ranging in size from a few residues to larger fragments of several hundred residues that retain structure and the ability to bind ligands. In general, less specific reagents, or exhaustive treatment with a protease, will produce smaller peptides. Larger fragments will be generated by chemical cleavage at rare amino acid residues, or by limited proteolysis at only the most accessible sites. Harsh treatments such as general acid hydrolysis or attack by vigorous non-specific proteases are not useful in peptide mapping, nor are the exolytic protease enzymes (amino-peptidases and carboxypeptidases); consequently they are not discussed in this chapter.

Cleavage at a specific amino acid residue, whether chemical or enzymatic, will usually be prevented if the residue or its neighbour is modified (for example, phosphorylation of a serine near to an arginine will prevent cleavage by trypsin) or if the protein is a mutant in which the residue has been replaced. A new site might also be generated by mutation, of course. Compare the products with the expected pattern to find the altered peptide.

2.1 Chemical methods

A chemical cleavage is often an early step in the treatment of proteins for sequencing. The reagent is chosen to break the polypeptide at sites that will overlap the specificity of proteolytic enzymes, preferably next to uncommon amino acid residues so that long fragments are obtained. It has been estimated that aspartate–proline bonds occur once in every 400 residues, and methionine about once per 100 residues. The presence of these residues can be confirmed by amino acid analysis of the target protein. Some reactions do not go to completion because of competition from nearby side chains, or because residual folded structures restrict the access of the reagent to the susceptible bond, even though the reagent is present in substantial molar excess over the target residue. Nevertheless, a combination of two or more cleavage methods will usually provide sufficient overlapping peptides to generate an unambiguous protein sequence. Several potentially useful reagents for chemical cleavage are described in *Table 1*.

A novel, specialized use of chemical reagents to cut proteins makes use of the oxidizing properties of molecular oxygen and the Fe (II) cation at the site where a nucleotide or other multivalent anion is bound to the protein. Cleavage by this system has been used to locate the metal–isocitrate binding site of pig heart isocitrate dehydrogenase (12). In similar experiments the amino acid residues involved in binding ATP to carbamoyl phosphate synthetase I (13) were located. The multienzyme CAD polypeptide contains this enzyme activity, but it cannot be cut in this way, presumably reflecting a different arrangement of side chains at the binding site.

Table 1. Chemical cleavage of specific peptide bonds

Reagent	Solvent, conditions	Specificity[b]	References
Mild acid cleavage	70% Formic acid; 37–40°C, 24 h	Asp–Pro	5,6
Cyanogen bromide	30- to 100-fold molar excess over Met; 70% formic acid or 0.1 M HCl; 25°C, 24 h	Met–Y	7,8
Hydroxylamine	2 M in 6 M GdmCl[a] pH 9–10; 37°C, < 4 h	Asn–Gly	9
Iodosobenzoic acid	Twofold molar excess over Trp; 80% acetic acid, 4 M GdmCl pH 8; 25°C, 24 h	Trp–Y	10
2-Nitro-5-thiocyanobenzoate followed by alkaline cleavage	Fivefold molar excess over thiol groups; 6 M GdmCl pH 8; 37°C, 15 min; separate from reagent, then expose to 6 M GdmCl pH 8; 37°C, 16 h	X–Cys	11

[a] GdmCl: guanidinium chloride.
[b] Cleavage is C terminal to residue X, N terminal to residue Y.

119

Protocol 1. General method for chemical cleavage

Equipment and reagents

- Substrate protein (usually in aqueous solution)
- Chosen chemical reagent(s) in safe container (*Table 1*)
- Acid or denaturant solution
- Centrifugal evaporator (e.g. SpeedVac by Savant)

Method

1. Set-up two parallel incubations in similar conditions of pH and temperature, omitting the reagent from one, especially if the conditions are acid. Cleavage at aspartate–proline bonds can occur in acid impregnated gel slices, accidentally or intentionally (14), and in acidic solutions of proteins.

2. Use the conditions summarized in *Table 1*. Take appropriate safety precautions when weighing or dispensing the reagent, and handle it carefully. Fume-cupboards, room ventilation, and cold traps for vacuum pumps must all be adequate.

3. At the end of the reaction, remove volatile reagents and solvents by lyophilization or by centrifugal evaporation, effectively quenching the reaction.

4. Add some water and check the pH of the reaction mixture; dry down again if necessary to remove excess acid. Resuspend the peptide products in the appropriate solvent for separation by gel electrophoresis or chromatography (see Section 3).

2.2 Proteolytic enzymes

Many of the endolytic proteases that are useful in peptide mapping hydrolyse the peptide bond next to specific side chains or classes of side chains, generating a large number of peptides from most substrate proteins. Other enzymes, such as plasmin and thrombin, have very specialized substrates *in vivo* but have nevertheless been valuable in producing defined fragments by digesting between domains in large proteins. Even proteases with very broad specificities can be used in controlled proteolysis experiments (see Section 2.3.1).

Table 2 summarizes relevant information on endolytic proteases that are currently commercially available from Worthington Enzymes and Boehringer Mannheim. Specialist companies such as these publish useful catalogues and newsletters describing applications for their enzymes. For general peptide mapping, buy good quality proteases from specialist suppliers; sequencing grade enzymes are not usually necessary.

Complete proteolysis at all the relevant residues in a protein will generate a distinct mixture of peptide fragments according to the enzyme that is used. Many of the products may be digested by a second protease with a different specificity, either in the same incubation or after separation of the fragments. Thermolysin, for example, may be used at high temperatures to digest a trypsin-resistant core caused by disulfide bridges in the substrate protein. Trypsin and chymotrypsin may be used together to obtain very short peptides with a restricted variety of C terminal residues. Alkylated ('modified') trypsin is now used for long incubations and for protein sequencing applications, since it is resistant to the autolysis that would otherwise reduce the efficacy of the enzyme and produce spurious peptides from trypsin.

Proteases that remain active when the substrate protein is denatured will ensure extensive proteolysis: for example, pepsin is active *in vitro* and *in vivo* at very acid pH, and thermolysin remains active at high temperatures. Many proteases remain active in the presence of chaotropic agents that may unfold the substrate protein, and the activity of proteases in SDS solution is particularly useful in the 'Cleveland mapping' strategy (15) (Section 2.4). Conversely, the activity of thermolysin can be decreased by lowering the temperature, and a high salt concentration (16) will inactivate chymotrypsin.

At the end of a digestion incubation the protease must usually be inactivated. Immobilized inhibitors can be used to remove some proteases from the incubation mixture. Natural protease inhibitors are polypeptides and proteins such as pancreatic trypsin inhibitor (BPTI, 6.5 kDa) and α2-macroglobulin (725 kDa). To avoid introducing a new peptide or protein into a mixture that may be processed for sequencing, either use a non-peptide inhibitor or rely on the boiling in SDS solution that precedes electrophoresis to inactivate the protease. *Tables 2* and *3* include information on such inhibitors and on active site titrants that can be used to quench the proteolytic activity. Choose a non-reversible reagent, since a reversible inhibitor will dissociate if the mixture is to be diluted for a further process.

Most digestions will proceed at neutral or slightly alkaline pH, for example in buffers based on 50–100 mM Tris–HCl, Hepes–KOH, or sodium phosphate pH 7–9. For some applications, a volatile buffer such as one of the following is useful:

- ammonium bicarbonate pH 7.8–9.0

- ammonium acetate pH 4.0

- *N*-ethyl(/methyl) morpholine-acetic acid pH 8.0–8.5

2.3 Controlled proteolysis and complete proteolysis

The two approaches described in this section give different but complementary information: controlled proteolysis depends greatly on the conformation of the substrate protein, while complete proteolysis reflects the sequence of the target.

Table 2. Commercially available proteases for use in peptide mapping

Enzyme	Sources	Mol. wt ($\times 10^{-3}$)	Specificity[a] optimum	pH conditions	Denaturing inhibitors[b] tolerated	Non-peptide conditions	Storage information[c]	Other
Chymotrypsin A$_4$[d] (α-chymotrypsin) EC 3.4.21.1	Bovine pancreas	25	Tyr–Y, Phe–Y, Trp–Y (Leu–Y, Met–Y, Ala–Y)	7.5–8.5	0.1% SDS; 5 M NaCl	PMSF, TPCK, Ag$^+$, Hg$^+$	Store dry at 5°C, stable for days at pH 3	A$_{280}$ = 20.4, stabilized by Ca^{2+}
Clostripain (clostridiopeptidase B) EC 2.3.22.8	Clostridium histolyticum	50	Arg–Y	7.4–7.8		Thiol reactants, PMSF	Store dry at 5°C, activate overnight	Requires 2.5 mM DTT, 1 mM Ca^{2+}
Elastase[d] EC 3.4.21.36	Porcine pancreas	26	Ala–Y, Gly–Y	8.5	0.1% SDS	PMSF	Stable at pH 3–6 for 6–12 months	A$_{280}$ (0.1 M NaOH) = 22, autolyses at pH ~ 8
Endoproteinase Arg-C (submaxillary protease) EC 3.4.21.40	Mouse submaxillary glands	30	Arg–Y	8–8.5	0.1% SDS	Acid (pH 2)	Stable in 1 mM HCl	Specificity broadens in long incubation; use at high ratio for short periods
Endoproteinase Glu-C (V8 protease) EC 3.4.21.19	Staphylococcus aureus V8	30	Glu–Y (Asp–Y)	7.8 & 4.0	0.5% SDS, 6 M urea, 5.5 M GdmCl	Monovalent anions	Stable to freeze–thawing in distilled water; precipitates at pH < 4	A$_{280}$ = 3.8, specific to Glu in NH$_4$HCO$_3$ buffers (pH 7.8, pH 4.0); in phosphate buffer pH 7.8, cleaves also at Asp

Enzyme / EC	Source		Cleavage specificity[a]	pH	Denaturants / conditions	Inhibitors	Storage / stability	Notes
Endoproteinase Lys-C EC 3.4.99.30	Lysobacter enzymogenes	33	Lys–Y	8.5–8.8	0.1% SDS, 5 M urea	TLCK	Stable one day at 4°C in pH 5–12	
Papain[d] EC 3.4.22.2	Papaya latex	23	Arg–Y, Lys–Y, Phe–X–Y	6.0–7.0	8 M Urea, 50% methanol	Hg^+, thiol reactants, heavy metals, PMSF, H_2O_2	Labile at acid pH, activate in reducing agents	$A_{280} = 25.0$
Pepsin[d] EC 3.4.23.1	Porcine stomach	35	Phe–Y, Leu–Y, pairs of non-polar residues	1.0	Acid pH denatures substrates	Neutral pH	Store in 10 mM HCl at 5°C	$A_{280} = 14.7$, unstable above pH 6
Thermolysin[d] EC 3.4.24.4	Bacillus thermoproteolyticus	38	X–Leu, X–Phe, (other non-polar residues)	7–9	50% Active after 1 h at 80°C; 1% SDS, 8 M urea, 2 M NaCl	Hg^+, Ag^+, 5 mM EDTA, 1,10-phenanthroline		Zn^{2+} essential for activity; stabilized by 2 mM Ca^{2+}
Trypsin[e] EC 3.4.21.4	Bovine pancreas	24	Arg–Y, Lys–Y	7.5–8.5	0.1% SDS, 6 M urea	TLCK, PMSF Ag^+	10 mg/ml in 1 mM HCl, 5°C	$A_{280} = 14.3$, inactivate contaminating chymotrypsin with TPCK

[a] The peptide bond cleaved by the protease is indicated by the dash C terminal to residue X, N terminal to residue Y. Secondary specificity is shown in brackets.
[b] See Table 3.
[c] The A_{280} given is the absorbance at 280 nm of a 1% (w/v) (10 mg/ml) solution in water unless otherwise stated.
[d] Proteases with a broader specificity, but which may be useful in extreme conditions (thermolysin, pepsin) or for controlled proteolysis.
[e] 'Modified trypsin' (Promega, Boehringer) is resistant to autolysis during a long incubation.

Table 3. Non-peptide inhibitors for proteases used in mapping

Inhibitor	Specificity	Stock solution[a]	Final (working) concentration[b]
4-(2-Aminoethyl) benzene sulfonyl fluoride (AEBSF)[c]	Serine proteases, irreversible	50 mg/21 ml water = 10 mM, stable at 4°C	1 mM or less, hydrolysed at pH > 8; use fresh if possible
(4-Amidophenyl) methane sulfonyl fluoride (APMSF)[d]	Trypsin-like serine proteases, irreversible binding to active site	5 mg/346 μl water = 50 mM, stable at −20°C	10–100 μM—dilute stock 1:500 or 1:5000 into mixture; hydrolysed rapidly in pH 7 buffers
3,4-Dichloro-isocoumarin[c,d]	Serine proteases, irreversible	10 mg/4.65 ml dimethyl sulfoxide or dimethyl formamide = 10 mM, stable at −20°C	5–100 μM—dilute stock 1:100 into mixture; hydrolysed rapidly in pH 7 buffers
E-64[c,d]	Thiol proteases; irreversibly forms thioether at active site	5 mg/13.99 ml = 1 mM (use water or 50% ethanol), store at 4°C	Dilute 1:100 to give final 10 μM. Stable at pH 2–10
EDTA-Na₂	Metalloproteases, chelates Ca²⁺, Zn²⁺	0.5 M at pH 8.9, 0.1 M at pH 7.0 (adjust pH with NaOH), store at 4°C	1–10 mM
Heavy metals: HgCl₂, AgNO₃ Phosphoramidon	Thiol proteases Thermolysin and other metallo-proteases from micro-organisms	Aqueous solutions 1 mg/ml water; also soluble in methanol and dimethyl sulfoxide	1 mM; remove EDTA and thiol reagents Dilute 1:200 to give final 8.5 μM
Phenylmethylsulfonyl fluoride (PMSF)[e]	Serine proteases, papain; irreversibly forms sulfonate at active site	1 g/28.7 ml 2-propanol = 200 mM, stable at 25°C	Dilute directly 1:200 for final 1 mM; add fresh to aqueous solutions
L-1-Chloro-3-(4-tosylamido)-7-amino-2-heptanone-hydrochloride (TLCK)	Trypsin-like serine proteases, some thiol proteases (not chymotrypsin); irreversible	50 mg/13.54 ml = 10 mM, stable at or below pH 6, e.g. 1 mM HCl, store at 4°C	100 μM; dilute stock 1:100
L-1-Chloro-3-(4-tosylamido)-4-phenyl-2-butanone (TPCK)	Chymotrypsin and papain; irreversible	250 mg/71 ml = 10 mM, store in ethanol, 4°C	100 μM; dilute stock 1:100

[a] Stock solutions made up to these concentrations will be stable for at least six months at the specified temperature.
[b] Working concentrations should be adjusted if a particularly high concentration of enzyme is used.
[c] Obtainable from Calbiochem.
[d] Obtainable from Boehringer.
[e] PMSF is the most toxic of these enzyme inhibitors: however, sensible precautions should be taken with all of them.

124

2.3.1 Controlled (limited) proteolysis

In their native conformation, most proteins are almost entirely resistant to attack by low concentrations of proteases. Proteolysis would be expected only at the most exposed or disordered region of the polypeptide backbone when a protease, however broad its specificity, is given only limited opportunity to attack the substrate protein. Many large proteins are composed of several discretely folded domains, some linked by disordered or exposed segments of polypeptide. Thus the mixture of peptides produced by controlled proteolysis can give valuable information on the accessibility of sites in the substrate, and on the changes that may follow a covalent modification or the binding of ligands (see *Figure 1*).

The following protocol may be used as an initial screening of proteases

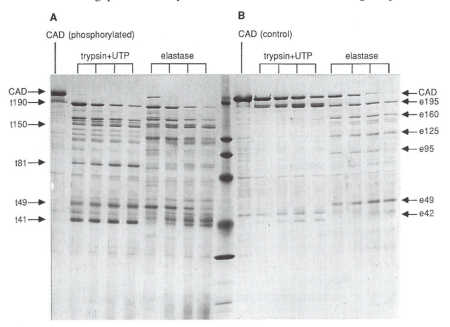

Figure 1. Controlled proteolysis: altered cleavage of ligand-bound or phosphorylated protein. The multienzyme polypeptide CAD was phosphorylated *in vitro* using ATP and the catalytic subunit of cAMP-dependent protein kinase (2,18). Both phosphorylated CAD and a control sample were incubated for 10 min at 37°C with 0.2 mM UTP before addition of trypsin, or for 10 min in buffer before addition of elastase, giving final weight ratios as follows: trypsin/phosphorylated CAD, 1/770; elastate/phosphorylated CAD, 1/115; trypsin/native CAD, 1/470; elastase/native CAD, 1/70. Samples were removed from the incubation immediately before the addition of proteases (undigested control), and after 10, 20, 40, and 60 min digestion. A stained SDS–PAGE gel of 5–15% acrylamide concentration is shown. (A) Illustrates that phosphorylated CAD is more readily digested by both elastase and trypsin than the control CAD. (B) Illustrates the restriction on tryptic cleavage caused by binding of UTP to native CAD. Note that this effect is abolished by phosphorylation of CAD. *Arrows* indicate the apparent molecular sizes of the major proteolytic products: on the left side, trypsin products; on the right, elastase products.

with a given substrate protein. Introduce variations by denaturing the substrate slightly, for example by heating or altering the pH of the solution.

Protocol 2.   Screening of proteases for controlled proteolysis

Equipment and reagents

- Substrate protein
- Digestion buffer (e.g. 50 mM Tris–HCl, or a volatile buffer—see text above)
- Protease stock solution (*Table 2*)
- Microcentrifuge tubes; screw-top microcentrifuge tubes
- Water-bath; boiling water-bath or heater block
- SDS–PAGE gel apparatus and slab gel (see Chapter 1)
- Molecular weight markers for gel system

Method

1. Dissolve the substrate protein (about 25 μg per tube) at 0.5–1.0 mg/ml in a buffer[a] that stabilizes its native structure. Allow solutions to reach the temperature of the incubation in a water-bath.

2. Screen several proteases of different specificities, including those that cleave at common residues (trypsin, chymotrypsin) and enzymes with broader specificities (elastase, thermolysin). Dilute the stock solutions, or dissolve directly, into the incubation buffer. Add to the substrate protein to give weight ratios of protease/substrate in the range 1/100–1/1000. Adjust the contents of all the incubation tubes to the same final volumes.

3. Return tubes to water-bath to begin incubation. At timed intervals, such as 15, 30, 45, 60 min after the start, remove portions containing 5 μg starting material from the incubation tubes into screw-top microcentrifuge tubes; mix with SDS sample buffer,[b] and immediately boil the sample for 2 min to ensure that the protease is denatured.

4. Apply the samples to an SDS slab gel to ensure the same running conditions (see Chapter 1). A full-size gel (14 cm wide) can accommodate 20 lanes, each requiring at least 3 μg peptide material. Mini-gels (8 cm wide), or the even smaller Phast Gel System from LKB, will require less material and a shorter running time. Run a lane containing molecular weight markers, and control lanes with protease alone and substrate alone. Significant autolysis of the protease should not occur in a short incubation, so it should be present as a single band in the digestion mixture.

5. Fix and stain the gel in the normal way (see Chapter 1), or transfer peptides to nitrocellulose for further treatment of the electroblot (see Section 4).

[a] Avoid buffers containing potassium ions since the resulting precipitate of potassium dodecyl sulfate in the sample buffer will cause peptides to aggregate in the wells before electrophoresis.
[b] Use the standard recipes for the Laemmli (17) system (see Section 3.1 and Chapter 1).

2.3.2 Complete (exhaustive or total) proteolysis

Total proteolysis at individual specific residues generates a number of peptide products that can often be predicted from the amino acid analysis of the protein. The method is used for comparative mapping of proteins with similar sequences, or to obtain short peptides whose sequence or degree of post-translational modification can be analysed (see Section 5 and Chapter 4). Complete digestion at all the relevant sites is ensured by exposing the substrate to a high ratio of protease and extending the incubation time. Most enzymes will also autolyse under these conditions, giving rise to spurious peptides that may complicate any efforts at sequencing fractions from the digestion mixture, so a modified trypsin is recommended.

If the substrate protein has been chemically modified and then precipitated by trichloroacetic acid (TCA), the pellet should be washed several times with cold water. Ammonium sulfate pellets can usually be resuspended, and desalted by gel filtration if necessary. Proteins, or large fragments resulting from limited proteolysis, can be separated by SDS electrophoresis, then completely digested from the surface of electroblots or from a slice cut out of a polyacrylamide gel (Sections 5.1, 5.2, and Chapter 1).

Protocol 3.   Complete proteolysis

Equipment and reagents
- Substrate protein and buffers as in *Protocol 2*
- Centrifugal evaporator
- 'Modified' trypsin (Promega, Boehringer)

Method

1. Resuspend the substrate and protease in an appropriate buffer; a volatile buffer is usually chosen where the digest will be concentrated or dried down at the end of the incubation.

2. Add the protease at a relatively high ratio to the substrate, such as 1/10 (w/w). Incubate at 25–30°C for an overnight period (16–24 h). 'Modified' trypsin is a good first choice for this step, since it will remain active during the incubation; otherwise, add additional aliquots of the protease during the long incubation.

3. At the end of the incubation, stop the reaction by lyophilization or freeze the samples and concentrate them by centrifugal evaporation to remove the buffer. Resuspend the products and separate them by high-pressure liquid chromatography (HPLC). Each peak fraction should correspond to a peptide with a different sequence (see Section 5.3).

2.4 Proteolysis in SDS solutions

This procedure is the basis for the popular 'Cleveland' two-dimensional map (15), widely used for comparison of proteins from different sources. An SDS-containing solution may also be used as a mild denaturant before digestion of a substrate as described in the first protocol. Proteolysis is extensive but not total, probably because the protease itself is partly denatured in the presence of SDS, and the peptides are usually large enough to be separated effectively by SDS electrophoresis in a 15% acrylamide gel (see Chapter 1).

The protease from *Staphylococcus aureus V8* (endoproteinase Glu-C) is widely used in 'Cleveland' mapping. Several other proteases, shown in *Table 2*, are also stable in low concentrations of SDS. These include elastase, chymotrypsin, papain, and subtilisin. A screening experiment, similar to that described in *Protocol 2*, can be set-up to identify the most useful protease. Combinations of these enzymes could be used in SDS-containing solutions to increase the extent of cleavage. Ions or chelating agents required by some proteases can be included in the stacking gel solution or the running buffer.

Protocol 4.   Digestion of purified proteins in the presence of SDS

Equipment and reagents

- Substrate protein
- Digestion buffer: 0.1% SDS, 10% (v/v) glycerol, 1 mM dithiothreitol (DTT), 25% (v/v) stacking gel buffer[a]
- Bromphenol blue 0.1% (w/v) solution
- Protease (*Table 2*)
- SDS–PAGE equipment and slab gel (see Chapter 1)

Method

1. Resuspend the substrate protein in the digestion buffer at 0.5–1.0 mg/ml.

2. Boil the solution for 2 min to denature the protein.

3. After cooling, add protease to digest the substrate protein. Boil the sample again (and add bromphenol blue to a final concentration of 0.01% if desired) before loading into the well of the SDS electrophoresis gel. Run the gel in the standard way.

3. Alternatively, load the boiled protein from step 2 into the sample well of the polyacrylamide gel, overlay with up to 1 µg of protease in the same buffer (not boiled, no bromphenol blue), and begin electrophoresis. Turn off the current when the samples have stacked in the upper gel, and allow proteolysis to take place for 15–30 min. The protease is effectively removed from the substrate by restarting electrophoresis and continuing in the standard way.

[a] Laemmli gel (17) recipe, Chapter 1.

---

3. Separation of polypeptide fragments

Polypeptide fragments may be separated by any of the widely-used methods for protein purification, paying attention to the likely size of fragments resulting from digestion or chemical fragmentation. The methods usually depend on differences in size, charge, or ability to bind ligands; these differences can be exploited to identify sites at which the protein is modified or where ligands bind, or to obtain specific peptide fragments for sequencing.

The trend towards miniaturization in hardware leads to faster run times for most of these separation methods, and sensitivity in many of the subsequent analytical steps is now adequate for the small samples that will be processed.

3.1 Polyacrylamide gel electrophoresis and isoelectric focusing

3.1.1 Separation by size

In the presence of SDS and reducing agents, polypeptides are separated by PAGE according to the length of the polypeptide chain, which can be determined with molecular weight markers run in the same conditions. The recipes and methods for this technique can be found in Chapter 1. Some points that are particularly relevant to peptide mapping are as follows:

(a) The polyacrylamide gel has a molecular sieving effect, thus restricting the size range of peptides that can be resolved. Acrylamide of 5% or 7.5% forms a soft gel that will resolve proteins of 60–200 kDa, so a higher concentration is required for smaller peptides. The limit is around 20% acrylamide, which forms a brittle gel that is difficult to handle when removed from the glass plates after the run. The resolution is best when fresh buffers and acrylamide solutions are used. The range of separable peptides now extends down to about 2 kDa in the Tris–Tricine system (19); gels and buffers for this method can be purchased ready-made from Bio-Rad.

(b) The molecular weight markers must be selected carefully, according to the gel system that is used. Sigma sell a number of kits, including a mixture of proteins that is particularly suitable for the conditions of the Laemmli gel (SDS-6H), and a mixture of polypeptides for Tris–Tricine gels (MW-SDS-17S).

(c) Slices excised from SDS–polyacrylamide gels, whether from the first or second dimension, are a convenient source of purified, immobilized polypeptide that can be exposed to chemical or enzymic hydrolysis. Some acid hydrolysis may have taken place during staining/destaining in acetic acid/methanol solution (see Section 2.1). Reversible cross-linking reagents are available (Bio-Rad, Pierce) that allow the gel to be dissolved after electrophoresis. A protocol is given for digesting polypeptide in slices from the Laemmli gels *in situ* (*Protocol 5*). Another protocol appropriate for MS analysis is given in Chapter 2, *Protocol 6*.

This method and variations on this theme have been widely used in our unit to generate sequence information either from protein subunits that can only be resolved by SDS–PAGE, or from proteins that are difficult to purify to homogeneity by conventional chromatographic methods. The method was used to identify the regulatory subunits of the skeletal muscle M-subunit and liver G-subunits of protein phosphatase 1 (20,21), the identity of a novel liver protein kinase (22), and the identity of the proteins associated with c-raf, such as the 14-3-3 proteins and heat shock protein Hsp 90 (unpublished work). Large and hydrophobic peptides do prove difficult to extract from the gel piece during digestion, but extractions with strong denaturants or acids/bases have been successful. However, by modifying this technique for 'in-gel' CNBr digestion in 70% formic acid, some of these problems can be overcome, and this method was used successfully to determine the carboxy terminus of a recombinant variant of MAPKAP kinase-2 that had been C terminally truncated in the expression system and was no longer a substrate for its 'upstream' activating protein kinases p38 RK or MAP kinase (23). The possible combinations of proteolytic or chemical digestions of proteins in gels is wide and varied and relies purely on the use of PAGE to separate the protein(s) of interest.

Protocol 5. Digestion of polypeptides *in situ* in gel slices

Equipment and reagents

- Stained/destained gel[a]
- Protease (alkylated trypsin)
- Microcentrifuge tubes, water-bath, other equipment as in *Protocol 3*

Method

1. Place gel on a clean glass plate and cut out the relevant band(s). Cut the bands into small cubes and place into a microcentrifuge tube. Prepare a blank digestion by using a similar sized piece of the gel which is devoid of protein and treat as in steps 1–5.

2. Wash the gel slices in Milli-Q water (five changes of 1 ml) for about 1 h to remove the acetic acid.

3. Take the gel pieces to near dryness by centrifugal evaporation.

4. Suspend the gel pieces in 0.2–0.5 ml of 50 mM Tris pH 8, 0.01% (v/v) alkylated Triton X-100 (used to assist the extraction of the peptides whilst being nearly undetectable at 214 nm during HPLC analysis) containing 1 μg of alkylated trypsin or other protease, and incubate at 30°C for 20 h.

5. Remove the supernatant, extract the gel piece with 0.3 ml buffer without trypsin, and combine extracts. The extracts can separated by HPLC immediately or stored frozen until required. Comparative maps of the blank digestion versus the protein samples can identify artefact or protease-derived peaks from the peptide map.

[a] Stain gel with Coomassie blue as normal (see Chapter 1, *Protocol 6*), but avoid touching the gel with bare fingers, then destain. For > 5 μg/band, stain for 5 min then destain only long enough for the bands to become visible.

3.1.2 Separation by charge

In the absence of SDS, polyacrylamide gel electrophoresis separates polypeptides according to their charge in the prevailing buffer system, with some molecular sieving by the gel (see Chapter 8). Native gels can be run cheaply, with the same equipment as SDS–PAGE, and the pH can be chosen from a wide range (24). After electrophoresis, some enzymes can be assayed *in situ* by diffusing the substrate into a native gel (25) (see Chapter 1), and the protein itself can be digested in a similar protocol to that given for SDS gel slices (*Protocol 5*). Low concentrations of polyacrylamide may be necessary if polypeptides remain associated after controlled proteolysis of a substrate protein; the corollary is that peptides that do associate can be identified by subsequent electrophoresis in dissociating conditions (see Section 3.5). Native gels are not widely used in peptide mapping because similar information about charge can be obtained by isoelectric focusing (IEF).

IEF uses very low concentrations of polyacrylamide to support proteins that are separated in a pH gradient generated by the current passing through a mixture of ampholytes. The proteins move along the pH gradient until they reach their isoelectric point; large molecules may take several hours to reach equilibrium. Proteins or peptides differing in a single charge, for example caused by post-translational modification, can be separated by IEF before the site(s) for modification are identified by comparative mapping of digestion fragments from each. IEF can be carried out in the presence of 9 M urea or SDS, if non-ionic detergents are also present (26,27), which will dissociate any non-covalent complexes between fragments. The use of analytical IEF to compare the final charge on small radiolabelled fragments resulting from complete proteolysis of fragments from a larger protein is described in *Protocol 9*.

3.2 Chromatography

3.2.1 Gel filtration

Gel filtration is a familiar method for resolving proteins according to size (see Chapter 9). In non-denaturing buffer, structured fragments or domains that associate to form oligomers or intersubunit complexes will elute from the column at positions corresponding to larger sizes than would be expected from the length of their polypeptides. In denaturing solutions such as urea, polypeptides will elute according to their individual size. The fractions can be assayed for enzyme activity or other properties to identify the interacting fragments.

3.2.2 HPLC

Liquid chromatography at high pressure (HPLC) and sensitive detection methods allow rapid and reproducible separation of complex mixtures of peptides. Reversed-phase C_{18} columns separate peptides of mass up to 50 kDa according to their hydrophobicity in a gradient of TFA/acetonitrile. Resolution depends on the steepness of the increase in acetonitrile concentration, so the gradient should be programmed to be more shallow around the concentration where more than one of the peptides of interest elute. The elution of short peptides should be monitored by absorbance at a wavelength in the far-UV (214 nm), since few will contain the aromatic side chains that contribute to the absorbance at 280 nm. Complete proteolysis of a ^{32}P-phosphorylated protein will liberate a small number of labelled peptides, which can be detected by continuous monitoring of Cerenkov radiation from the eluate as shown in *Figure 2*, or by sampling the ^{32}P-labelled fractions. The excellent reproducibility of HPLC runs means that several similar samples can be compared directly in parallel conditions; for example, the extent of phosphorylation at a given site (defined by the elution position of a small peptide after complete proteolysis) can be compared according to the hormone treatment of the cells from which the protein was purified (28).

3.3 Capillary zonal electrophoresis (CZE)

This technique has evolved into one of the most powerful techniques for separating peptides and proteins. Samples as small as 1 nl can be loaded into a fused silica capillary (25–140 cm long with internal diameter of 50 μm) and then electrophoresed across a very high potential (15–30 kV). Peptides are detected by their UV absorption at 200 nm with the capillary itself being the flow cell. The polyimide coating of the fused silica is burnt off, and the clear silica capillary runs between the light source and the photomultiplier tube of a UV detector. In 'classical' free solution CZE, peptides are separated according to their charge, usually in acidic citrate or phosphate buffers, with the detector end of the capillary being at the cathode and the injector end at the anode. In this configuration, basic peptides migrate faster than neutral or acidic peptides. Changing the pH of the solution and the polarity of the electrodes reverses the separation characteristics. New coated capillaries and gel-filled capillaries allow the separation of large polypeptides, which previously adsorbed to the charged silanols of uncoated fused silica capillaries. The examples of separation strategies are far too numerous to list here. How-

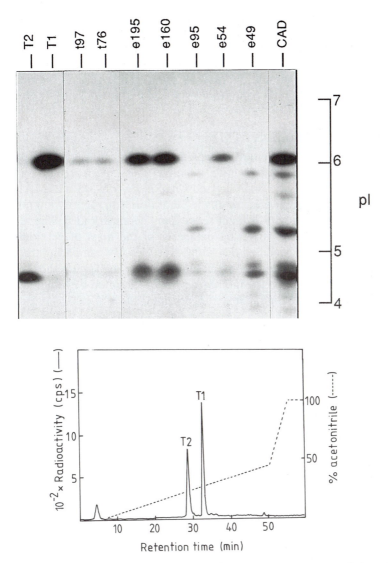

Figure 2. Use of HPLC and analytical IEF to identify radiolabelled peptides. Selected ^{32}P-labelled bands were cut from the electroblot shown in *Figure 3* and exposed overnight to concentrations of trypsin as described in *Protocol 9*. The resulting short radiolabelled peptides were analysed by thin-layer IEF. (A) Autoradiograph of the dried IEF gel, showing the lanes identified according to the proteolytic fragments that were exhaustively digested. Coloured proteins were used as markers to estimate the pI values. (B) Authentic peptides T1 and T2, which were sequenced and also used as IEF markers, were obtained from exhaustive trypsin digestion of TCA precipitated phosphorylated CAD. The resulting short peptides were separated by reverse-phase HPLC on a C_{18} column in 0.1% (v/v) TFA using a linear gradient from 0–40% (v/v) acetonitrile. The radioactivity in the eluate was monitored continuously.

ever, the manufacturers of CZE units, such as Applied Biosystems, Bio-Rad, Beckman, and Waters provide application notes for a wide range of peptide separations. The Beckman P/ACE system has introduced very sensitive laser-induced fluorescence detection. A recent publication separated fluorescein-derivatized phosphopeptides in this way, measuring less than 75 attomoles of phosphoserine-containing peptides (29).

3.4 Mass spectrometry (MS)

Modern technology for peptide mapping of low abundance proteins has been complemented by the development of novel mass spectroscopic techniques ideally suited to peptide analysis (see Chapter 2). ES MS and MALDI MS have been used to great effect in many laboratories to identify unknown proteins in combination with Edman sequencing chemistry (30). The technique has recently been used to set-up databases of the mass of peptides arising from a protein after specified digestion strategies. By comparing the peptide mass fingerprint obtained from an unknown protein with those in mass databases, the identity of the unknown protein can often be assigned, as each protein should have a unique mass fingerprint (31).

3.5 Combining techniques to generate maps

Several of the methods that have been described already could be combined to generate a two-dimensional map of peptide fragments. For example, a purified protein could be incubated with proteases, the fragments separated by HPLC, and fractions applied consecutively to the wells of a slab gel; the resulting spectrum of fragments would have been separated according to hydrophobicity in the first dimension, by size in the second dimension. In another combination, non-covalently associated proteins or large proteolytic fragments may run together in a 'first dimension' native gel; they would then be separated by the SDS in a perpendicular run using the Laemmli gel system. The map in turn may be analysed further, by antibody cross-reaction on blotted proteins, or by excising bands for sequencing: these approaches will be covered in the final sections.

3.5.1 Two-dimensional mapping

The two-dimensional method developed by O'Farrell (32) for the comparative mapping of proteins obtained from cell extracts has been widely used in peptide mapping. Very small quantities of proteins could be detected because the cell extract had been biosynthetically labelled with ^{35}S or radioiodinated (33). Sensitive staining methods or ^{32}P-labelling can be used to detect peptide products from digestions. The method, which resolves proteins according to charge and size, is described fully in another volume of this series (34). The following scheme could be followed to map peptide products.

Protocol 6. 'Two-dimensional' mapping of peptides

Equipment and reagents
- Substrate protein, protease, volatile buffer (*Protocols 2* and *3*)
- IEF equipment and rod gel, or Multiphor (Pharmacia) and Immobiline dry strips of appropriate pH range[a]
- SDS–PAGE equipment and slab gel
- Coloured proteins for pI markers (Merck)
- Molecular weight markers for SDS–PAGE
- Silver stain (see Chapter 1, *Protocol 7*)

Method

1. Incubate substrate protein with protease in a volatile buffer. Dry down at the end of digestion.

2. Resuspend the sample in water and apply to the top of an IEF rod gel, or to an Immobiline strip (Pharmacia). Use a broad range of ampholytes to cover the likely isoelectric points of the peptides. Run until the peptides have reached equilibrium. The position of coloured proteins used as markers will indicate whether the peptides have reached their final positions.

3. Place the IEF rod gel or strip between the glass sheets used to make a slab gel, positioning the IEF gel to lie at the top of the separating gel. Choose an appropriate concentration of acrylamide for the SDS–PAGE gel, or use a gradient of 7.5–15% acrylamide for good separation of a wide size range of peptides. Pour the separating gel, and when it is set, pour the stacking gel to polymerize around the IEF gel. Use a one-toothed comb or a nylon spacer strip to form a well for molecular weight marker mixtures.

4. Run the gel as usual. Stain with Coomassie blue, or with a silver stain to detect smaller quantities of peptide material (see Chapter 1). The mapped proteins may also be transferred to an electroblot.

[a] Ensure that mercaptoethanol is not used as a reducing agent in the IEF gel, as it may inhibit polymerization of the acrylamide in the second dimension (34).

3.5.2 Off-diagonal maps

The concept of diagonal mapping is slightly different from the two-dimensional methods discussed previously, and it could be better named off-diagonal mapping. Instead of separating a mixture of peptides according to two different criteria, the same separation method is used on a sample before and after modification to alter the properties of the selected peptides. Unmodified peptides will travel to the same position as before, but modified peptides will fall, actually or notionally, 'off-diagonal'.

The concept was first used (35) to identify, by paper electrophoresis, peptides whose charge had been altered by chemical treatments such as maleylation of lysine residues or cleavage of disulfide bonds (to give two peptides of different charges). A strip of the electrophoretogram was exposed to the chemical treatment before being sewn in place on a second sheet of paper for electrophoresis in the perpendicular direction. Clearly a similar principle could be applied to separation on gels, such as IEF or electrophoresis, which are now used in preference to paper separations. Many reagents will be able to permeate a gel of low acrylamide concentration (5–7.5%), which could then be applied to the origin of a second gel run under the same conditions.

An HPLC-based application identified peptides containing phosphoserine residues by comparing their elution in reversed-phase chromatography before and after chemical modification of the phosphoserine groups (36); elution of the modified peptides was retarded.

Alterations in size can be observed by electrophoresis before and after cleavage of covalent bonds. The most obvious application is the digestion of large polypeptide fragments *in situ* in a polyacrylamide gel, followed by a second SDS–PAGE step to identify fragments that are partial hydrolysis products of the same original protein.

Protocol 7. Off-diagonal 'Cleveland' mapping on SDS–PAGE

Equipment and reagents
- Stained or unstained SDS slab gel[a]
- Protease and digestion buffer (*Protocol 4*)
- Molecular weight markers

- SDS–PAGE equipment and slab gel (see Chapter 1)

Method

1. Cut an entire lane from a stained or unstained SDS slab gel in which proteins or large peptide fragments have been separated (37). Alternatively, use individual bands cut either from wet gels or from dried gels (for example, those used for autoradiography).

2. Equilibrate the separate bands or the long slice thoroughly in digestion buffer containing 0.1% SDS, being certain to remove acetic acid from the slice. Unfixed gels that have been dried between sheets of cellophane may be successfully rehydrated in this buffer.

3. Carefully lay the gel slice along the top of a slab gel. Bio-Rad sell a '2D comb' that can be used to form a stacking gel with one long section, suitable for this application, and an adjacent lane for molecular weight markers. Alternatively, push the hydrated gel pieces into the wells of a second gel. Run a lane with molecular weight markers.[b]

4. Overlay with buffer and protease solution and proceed with electrophoresis and digestion as described in *Protocol 4*.

5. Stain and destain the slab gel, plus an equivalent slice from the original

---

gel. You will find families of fragments, running down the second slab gel, that derive from the larger fragments or proteins in the first dimension. Large proteolytic fragments will generate a spectrum of smaller fragments that are found in related fragments or the parent protein.

[a] Mercaptoethanol, which may be present in some unstained gel slices, and which is also required by papain, will inhibit polymerization of the acrylamide in the second dimension (34,38); dithiothreitol (DTT) should be used instead because it is effective as a reducing agent at lower concentrations (e.g. 1 mM).
[b] Remember to use appropriate concentrations of acrylamide and the correct range of molecular weight markers to cover the different size range of the peptides in the two dimensions.

Peptides linked by disulfide bridges will remain associated in a first dimension run without reducing agents, but they would be separated by the mercaptoethanol in the sample buffer used in the Laemmli system. An extension of that concept has been used to follow the cross-linking within a multi-subunit enzyme, pyruvate dehydrogenase (39), by separating in the first dimension by SDS–PAGE, followed by treatment of the tube gel (or slice from the slab) with a reagent that will break the cross-link. In this case, the *bis*(imidoester) link was broken by methylamine/acetonitrile treatment. In theory, the cross-linked proteins could have been digested to yield a population of fragments, pairs of which would have remained cross-linked. These peptides, having been isolated on an off-diagonal map as above, could be sequenced to identify the residues at which the protein surfaces interact in the complex.

4. Electroblots as the second or third dimension

The first step of the 'Western' blotting procedure is the transfer of proteins and polypeptides from gel slabs to blotting membranes (40): whether the proteins have been run in parallel lanes or separated in two dimensions, the blot (electroblot) will accurately represent the positions reached in the gel by the different species, which can now be identified through their antibody-binding ability or through sequencing of the polypeptides from the surface of the gel. Because the polypeptide is adsorbed at or near the surface of the membrane, it remains accessible to chemical or enzymic attack.

4.1 Electroblotting procedure

A procedure for electroblotting is given in Chapter 3, *Protocol 12*. While several hours are usually required to stain and destain a full-size gel, an electroblot can be obtained in less than one hour using a semi-dry blotting system such as the Novablot (Pharmacia), the SemiPhor (Hoefer), Trans-Blot and Mini Trans-Blot (Bio-Rad), and the Fast Blot (Biometra). Gels and blotting membranes may be stacked between paper in the semi-dry blotter to allow the transfer of proteins from several gels under identical conditions. Pharmacia and Bio-Rad also sell 'wet' transfer equipment that uses the same tank as the slab gel apparatus. Because the electrodes are farther apart, the latter type of transfer takes longer than semi-dry blotting, but in some mapping protocols an overnight transfer in the wet system may be more convenient. Alternatively, if a higher voltage is used, most proteins will be transferred quantitatively to nitrocellulose in 4–6 h, allowing time to begin an overnight incubation of the electroblot with specific antibody. The wet system is considerably cheaper to set-up than the semi-dry equipment, although it uses much larger volumes of buffer for each transfer.

The choice of membrane is an important factor in peptide mapping. Small fragments, such as the 12 kDa peptides that could be resolved in a 15% acrylamide SDS–PAGE gel, might be 'lost' during transfer to a nitrocellulose membrane; therefore small pore size, 0.22 μm rather than 0.45 μm, should be chosen. Even after transfer, small peptides that are visible in autoradiographs may be lost during the washing and blocking steps of the immunostaining procedures, as can be seen in *Figure 2*.

Membranes based on nitrocellulose, such as Hybond (Amersham), are useful for analysis of proteins and polypeptides. Hybond C is a more robust membrane, which will allow reprobing (washing to remove an antibody–enzyme–substrate mixture followed by a new series of steps); Hybond ECL Western is more fragile but gives excellent results in the chemiluminescent detection methods. Other companies that supply nitrocellulose membranes are Bio-Rad, Pharmacia, Millipore, Schleicher and Schüll. Nitrocellulose membranes are wettable in aqueous buffers. The PVDF (polyvinylidene difluoride) membranes available from several suppliers are more robust and have a higher binding capacity and a higher affinity for small proteins and peptides.

4.2 Identification of polypeptide fragments

4.2.1 Protein staining

It is useful to locate proteins and polypeptides on the blotting membranes, for example before cutting parallel lanes for probing with different antisera, and a reversible stain is preferred. After removing the electroblot from the transfer apparatus, rinse nitrocellulose membranes thoroughly in water, and PVDF membranes in methanol. Transfer nitrocellulose membranes to a solution of 0.2% (w/v) Ponceau S in 1% (v/v) acetic acid and agitate gently; pink stain should be visible within a minute (41). Rinse rapidly in water to remove the background stain, and allow the blot to dry in air. Nitrocellulose can be restained if rinsing has gone too far. PVDF is wettable in 100% methanol, so it can be reversibly stained with Coomassie blue R-250 (0.1% (w/v) in 50% methanol) to identify interesting protein bands (42). Rinse in

50% methanol; rewet the dried blot with 100% methanol to remove the stain from bands, and proceed to immunostaining (*Protocol 8*), or to sequencing directly from the blot (see Section 5.1).

Record the position of the stained bands on a Polaroid photograph or by tracing through a lightweight acetate sheet with a permanent overhead projector pen. Mark molecular weight marker bands in pencil on the membrane itself. If pre-stained marker proteins are used, these will still be visible on the electroblot.

4.2.2 Specific antibodies

Antiserum raised against an intact protein will cross-react against nearly all the fragments from limited proteolysis, and with common epitopes on related proteins (see Chapter 3). Antisera with limited specificity will identify only fragments derived from the domain against which they were raised. To raise antibodies against defined fragments, the products of limited proteolysis may be separated chromatographically and the fraction injected into rabbits with an adjuvant. This method was used to raise the sera that demonstrated (see *Figure 3*) that a large fragment from the elastase digestion of CAD, a multi-enzyme polypeptide, contained both dihydroorotase and aspartate transcarbamoylase domains (3,43). If the immunogen is scarce or difficult to purify, the appropriate slice from a gel or from an electroblot may be macerated (44) and used directly for inoculation. Fragments of 12 kDa can readily be resolved in electrophoresis in SDS–PAGE and will be large enough to acquire structure in solution. It may be preferable to extract peptides from the gel by macerating it in a solvent and then removing the gel pieces from the solution (for example, by concentrating the solution in a Microcon tube with a Micropure insert—both available from Amicon). Nitrocellulose is soluble in dimethyl sulfoxide; small quantities can be emulsified with adjuvant, but glass syringes should be used.

Immunostaining methods rely upon a primary antibody specific to the protein of interest, and then upon a detection system based on a secondary antibody and an amplification step that marks the presence of the antibodies, and hence the antigen, at that position on the electroblot. Enzyme-linked systems have now almost completely replaced radioiodinated antibodies as the detection method. The enzymes horse-radish peroxidase and alkaline phosphatase have been widely used: both colorimetric and chemiluminescent substrates (e.g. the system from Amersham) can be used with each. The colorimetric substrates form a precipitate that remains on the membrane after the substrate solution has been washed away, but the blot must be protected from light to prevent fading. The blot should be photographed as soon as possible. The chemiluminescent substrate emits light when it is hydrolysed, so its detection requires X-ray film and dark-room facilities. The film gives an exact record of the position of the enzyme labelled antibody, and by adjusting the time of exposure (30 sec to 2 min), a quantitative measure of the amount of

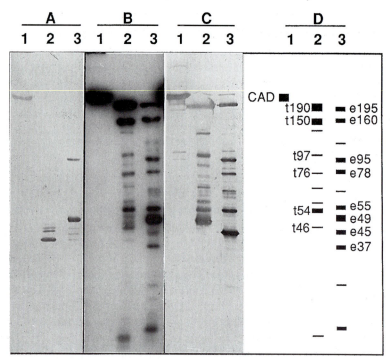

Figure 3. Use of the immunoblot to map immunoreactive and radiolabelled products of controlled proteolysis. CAD that had been phosphorylated using [γ-^{32}P]ATP was incubated without protease (lane 1), with trypsin(1/1000 for 10 min, lane 2), or with elastase (1/150 for 15 min, lane 3). Identical aliquots of each mixture were separated in parallel by SDS–PAGE and electroblotted on to nitrocellulose. Autoradiography was used to detect ^{32}P-labelled peptides, and the same electroblots were then used for immunostaining, followed by tryptic digestion of selected regions (*Figure 2*). Use of the same blots allowed accurate alignment of radioactive and immunostained peptides. (A) Immunostained using antiserum raised against the fragment of CAD corresponding to aspartate transcarbamoylase activity. (B) Autoradiograph of electroblot used in (A). (C) Immunostained using antiserum raised against the dihydro-orotase fragment of CAD. (D) Sketch of panel B showing ^{32}P-labelled proteolytic fragments, identified according to the protease used and the molecular size of the peptide.

bound antibody can be obtained. Usually the signal decays within an hour of adding the substrate to the immunostained blot, but Pierce have recently advertised a system whose signal remains high for 2–4 h.

Most manufacturers who sell membranes and blotting systems also sell the immunostaining reagents in kit form. These contain excellent instruction sheets that have been optimized for the reagents in the kits, so a very brief protocol is given here. The specific reagents can then be purchased individually and the buffers made up in your own laboratory. A general method of immunostaining is given in *Protocol 8*; another is given in Chapter 3, *Protocol 12*.

Protocol 8. General method for immunostaining

Equipment and reagents[a]

- Flat dishes or plastic boxes a little larger than the electroblot
- Shaking platform
- Enzyme-linked secondary antiserum (raised against IgG from the species that provided the primary antiserum)
- Enzyme substrate (and hydrogen peroxide if peroxidase is used)

- Primary antiserum
- Tris-buffered salt solution: 25 mM Tris–HCl pH 7.5, 0.5 M NaCl
- Detergent solution: Tween 20 or Triton X-100, 0.1% (v/v) in the Tris–NaCl solution
- Blocking protein: bovine serum albumin or dried low-fat milk

Method

1. Wet the blot in water (nitrocellulose) or methanol (PVDF membrane), after marking the molecular weight standards and other features, such as lane numbers.

2. Transfer the wet membrane immediately to the Tris-buffered salt solution and agitate gently for 1 h in a 3% (w/v) solution of blocking protein in the same solution.

3. Wash the membrane, and place it in fresh Tris–NaCl solution containing the primary antibody (1:100 to 1:2000 dilution by vol.). Agitate gently for 1 h at room temperature.

4. Wash the membrane with several changes of the detergent solution, 3 × 10 min.

5. Agitate gently in fresh detergent solution containing the secondary antibody (1:2000 by vol.) for 45–60 min at room temperature.

6. Wash the membrane with several changes of the detergent solution.

7. Soak the membrane in a solution of the substrate. Have ready a dish of ice-cold water to quench colorimetric reactions, or film and developing reagents for chemiluminescent substrates.

8. Dry blots between sheets of filter paper. Record stained bands.

[a] All except the first three items are usually available as a kit.

4.2.3 Radioactive labels

Radiolabelled peptides should be found after digestion of proteins whose side chains have been labelled with ^{32}P by protein kinase, or chemically with radioactive active site analogues. Any covalently-bound radioactive or fluorescent group, if it is not hydrolysed during the digestion and separation of the protein, may be used to indicate the site at which it is attached. The following protocol was used to identify ^{32}P-labelled peptides by analytical IEF (3,4) as seen in *Figure 3*.

Protocol 9. Identification of radiolabelled peptides by analytical IEF

Equipment and reagents

- Reversibly-stained electroblot (see Section 4.2.2)
- X-ray film and cassette, or Phosphor Imager screen and cassette (Molecular Dynamics)
- Volatile buffer: 0.1 M *N*-ethyl-morpholine/ acetic acid pH 8.3

- Blocking solution (see *Protocol 8*) or use 0.5% (w/v) polyvinylpyrrolidone in 0.1 M acetic acid (45).
- Protease: alkylated trypsin (see *Protocol 3*)
- Flat-bed IEF gel and apparatus (e.g. Bio-Rad Mini IEF cell)
- Coloured pI marker proteins

A. *Location of radioactive bands on the electroblot*

1. (a) If weak β-emitters are to be located, mark the corners of the membrane asymmetrically with tiny dots of India ink mixed with waste ^{14}C-labelled reagent, if necessary, and wrap carefully in Saran Wrap or cling film. Place the membrane against a PhosphorImager screen in a cassette for at least three days.

   (b) If the peptide is labelled with ^{32}P, proceed under safe-light in a dark-room. Tape the membrane to an acetate sheet cut exactly to the size of an X-ray cassette, making certain that the top right corner (for example) corresponds to one unique position on the X-ray film. Insert into the cassette, with a sheet of cling film between the membrane and the film emulsion, and keep in the deep freeze for several days.

2. Develop the X-ray film or scan the Imager screen.

3. Align the membrane with the image; a light box will be useful whether it is on film or on paper.

4. Cut out the bands of interest with a new scalpel blade, and place each into a labelled microcentrifuge tube. The Cerenkov radiation from ^{32}P-labelled fragments can be detected from the dry pieces of membrane. Fragments containing as little as 500 c.p.m. can be used in the next step.

B. *Exhaustive proteolysis of electroblotted proteins*

1. 'Block' the membrane surface, to prevent adsorption of the protease, by immersion in the blocking solution for at least 1 h at room temperature.

2. Rinse the membrane pieces with several changes of the volatile buffer.

3. Incubate each piece in a minimum volume of the same buffer, and add alkylated trypsin to a final protease/substrate ratio of 1:10 (w/w).

4. After 16–24 h digestion at 37°C, remove the supernatants into clean

microcentrifuge tubes; wash the pieces with buffer, and add the wash-ings to the supernatants.

5. Dry the peptide solutions by centrifugal evaporation; count the Cerenkov radiation in the dried supernatants, plus that remaining on the membrane to confirm that the radiolabelled peptides have been released by the protease.

C. *Analytical IEF*

1. Prepare a slab gel on which several samples can be analysed by flat-bed IEF, or use ready-made gels on plastic backing film. Phosphory-lated peptides tend to be acidic, so the labelled bands may be found in the low pH region of the gel.

2. Resuspend the dried samples and apply to the gel; apply coloured pI marker proteins. Run according to the instructions supplied with the flat-bed equipment.

3. Mark the positions of the coloured markers on the backing sheet, as the colours fade during drying. Measure and plot the distance travelled against the pI of the standards. Dry the gel.[a]

4. Place the dried gel into a polythene bag and expose to X-ray film or to the PhosphorImager.

5. Read off the pI values of the labelled peptides against the plot of the marker proteins. A series of radiolabelled bands in one IEF sample indicates a population of peptides labelled to different extents at more than one site.

[a]Gels containing glycerol will not dry fully.

5. Sequencing strategies

In many cases, sequencing is the final stage of a modern peptide mapping procedure. Gas phase sequencers are now so sensitive that less than a 10 pmol of protein can be used for a partial sequence, sufficient to place the peptide within the sequence of the protein. The use of sequencing to identify the residue at which a protein is phosphorylated is described in the final section.

5.1 Use of the electroblot

Internal amino acid sequences can be obtained by digesting the proteins on the surface of the blot, as described in *Protocol 9*, using 5% acetonitrile as the solvent (45). Short peptides liberated into solution are separated by HPLC, and selected fractions may be used in the gas phase sequencer. Electroblots may also be used as supports for sequencing from the N terminal residue of a fragment. The membranes used in this application have a high binding capacity, and must be robust. Immobilon (Millipore) and ProBlott (Applied Biosystems) are excellent. The general procedure for running gels is followed, with the precaution of adding thioglycolate in the SDS–polyacrylamide gel to prevent oxidative blocking of the amino terminus during the run. The blotting procedure uses buffers such as Caps (46), *N*-ethyl-morpholine (47), or borate, which will not interfere with the sequencing steps or provide spurious glycine residues.

5.2 Sequencing from gel slices

Direct protein sequencing from gel slices is difficult, but the protein in the gel slice can be eluted into conditions suitable for protein sequence analysis. The gel slice, preferably unstained and not fixed, can be electroeluted or passively eluted in buffers with or without SDS (the larger the protein, the greater the need for SDS). The sample is now in a very dirty condition for sensitive protein sequencing, but protein can be precipitated with either TCA or cold methanol/ethanol to remove buffer/detergent contaminants. However, low levels of protein (0.5–5 μg) or protein in relatively large volumes (0.5–5 ml) may not be quantitatively precipitated or resolubilized.

Other possibilities are the inverse chromatography strategies devised by Simpson and co-workers (48) and the concentration of the eluted protein on a PVDF membrane in a microconcentrator as marketed by Applied Biosystems under the trade name ProSpin. This technique relies on the same blotting membrane described previously (Section 5.1): the protein adsorbs to the polymer surface, and the contaminants are washed off the membrane with methanolic mixtures such as SDS–PAGE destaining solution. The membrane is 'punched out' of the microconcentrator and can be directly placed in the gas phase protein sequencer to generate contaminant-free sequence analysis.

A mixed hydrophobic/hydrophilic peptide support has recently been developed for use in the Hewlett-Packard protein sequencer. This support can be used on a work station to load peptides and proteins from previously unsequencable matrices and to remove the contaminants before sequence analysis. This novel matrix for peptide support and its combination with bidirectional flow protein sequence chemistry is a major advance in the field; further information should be obtained from the company.

5.3 Sequencing from HPLC fractions

It is crucial to emphasize that the quality of the solvents used in the HPLC determines the quality of the peptide sequence obtained, especially when there is less than 10 pmol of peptide. We recommend the use of HPLC grade acetonitrile, Aristar or sequencing grade ion-pairing agents such as TFA (Pierce sequencing grade is highly recommended) and, most importantly, the purest water available, and the use of dedicated glassware. Peaks should be collected manually, by inspection of the elution profile, wherever possible rather than using a fraction collector set on time parameters. Fractions containing two or more peaks will of course result in multiple sequences that may be impossible to decipher if the peptides are present in equimolar amounts.

Where possible, mini (2.1 mm i.d.) or microbore (< 1 mm i.d.) HPLC should be used for low abundance peptides, as this maximizes the ratio of peptide to HPLC solvent and usually means that the sample need not be concentrated before applying it to the protein sequencer support. This may appear a trivial point, but it is a common consensus amongst protein chemists that vacuum concentration of < 10 pmol of a peptide, especially a large or hydrophobic peptide, will usually result in the peptide 'vanishing' to the walls of the microcentrifuge tube, to remain insoluble. In turn, modern protein sequencer supports usually hold no more than 15–30 μl of sample at any one application, which is then dried under a stream of argon. Thus a 1 ml fraction from a 4.6 mm i.d. HPLC column will take hours to load and will have ten times the buffer contaminants of a 0.1 ml fraction from a mini-HPLC column, making initial amino acid assignments difficult, if not impossible.

If it is available, MS (Section 3.4 and Chapter 2) will indicate the purity of the peptide fraction, plus its size, and possibly the presence of post-translational modifications not detectable by Edman chemistry. If the peptide fraction is impure, rechromatography of the peptide with either a modified gradient or a different ion-pairing agent will usually result in a pure peptide fraction being obtained. If TFA is used in the first dimension, a change to phosphate, NaCl, or ammonium acetate in a second dimension will often prove successful, and these buffers are compatible with modern protein sequencers.

5.4 Phosphorylated peptides (see also Chapter 4)

5.4.1 Identification of phosphopeptides

Substrates for a protein kinase can be labelled with ^{32}P, either by incubating the purified protein and kinase together *in vitro*, or by equilibrating labelled phosphate into cellular ATP, where it can be used by endogenous kinases to label the substrate *in situ*. The stoichiometry of incorporation can be calculated by precipitating a known amount of substrate with TCA, if the specific radioactivity of the ATP present *in vitro* or within the cells is known (49).

Radiolabelled peptides released by controlled proteolysis and then separated by SDS–PAGE can be located on dried gels or electroblots by autoradiography or imaging as described in *Protocol 9*. Subsequent complete proteolysis of a radioactive fragment will generate one or more radiolabelled peptides and a larger number of unmodified peptides that can be separated by HPLC. The radioactive fractions may be directly applied to the gas phase sequencer.

Peptides that have been labelled using non-radioactive ATP are distinguishable from the unmodified equivalent peptides through their increased negative charge, leading to a different pI in IEF separation and a different elution from reverse-phase and ion exchange chromatography, and through the increase of 80 mass units per phosphate group, which can be measured using MS. Using MALDI MS, as little as 10 fmol of phosphopeptide can be analysed; this method was used to confirm the presence of a 'cold' phosphorylation site in a novel protein kinase related to the protein kinase C family (22).

5.4.2 Identification of the phosphorylated residue

In gas/pulsed liquid phase protein sequencers, phosphorylated serine and threonine residues undergo β-elimination of the phosphate to generate the dehydro derivatives of these amino acids. The liberated phosphate is poorly soluble in the non-polar solvents used to extract the ATZ-amino acids during the Edman chemistry, and little ^{32}P phosphate is released from the reaction cartridge. Aqueous methanol has been successfully used to extract the liberated phosphate without washing out the peptide (50), but the reagents used to make the derivatized glass fibre are not available commercially.

A derivatized PVDF system has subsequently been devised, to be used in conjunction with the Milligen protein sequencer, and available under the trade name of Sequelon-AA (51). Peptides are coupled to a disc of Sequelon-AA in MES buffer pH 4.5–5 with 10 mg/ml *N*-ethyl-*N'*-(3-dimethylaminopropyl) carbodiimide in 10–20% acetonitrile. The disc is washed in aqueous acetonitrile; the Cerenkov counts provide an estimate of the degree of coupling with the labelled peptide. The extraction solvent in the gas/pulsed liquid phase sequencers is modified to aqueous methanol (52), acetonitrile (53), or anhydrous TFA (54). The peptide is usually sequenced by conventional procedures to determine the primary structure of the peptide, followed by a parallel solid phase process, collecting the ATZ-amino acids as they elute, and counting the fractions in a scintillation counter to determine the site(s) of phosphorylation. Alternatively, automated analysis of the sequence and of the phosphorylation sites can be performed by analysing a fraction of each PTH-amino acid by HPLC, and collecting the rest in fractions for scintillation counting.

This method can be applied to phosphorylated serine, threonine, and tyrosine residues, and to other post-translational modifications such as glycosylation. The ability to determine multiple phosphorylation sites is shown in *Figure 4*. This peptide was generated from *in vivo* phosphorylation of ribosomal protein S6 and shows five phosphorylation sites within a 19 residue peptide generated by digestion with Lys-C protease (see *Table 2*). The order of labelling at multiple sites can also be determined using this method (55).

A non-radioactive peptide containing a phosphorylated residue can be identified by its reaction with ethanethiol (36,56). The peptide, now contain-

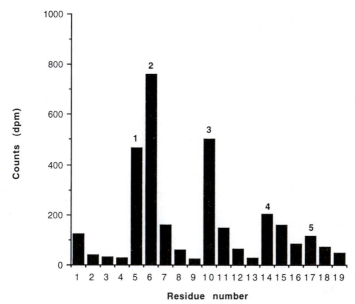

Figure 4. Solid phase sequence analysis of a phosphopeptide. Ribosomal protein S6 (which had been ^{32}P-labelled *in vivo* in Swiss 3T3 cells after stimulation by epidermal growth factor) was separated from other 40S ribosomal proteins by 2D SDS–PAGE, and the gel piece digested with endoproteinase Lys-C (*Protocol 5*). The peptides were resolved by HPLC and the radiolabelled peptide (residues 231–249: RRRLSSLRAST-SKSESSQK) coupled to Sequelon-AA and sequenced on an Applied Biosystms 477A sequencer with 50% acetonitrile/water as the extraction buffer. ATZ-amino acid fractions were collected from each cycle of Edman chemistry and subjected to scintillation counting. The phosphorylation sites 1–5 correspond to serine residues 235, 236, 240, 244, and 247.

ing *S*-ethyl cysteine instead of phosphoserine, elutes later from reverse-phase HPLC than the unmodified equivalent, and it can thus be identified by comparison with the HPLC profile of the mixture before chemical treatment.

In conventional Edman degradation, the phosphorylation site in non-radioactive phosphorylated peptides can be determined by the characteristic dehydration products of phosphoserine and phosphothreonine. The expected PTH derivative will be missing, and a dehydro derivative will form a new peak detectable by absorbance at 330 nm and a DTT adduct (57).

Acknowledgements

This work was supported by the Medical Research Council: by project grants to E. A. C., and through the MRC Protein Phosphorylation Unit (N. A. M.). *Figures 1, 2,* and *3* were reproduced with permission from the *European Journal of Biochemistry* and *EMBO Journal*, and material in *Tables 2* and *3* with permission from Boehringer Mannheim and Calbiochem.

References

1. Davidson, J. N., Chen, K. C., Jamison, R. S., Musmanno, L. A., and Kern, C. B. (1993). *BioEssays*, **3**, 157.
2. Carrey, E. A. (1986). *Biochem. J.*, **236**, 327.
3. Carrey, E. A. and Hardie, D. G. (1988). *Eur. J. Biochem.*, **171**, 583.
4. Carrey, E. A. and Hardie, D. G. (1986). *Anal. Biochem.*, **158**, 431.
5. Piszkiewicz, D., Landon, M., and Smith, E. L. (1970). *Biochem. Biophys. Res. Commun.*, **40**, 1173.
6. Landon, M. (1977). In *Methods in enzymology* (ed. C. H. W. Hirs and S. N. Timasheff), Vol. 47, p. 145. Academic Press, New York.
7. Spande, T. F., Witkop, B., Degani, Y., and Patchornik, A. (1970). *Adv. Protein Chem.*, **24**, 97.
8. Gross, E. and Witkop, B. (1962). *J. Biol. Chem.*, **237**, 1856.
9. Bornstein, P. and Balian, G. (1977). In *Methods in enzymology* (ed. C. H. W. Hirs and S. N. Timasheff), Vol. 47, p. 132. Academic Press, New York.
10. Mahoney, W. C., Smith, P. K., and Hermodson, M. A. (1981). *Biochemistry*, **20**, 443.
11. Stark, G.R. (1977). In *Methods in enzymology* (ed. C. H. W. Hirs and S. N. Timasheff), Vol. 47, p. 129. Academic Press, New York.
12. Soundar, S. and Colman, R. (1993). *J. Biol. Chem.*, **268**, 5264.
13. Alonso, E. and Rubio, V. (1995). *Eur. J. Biochem.*, **229**, 377.
14. Rittenhouse, J. and Marcus, F. (1984). *Anal. Biochem.*, **138**, 442.
15. Cleveland, D. W. (1983). In *Methods in enzymology* (ed. S. Fleischer and B. Fleischer), Vol. 96, p. 222. Academic Press, New York.
16. Carles, C., Huet, J.-C., and Ribadeau-Dumas, B. (1988). *FEBS Lett.*, **229**, 265.
17. Laemmli, U. K. (1970). *Nature*, **227**, 680.
18. Carrey, E. A., Campbell, D. G., and Hardie, D. G. (1985). *EMBO J.*, **4**, 3635.
19. Schägger, H. and von Jagow, G. (1987). *Anal. Biochem.*, **166**, 368.
20. Moorhead, G., MacKintosh, R. W., Morrice, N. A., Gallagher, T., and MacKintosh, C. (1994). *FEBS Lett.*, **356**, 46.
21. Moorhead, G., MacKintosh, C., Morrice, N. A., and Cohen, P. (1995). *FEBS Lett.*, **362**, 101.
22. Morrice, N. A., Gabrielli, B., Kemp, B. E., and Wettenhall, R. E. H. (1994). *J. Biol. Chem.*, **269**, 20040.
23. Ben-Levy, R., Leighton, I. A., Doza, Y. N., Attwood, P., Morrice, N. A., Marshall, C. J., *et al.* (1995). *EMBO. J.*, **14**, 101.
24. Hames, B. D. (1981). In *Gel electrophoresis of proteins: a practical approach* (ed. B. D. Hames and D. Rickwood), p. 1. IRL Press, Oxford.
25. Nimmo, H. G. and Nimmo, G. A. (1982). *Anal. Biochem.*, **121**, 17.
26. An der Lan, B. and Chrambach, A. (1981). In *Gel electrophoresis of proteins: a practical approach* (ed. B. D. Hames and D. Rickwood), p. 157. IRL Press, Oxford.
27. Reinhart, M. P. and Malamud, D. (1982). *Anal. Biochem.*, **123,** 229.
28. Munday, M. R., Campbell, D. G., Carling, D., and Hardie, D. G. (1988). *Eur. J. Biochem.*, **175**, 331.
29. Fadden, P. and Haystead, T. A. J. (1995). *Anal. Biochem.*, **225**, 81.

30. Elicone, C., Lui, M., Geromanos, S., Erdjument-Bromage, H., and Tempst, P. (1994). *J. Chromatogr. A.*, **676**, 121.
31. Sutton, C. W., Pemberton, K. S., Cottrell, J. S., Corbett, J. M., Wheeler, C. H., Dunn, M. J., *et al.* (1995). *Electrophoresis*, **16**, 308.
32. O'Farrell, P. H. (1975). *J. Biol. Chem.*, **250**, 4007.
33. Garrels, J. I. (1983). In *Methods in enzymology* (ed. R. Wu, L. Grossman, and K. Moldave), Vol. 100, p. 411. Academic Press, New York.
34. Sinclair, J. and Rickwood, D. (1981). In *Gel electrophoresis of proteins: a practical approach* (ed. B. D. Hames and D. Rickwood), p. 189. IRL Press, Oxford.
35. Brown, J. R. and Hartley, B. S. (1966). *Biochem. J.*, **101**, 214.
36. Holmes, C. F. B. (1987). *FEBS Lett.*, **215**, 21.
37. Fischer, S. G. (1983). In *Methods in enzymology* (ed. R. Wu, L. Grossman, and K. Moldave), Vol. 100, p. 424. Academic Press, New York.
38. Hames, B. D. (1981). In *Gel electrophoresis of proteins: a practical approach* (ed. B. D. Hames and D. Rickwood), p. 219. IRL Press, Oxford.
39. Packman, L. C. and Perham, R. N. (1982). *Biochemistry*, **21**, 5171.
40. Towbin, H., Staehelin, T., and Gordon, J. (1979). *Proc. Natl. Acad. Sci. USA*, **76**, 4350.
41. Salinovich, O. and Montelaro, R. C. (1986). *Anal. Biochem.*, **156**, 341.
42. Applied Biosystems (Nov. 1986). User Bulletin issue no. 25, Protein Sequencer Model 477A.
43. Grayson, D. R., Lee, L., and Evans, D. R. (1985). *J. Biol. Chem.*, **260**, 15840.
44. Diano, M., Le Bivic, A., and Hirn, M. (1987). *Anal. Biochem.*, **166**, 224.
45. Aebersold, R. H., Leavitt, J., Saavedra, R. A., Hood, L. E., and Kent, S. B. H. (1986). *Proc. Natl. Acad. Sci. USA*, **84**, 6970.
46. Matsudaira, P. (1987). *J. Biol. Chem.*, **262**, 10035.
47. Aebersold, R. H., Teplow, D. B., Hood, L. E., and Kent, S. B. H. (1986). *J. Biol. Chem.*, **261**, 4229.
48. Moritz, R. L. and Simpson, R. J. (1992). *J. Chromatogr*, **599**, 119.
49. Haystead, T. A. J. and Hardie, D. G. (1986). *Eur. J. Biochem.*, **240**, 99.
50. Wettenhall, R. E. H., Aebersold, R. H., and Hood, L. E. (1991). In *Methods in enzymology* (ed. T. Hunter and B. M. Sefton), Vol. 201, p. 186. Academic Press, New York.
51. Bodwell, J. E., Orti, E., Coull, J. M., Pappin, D. J. C., and Smith, L.I. (1991). *J. Biol. Chem.*, **266**, 7549.
52. Stokoe, D., Campbell, D. G., Nakielny, S., Hidaka, H., Leevers, S. J., Marshall, C., *et al.* (1992). *EMBO J.*, **11**, 3985.
53. Gerondakis, S., Morrice, N. A., Richardson, I. B., Wettenhall, R. E. H., Fecondo, J., and Grumont, R. J. (1993). *Cell Growth Diff.*, **4**, 617.
54. Taylor, L. K., Marshak, D. R., and Landreth, G. E. (1993). *Proc. Natl. Acad. Sci. USA*, **90**, 368.
55. Wettenhall, R. E. H., Erikson, E., and Maller, J. L. (1992). *J. Biol. Chem.*, **267**, 9021.
56. Meyer, H. E., Hoffmann-Posorske, E., Korte, H., and Heilmeyer, L. M. G. (1986). *FEBS Lett.*, **204**, 61.
57. Polya, G. M., Morrice, N. A., and Wettenhall, R. E. H. (1989). *FEBS Lett.*, **253**, 137.

6

Counting integral numbers of residues by chemical modification

MICHELLE HOLLECKER

1. Introduction

Proteins have integral numbers of each of the 20 amino acids, but all the currently accepted methods of determining this number measure only a non-integral ratio of moles of amino acid per mole of protein. This value is rarely found to be close to an integer, due to experimental error and uncertainty about the molecular weight of the protein (1–3).

This chapter describes one method that gives integral values for the number of cysteine and lysine residues as well as disulfide bonds present in a protein, independently of any other property of the polypeptide chain, including its molecular weight. The general procedure of determining the number, N, of residues of a given type is to modify them gradually and specifically to generate a complete spectrum of molecules with 0, 1, 2... N groups modified. These species are then counted after separating them by a technique sensitive to only the number of groups modified.

Such an approach should be useful for other amino acids than cysteine and lysine residues, provided they have reactive groups on their side chains and the modification reaction is specific and stoichiometric. In the particular case of cysteine and lysine residues described here, the method relies on the charge differences introduced by specific chemical modification of the amino acid, using either a single reagent or the competition between two or more different reagents (4–8). As the modification alters the ionic charge, the separation of the species with different numbers of groups modified can be readily accomplished by electrophoresis (see Chapter 8), isoelectric focusing, or ion exchange chromatography.

1.1 Advantages of the method

The procedure developed here presents several advantages:

(a) It is simple and requires very little equipment. One needs only the usual glassware, a pH meter, a small Sephadex G25 column, and a gel electrophoresis apparatus.

(b) It is rapid. The experiment consists mainly of adding various amounts of reagent to the protein, taking a sample of each solution, and analysing it by electrophoresis. This method usually yields results in less than a day.

(c) It requires only limited amounts of protein. The preparation of protein needs only to be nearly homogeneous (which is also necessary, although not sufficient, for the determination of a correct value using currently accepted procedures).

(d) It can be applied to any protein, even if its molecular weight or concentration is not known. Furthermore, it can be used to determine the molecular weight of an unknown protein. Once the exact number of residues of one or more amino acids has been determined using the procedure described in this chapter, the relative amounts of all the amino acids determined by standard amino acid analysis can be converted directly to numbers of residues per polypeptide chain, and the molecular weight of the protein be calculated from the amino acid composition.

1.2 Limitations of the method

The procedure developed here will never replace the standard methods of amino acid analysis because it is unlikely that chemical reagents will be available for specifically modifying those amino acids with no reactive groups, for example glycine, proline, valine, leucine, etc.

2. Cysteine residues

2.1 Choice of reagents for modifying thiol groups specifically

Various reagents react with thiol groups, but iodoacetic acid and iodoacetamide are particularly suitable for such a study, for the following reasons:

(a) The reaction of thiols with iodoacetamide or iodoacetate is rapid and irreversible under essentially all conditions. So the modified species are trapped in a stable form.

(b) Carboxymethylation with iodoacetic acid introduces one new acidic group $-O_2CCH_2S-$ at each free cysteine residue, making the charge of the molecule more negative.

(c) Iodoacetamide, on the contrary, reacts with SH groups to give the neutral moiety H_2NCOCH_2S-, thereby not changing the net charge.

(d) With an excess of one of the reagents, all cysteine residues will have reacted either with iodoacetate or with iodoacetamide. With a mixture of the neutral and acidic reagents, there will be competition between them.

(e) Electrophoretic mobility through polyacrylamide gels depends upon both the net charge and the hydrodynamic volume of the protein (see

Chapter 8). As the hydrodynamic property of an unfolded protein can be the same whether iodoacetate or iodoacetamide has been used, the differences observed in the electrophoretic mobility of the species can be interpreted directly in terms of differences in charges.

The method described below consists of adding varying iodoacetamide to iodoacetate ratios to portions of the unfolded protein, in order to generate a complete spectrum of protein molecules with 0, 1, 2, ..., N acidic carboxymethyl groups, where N is the integral number of cysteine residues per protein molecule. In the case of unfolded, reduced bovine pancreatic trypsin inhibitor (BPTI), ribonuclease A (RNase A), and lysozyme (4), approximately equal competition between the two reagents was observed with a ratio of iodoacetamide to iodoacetate of $1:3$; this indicates that the former reacts three times more rapidly than the latter, as has also been observed by others (9). The entire spectrum of protein molecules modified at the thiol groups could be generated using ratios of $1:1$, $1:3$, and $1:9$ of neutral to acidic reagents. With other proteins with greater numbers of cysteine residues, it might also be necessary to use higher or lower ratios.

2.2 Experimental procedure

The method requires that all cysteine thiol groups have comparable reactivities, so the following considerations must be taken into account:

(a) Thiol groups buried in the interior of a molecule do not react readily with iodoacetate or iodoacetamide. Therefore the reaction must be carried out in both denaturing and reducing conditions, so that all thiol groups of the protein are chemically equivalent.

(b) Only ionized thiols are reactive; as the pK value of most thiols is generally in the region of 8–10, the pH of the protein solution should be neutral or slightly alkaline.

(c) High concentrations of iodoacetate or iodoacetamide slowly react with other groups of the molecule at high pH, especially amino groups and histidine residues. Therefore, the pH should not be much greater than 8, the protein should not be exposed to reagents unnecessarily long, and the solutions should be put on ice after reaction.

Protocols 1 and *2* are adequate to count up to eight cysteine residues in a protein. If only limited amounts of your protein are available, all volumes may be decreased tenfold.

Protocol 1. Denaturation and reduction of proteins

Reagents

- 1.0 M Tris–HCl pH 8.0
- 0.1 M EDTA pH 7.0
- 8.0 M urea (Sigma Ultra)

- 1.0 M dithiothreithol (DTT) (Cleland's reagent, from Sigma)

Method

1. Denature the protein and reduce all disulfide bonds by adding to 0.2 mg of protein 10 μl of each of the stock solutions of Tris–HCl, EDTA, DTT, and 1.0 ml of the urea solution.

2. Mix the solution carefully and leave it to incubate at 37 °C for at least 30 min.

[a] This protocol is appropriate for dried proteins, many variations are possible for proteins already in solution. The important parameters are that the pH be about 8, the DTT concentration be at least 10 mM, and the urea concentration be at least 8 M.

Protocol 2. Competitive alkylation of the thiol groups

Equipment and reagents

- Six Eppendorf tubes of 0.5 ml numbered 1–6
- Solution A: 46.5 mg of iodoacetamide in 1.0 ml of 0.25 M Tris–HCl pH 8.0
- Solution B: 46.5 mg of iodoacetic acid in 0.25 ml of 1 M KOH and 0.75 ml of 0.33 M Tris–HCl pH 8.0[a]

- Solution C: 50 μl of solution A and 50 μl of solution B
- Solution D: 150 μl of A with 50 μl of B
- Solution E: 450 μl of A with 50 μl of B

Method

1. Put 10 μl of solutions A, B, C, D, and E in tubes 1–5 respectively. Keep tube 6 aside.

2. Add 40 μl of the freshly unfolded reduced protein solution (obtained from *Protocol 1*) to each of the tubes 1–5.[b]

3. Gently mix the content of each tube once or twice with an Eppendorf pipette. Leave the solutions to incubate at room temperature for 15 min.

4. Transfer an aliquot of 10 μl from each of tubes 1–5 into tube 6. Mix gently.

5. Store all the tubes on ice, until the solutions are analysed on polyacrylamide gels (see Section 5).

[a] KOH solution is added to neutralize iodoacetic acid when it is in the free acid form. If you use iodoacetic acid in the sodium salt form, replace the volume of KOH by distilled water.
[b] Large quantities of alkylating reagents may need to be added if the reduced protein solution contains more than 20 mM dithiothreitol (DTT).

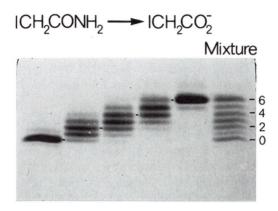

$$ICH_2CONH_2 \longrightarrow ICH_2CO_2^-$$

Figure 1. Electrophoretic separation of BPTI molecules with 0–6 acidic carboxymethyl groups on its six cysteine residues. The reduced protein was reacted with the neutral iodoacetamide (channel 1), acidic iodoacetate (channel 5), and 1:1, 1:3, and 1:9 ratios of neutral to acidic reagents in channels 2, 3, and 4 respectively. Channel 6 contains a mixture of equal portions of the samples applied to channels 1–5. Taken from ref. 4.

2.3 Counting cysteine residues

The cysteine contents of BPTI, RNase A, lysozyme, bovine α-lactalbumin, and β-lactoglobulin have been determined using the procedure described in *Protocols 1* and *2* (4). Molecules with varying numbers of acidic groups were separated by electrophoresis under conditions that are described in detail in Section 5.1. The separation obtained for one of these proteins is shown in *Figure 1*.

The reduced protein in which all thiol groups were blocked with iodoacetate migrated electrophoretically more slowly than that blocked with iodoacetamide, due to the acidic carboxymethyl groups (10). The reaction of the thiol groups with mixtures of neutral and acidic reagents generated an additional six bands expected with six cysteine residues in the case of BPTI, and eight additional bands corresponding to the eight cysteine residues of bovine pancreatic RNase A and hen egg lysozyme.

This procedure is sufficiently simple that it has been used to check amino acid compositions predicted by gene sequencing (11) and to correct errors in DNA sequencing (12).

3. Disulfide bonds

The procedures described below for counting disulfide bonds in a protein rely basically on the same strategy as the method for counting cysteine residues of Section 2, and the two methods present the same advantages of:

(a) Giving integral numbers of residues.

(b) Being suitable for the analysis of inhomogeneous protein solutions, as in the case of proteins synthesized in complex biological systems (7).

In practice, either of these two methods can be applied for the determination of disulfide content in a native protein, but the procedure of counting disulfide bonds presents the additional advantage that with a partially reduced protein, mixtures of protein species with different numbers of cleaved disulfides can be analysed separately (8).

3.1 Experimental procedures

The procedure of counting disulfide bonds is a two-step alkylation method. The protein is first submitted to partial reduction, and the free thiols thus generated trapped with an alkylating reagent. In the second step, the protein is then fully reduced and the remaining thiols blocked with a second alkylating reagent.

Two parameters are especially important in the process of partially unfolding a protein: the time of reaction and the concentration of reducing reagent. Two procedures can thus be applied for counting disulfide bonds, depending on the parameter that is set in the first step.

3.1.1 Disulfide bond counting as a function of reducing reagent concentration

The procedure developed by Hirose *et al.* (7) is especially adapted for proteins of unknown reactivity to reducing reagents. In the first step, the protein is submitted to increasing amounts of DTT, for a set time of reaction of 30 min. The concentrations of DTT should be chosen so as to range up to the concentration giving complete reduction and to give rise to a pattern of equally spaced bands between these two extremes. This procedure is exemplified in *Protocol 3*, for DTT concentrations of: 0, 50 μM, 100 μM, 200 μM, 500 μM, and 10 mM.

Protocol 3. Disulfide counting as a function of DTT concentration

Equipment and reagents

- Six Eppendorf tubes of 1 ml numbered 1–6
- A small Sephadex G25 column equilibrated with 0.1 M acetic acid
- Denaturing solution: 4.8 g urea (8 M) in 1.0 ml of 1.0 M Tris–HCl pH 8, 100 μl of 0.1 M EDTA, and water to 10 ml
- Solution A: 77 mg DTT in 1.0 ml H_2O
- Solution B: 50 μl of A and 950 μl H_2O

- Solution C: 200 μl of B and 800 μl H_2O
- Solution D: 100 μl of B and 900 μl H_2O
- Iodoacetamide solution: 185 mg of iodoacetamide in 1.0 ml of 0.5 M Tris–HCl pH 8
- Iodoacetate solution: 186 mg of iodoacetate in 0.5 ml of 0.5 M Tris–HCl pH 8 and 0.5 ml of 1.0 M KOH

A. *First-step alkylation*

1. Dissolve 3 mg of protein in 3 ml of denaturing solution. Transfer 0.5 ml of this protein solution to each tube 1–6.

---

2. Add 10 μl of solution A to tubes 1 and 6; 10 μl of solution D to tube 2; 10 μl of solution C to tube 3; 20 μl of solution C to tube 4; and 10 μl of solution B to tube 5.

3. Mix thoroughly and leave the samples to incubate 30 min at 37 °C.

4. Add 50 μl of the iodoacetate solution to tube 6 and 50 μl of the iodoacetamide solution to tubes 1–5.

5. Mix thoroughly and leave the samples to incubate at room temperature for 15 min.

6. Gel filter each sample on a Sephadex G25 column[a] and lyophilize the protein sample.

B. *Second-step alkylation*

1. Dissolve each sample in 0.5 ml of denaturing solution.

2. Add 10 μl of solution A to each tube 1–6 and repeat part A, step 3.

3. Add 50 μl of the iodoacetate solution to each tube and repeat part A, step 5. Store the samples on ice prior to gel electrophoresis (see Section 5).

[a] As an alternative method, you can also remove the small molecule reagents using a reversed-phase HPLC column to isolate the protein, or Centricon (Amicon) microdialysis tubes of the appropriate molecular weight cut-off can be used.

3.1.2 Disulfide bond counting as a function of time

The method developed by Ewbank and Creighton (8) can be applied for disulfide bond counting in the case of a native protein of known reactivity to DTT or to assess the numbers of disulfide bonds in mixtures of partially reduced proteins.

Protocol 4.   Disulfide bond counting as a function of time of reduction

Equipment and reagents

- Six Eppendorf tubes of 1 ml numbered 1–6
- Denaturing solution, solution A, iodoacetamide solution, and iodoacetate solution (see *Protocol 3*)

Method

1. Dissolve 3 mg of protein in 3.0 ml of denaturing solution.

2. Transfer 50 μl of iodoacetamide solution to tubes 1–6.

3. At time zero, add to the protein solution a suitable volume of solution A. Mix thoroughly and leave the solution to incubate at 37 °C.

Protocol 4. *Continued*

4. At different time intervals, withdraw 0.5 ml from the protein solution and transfer it to one of the tubes 1–6.

5. For each fraction removed, repeat *Protocol 3A*, steps 5 and 6.

6. For all six tubes, repeat the second-step alkylation, *Protocol 3B*.

3.2 Counting disulfide bonds

Following the procedure described in *Protocol 3*, Hirose *et al.* (7) analysed a series of proteins with disulfide bond contents ranging from 4–15. This method, used in combination with autoradiography, was especially useful in the analysis of disulfide bonds in proteins synthesized in complex biological systems.

The method described in *Protocol 4* was applied by Ewbank and Creighton (8) to follow the progressive unfolding of α-lactalbumin over time. An example of the results obtained is presented in *Figure 2*.

Electrophoresis was in the presence of 8 M urea, following the gel recipe given for acidic proteins in Chapter 8, *Protocol 1*. Using this procedure, it was possible not only to determine the total number of disulfide bonds in native α-lactalbumin, but also to assess the extent of reduction of the four disulfide bonds over time.

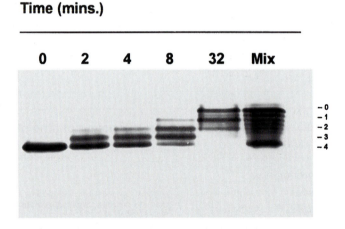

Figure 2. Counting disulfide bonds of bovine α-lactalbumin (α-LA). Bovine α-LA (15 μM) was reduced in 8 M urea by 100 μM DTT for different lengths of time ranging from 0 min (channel 1) to 32 min (channel 5). The free thiol groups were blocked with iodoacetamide, then the remaining disulfide bonds fully reduced, and the resulting free thiols blocked with iodoacetate. Channel 6 is a mixture of fully reduced α-LA blocked with different ratios of iodoacetamide and iodoacetic acid to generate molecules containing between zero and eight acidic carboxymethyl groups (see *Protocol 2*). Taken from ref. 8.

4. Lysine residues

4.1 Choice of reagent

Succinic anhydride is a non-volatile, stable, and easy to handle reagent. It reacts specifically with the ε-amino groups of lysine residues and the terminal α-amino group of proteins, in their non-protonated forms, converting them from basic to acidic groups. Succinylation of each amino group then alters the net charge of a protein by up to two unit charges.

A few precautions should be taken before starting the succinylation reaction.

(a) Succinic anhydride has been reported to react with free thiol groups (13). If the protein you want to study contains such thiol groups, first block them with a neutral reagent. As in *Protocol 2*, add 0.1 M iodoacetamide to your protein solution at about pH 8; after a few minutes at room temperature, remove the excess reagent by gel filtration on Sephadex G25.

(b) Lysine residues are usually on the surface of a folded protein, accessible to the solvent and the reagent. In the case of two small basic proteins RNase A and horse ferricytochrome *c*, it was not necessary to unfold the protein before reacting it with succinic anhydride. With an unknown protein, it may be prudent to do so, for example, by incubating it at 37 °C for 30 min in 8 M urea.

4.2 Experimental procedure

Gradual succinylation of amino groups is accomplished in *Protocol 5* simply by adding increasing amounts of reagent until all amino groups are modified. It is necessary that the portions added give a full spectrum of partly modified species, so it may be necessary to perform a second experiment with adjusted quantities of succinic anhydride.

Protocol 5. Succinylation of amino groups

Equipment and reagents
- Eight Eppendorf tubes of 0.5 ml, numbered 1–8
- Succinic anhydride (Sigma)

Method

1. Dissolve about 5 mg of protein in 5 ml of distilled water.[a]

2. While stirring, adjust the pH to 7 with NaOH.[b] Withdraw an aliquot of 100 μl of the solution, put it in tube 1, and keep it on ice.

3. Add 1 mg of succinic anhydride to the solution and readjust the pH to 7 with NaOH.[c]

4. When the pH has been constant for 10 min, withdraw 100 μl of the solution and put it in tube 2.

Protocol 5. *Continued*

5. Repeat steps 3 and 4 until you have added altogether 5 mg of succinic anhydride.

6. Add an excess of reagent, about 10 mg, to the remainder of the protein solution. Readjust the pH to 7, wait until it is constant for 15 min, and withdraw 100 μl for tube 7.

7. In tube 8, put 10 μl of each of the seven samples and mix the solution.

[a] 0.1 M NaCl or other salt may be included, if the protein is insoluble in distilled water, but do not include any amines.
[b] Succinylation of amino groups proceeds optimally between pH 7 and 9 (14). Very small quantities of reagent are required initially, however, so the amino groups of your protein may be too reactive. It might then be better to carry out the initial reactions at a lower pH (15).
[c] The pH should drop, due to the hydrolysis of the reagent to the dicarboxylic acid.

Samples 1 and 7 should give you the initial and final stages of the reaction, where none or all *N* amino groups are modified, respectively. Samples 2–6 should give an idea of the reactivity of your protein and enable you to design your own protocol, to generate a complete spectrum of molecules with 0 to *N* amino groups modified.

4.3 Counting amino groups

In the case of horse ferricytochrome *c* and RNase A (6), the reactions were performed with protein solutions of 1.0 mg/ml and portions of succinic anhydride varying from initially 10 μg up to 1 mg/ml of protein solution. With cytochrome *c*, its 19 lysine residues were resolved by adding 29 successive portions of reagent. With RNase A, which has only ten lysine residues and one α-amino group, the procedure required 16 additions of succinic anhydride.

The eight samples obtained for each protein were analysed by electrophoresis on a denaturing urea gel, so that succinylation of each of the amino groups has an equivalent effect on the electrophoretic mobility (see *Protocol 6*). The electrophoresis was run at 250 V (~ 10 mA/gel) for 2.5–3 h at room temperature. With the two small, basic proteins used here, containing relatively large numbers of lysine residues, the change in net charge produced by complete succinylation was very great, changing them from very basic to very acidic proteins. It was found preferable to suppress the ionization of the succinyl groups by separating the proteins at a very acidic pH (pH 3.6), where the proteins retained their net positive charge. The results obtained are illustrated in *Figure 3*.

As expected, succinylation of RNase A produced 11 additional bands, corresponding to the 10 lysine residues and one α-amino group it is known to contain. Horse cytochrome *c* gave 19 additional bands, corresponding to its 19 lysine residues. It has no α-amino group, as the amino terminus of the polypeptide chain is acetylated.

Ribonuclease A

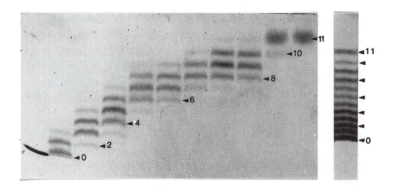

Cytochrome c

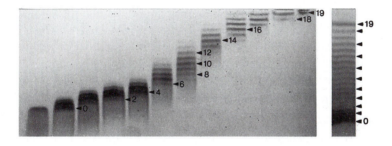

Figure 3. Counting amino groups of (*top*) bovine RNase A and (*bottom*) horse ferricyto-chrome *c* by electrophoretic resolution of the mixtures produced by progressive succiny-lation. The original unmodified protein is on the left and succinylation increases to the right. Electrophoresis at pH 3.6 was from top to bottom. The lanes on the far right are mixtures obtained by combining the individual samples and by carrying out the electro-phoresis at the lower pH of 3.45. Alternative bands are marked by *arrows*, with the number of succinyl groups indicated for a few of the bands. Taken from ref. 6.

5. Electrophoretic resolution of the modified species

The general method for counting amino acid residues described here for lysine and cysteine residues depends upon having a modification reaction that is sufficiently specific and stoichiometric, as well as a separation method able to resolve species differing by only one such modification. The method used will depend upon the specific protein and the modification, but usually can be accomplished either by electrophoresis (see Chapter 8), isoelectric focusing, or ion exchange chromatography.

A few considerations have to be taken into account when using this method:

(a) The crucial parameter for electrophoresis is the pH; the modified groups must differ in their net charge, yet the net charges of the molecules with 0 to N groups modified ideally should all be of the same sign, so that they may be analysed with a single electrophoretic run. For the carboxy-methyl cysteine groups and succinyl lysine groups described here, the pH should be greater than 3.5.

(b) It is important that the initial starting protein gives a single major band in the region of the gel where the modified molecules will migrate. The final protein sample should also give a single band; if it does not, the modification was either incomplete or not sufficiently specific, or the initial protein was heterogeneous. In between these two extremes of the initial and final samples, a complete spectrum of partially modified species should be generated. Ideally, the various samples should have overlapping spectra, so that no species are missed.

(c) The molecules with the same number of groups modified will usually differ in the identities of the groups. It is important that all modifications have about the same electrophoretic effect, irrespective of which groups have reacted, so the separation procedure must be carried out on unfolded molecules (e.g. in 8 M urea).

5.1 Counting cysteine residues and disulfide bonds

The samples with alkylated cysteine residues were analysed by denaturing polyacrylamide gel electrophoresis. The systems used were the discontinuous high pH electrophoresis system introduced by Davis *et al.* for acidic proteins or the low pH discontinuous buffer system of Reisfield *et al.* for basic proteins (for recipes, see Chapter 8, *Protocols 1* and *2*), both systems being modified to include 8 M urea in the separating gels.

5.2 Counting lysine residues

The electrophoretic system described in *Protocol 6* was found useful for separating the basic proteins cytochrome *c* and RNase A treated with succinic anhydride. Other electrophoretic systems that are sensitive to net charge on proteins are given in Chapter 8.

Protocol 6.   Recipe for 40 ml of 8.0 M urea, 11 % acrylamide gels

Reagents

- 1 M Tris–acetate buffer: 1 M acetic acid titrated to the given pH with solid Tris[a]
- Electrophoresis buffer: 1 M Tris–acetate buffer diluted to 0.05 M
- Acrylamide solution:[b] 60% (w/v) acrylamide, 0.4% (w/v) bisacrylamide
- TEMED solution: 1% (v/v) *N,N,N',N'* tetramethylethylenediamine adjusted to pH 4 with acetic acid
- Riboflavin solution: 4 mg riboflavin in 100 ml H_2O

Method

1. Mix together 7.3 ml of acrylamide solution, 5.0 ml of TEMED solution, 2.0 ml of 1 M Tris–acetate buffer, 19.6 g of urea, and adjust the volume to 35 ml with distilled water.

2. Degas the solution with a water pump to remove oxygen that inhibits polymerization.

3. Add 5.0 ml of the riboflavin solution.

4. Mix the solution thoroughly, then immediately pour it into the gel apparatus.

5. Polymerize with light.

6. Add 0.2 vol. of glycerol to your samples if they are not already in 8 M urea, as well as some methyl green dye.

7. Layer a volume of 20 μl of each sample into the wells of the gels.

8. Electrophoresis is usually at 250 V (~ 10 mA/gel) for 2.5–3 h at room temperature.

9. For staining and destaining of the gels, see the recipes given in Chapter 8, *Protocol 4*.

[a] The pH chosen depends on the protein studied; usually it ranges between 3.45 and 4.
[b] Acrylamide is toxic, so be careful not to come into contact with the monomer. Avoid inhaling its dust and wear gloves.

References

1. Light, A. and Smith, E. L. (1963). In *The proteins* (2nd edn) (ed. H. Neurath), Vol. I, p. 1. Academic Press, New York.
2. Schroeder, W. A. (1968). *The primary structure of proteins*. Harper and Row, New York.
3. Konigsberg, W. H. and Steinman, H. M. (1977). In *The proteins* (3rd edn) (ed. H. Neurath and R. L. Hill), Vol. III, p. 1. Academic Press, New York.
4. Creighton, T. E. (1980). *Nature*, **284**, 487.
5. Feinstein, A. (1966). *Nature*, **210**, 135.
6. Hollecker, M. and Creighton, T. E. (1980). *FEBS Lett.*, **119**, 187.

7. Hirose, M., Takahashi, N., Oe, H., and Doi, E. (1988). *Anal. Biochem.*, **168,** 193.
8. Ewbank, J. J. and Creighton, T. E. (1993). *Biochemistry*, **32,** 3677.
9. Webb, J. L. (1966). *Enzyme and metabolic inhibitors.* Vol. III. Academic Press, New York.
10. Creighton, T. E. (1974). *J. Mol. Biol.*, **87,** 579.
11. Stan-Lotter, H. and Bragg, P. D. (1985). *Arch. Biochem. Biophys.*, **239**, 280.
12. Stan-Lotter, H., Clarke, D. M., and Bragg, P. D. (1986). *FEBS Lett.*, **187**, 121.
13. Meighen, E. A. and Schachman, H. K. (1970). *Biochemistry*, **9**, 1163.
14. Klapper, M. H. and Klotz, I. M. (1972). In *Methods in enzymology* (ed. C. H. W. Hirs and S. N. Timasheff), Vol. 25, p. 531. Adademic Press, New York.
15. Habeeb, A. F. S. A., Cassidy, H. G., and Singer, S. J. (1958). *Biochem. Biophys. Acta*, **29**, 587.

7

Disulfide bonds between cysteine residues

WILLIAM R. GRAY

1. Introduction

Disulfide bonds between cysteine residues, to generate cystine, are a common post-translational modification of proteins, principally of those that operate in the extracellular milieu. A full description of the covalent structure demands that the connectivity of the bridged cysteines be analysed. Although disulfides are chemically rather simple, two factors can make the analysis difficult. First, the number of possible isomers grows rapidly as the number of bridges increases—1, 3, 15, 105, 945, etc. Second, base and reducing agents can catalyse exchange among partners, obscuring the original connectivity (1). Native proteins under physiological conditions are deceptively stable, but once they are denatured or nicked by proteases, disulfide exchange occurs when even a trace of thiol is present. Offsetting this complexity is one's prior knowledge of the amino acid sequence. This helps greatly in designing cleavage methods, and in simplifying peptide analysis—an accurate mass will often identify a peptide fragment, thereby defining a disulfide connection.

Whatever chemical approach is used, the final level of confidence depends on how well several lines of evidence converge. Complete rigour may be almost impossible with only small amounts of a protein or peptide, and one balances expenditure of material and effort against the potential consequences of error. If a protein is of exceptional interest, it may be more rewarding to obtain crystals suitable for X-ray analysis, which will provide independent proof of disulfide structure. Likewise, any really interesting peptide should be synthesized in a way that gives additional confirmation of its disulfide bridging.

The present article is addressed to those who have a particular structure to solve, but who are not technical specialists in one or another field of structural chemistry. It is assumed that they are experienced with HPLC purification of peptides, and that they have access to sequencing and/or mass spectrometry (MS) services (see Chapters 2 and 5). Methods for determining the number of disulfide bonds present are given in Chapter 6.

1.1 Planning the general approach

Choice of methodology is dictated by the length of protein, distribution of Cys residues, how much starting material can be spared, and what analytical facilities are available. In general it is useful to think in terms of two complementary approaches.

(a) Classical fragmentation strategy (2):
- *i.* digest protein, avoiding disulfide exchange
- *ii.* separate fragments
- *iii.* detect cystine-containing peptides
- *iv.* identify subfragments, and hence individual cysteine residues that are connected

(b) TCEP partial reduction strategy (3):
- *i.* reduce peptide incompletely using TCEP (*tris*-[2-carboxyethyl]-phosphine) at pH 3
- *ii.* separate full-length reduction intermediates
- *iii.* label thiols of one or more products
- *iv.* identify labelled cysteine residues

The choice between them depends on the problem to be solved, rather than sensitivity. Either may work with as little as 1 nmol, or require as much as 10–100 nmol, depending on the peptide. For proteins larger than 50 residues the classical strategy is almost always needed—but if some cysteine residues are clustered in the primary structure, some proteolytic fragments may have more than one disulfide, requiring another step to solve the connectivities. In such cases the fragment can be analysed with TCEP.

Conversely, TCEP reduction at acid pH may now be the preferred method for peptides less than 50 residues, especially if the cysteines are clustered. In this size range a single sequence run will often identify all labelled cysteines, even though a full sequence might not be obtained for an unknown peptide. In some cases, digestion of the labelled peptide and isolation of the fragments could still be simpler; this approach should also be used for terminally blocked peptides.

In either case, the problem of thiol–disulfide exchange must be faced before one can have confidence in the results. With the digestion strategy, finding a consistent set of disulfide-bridged fragments is not enough—every fragment in the digest must be consistent with the proposed structure. With the partial reduction approach, some exchange may occur during labelling—one must sequence the correct product.

Proteins vary so widely that no one set of recipes will work every time. Successful analysis will come by learning how to adapt the procedures, rather than from slavishly following them. This paper focuses on basic principles. It should be used with reference to journal articles that employ the various methods discussed in the text (3–6).

2. The classical fragmentation strategy

Sanger's original work on insulin (1,2) laid many of the foundations for disulfide analysis. It is still widely applicable, greatly enhanced by modern methods of peptide separation and analysis.

2.1 Fragmentation of protein

The aim is to produce a definitive set of peptide fragments by enzymatic or chemical digestion (see Chapter 5). With very big proteins it may be necessary to obtain large fragments by selective cleavage and to digest these further with general purpose proteases. The greatest risk of disulfide exchange occurs during digestion: even though the protein, enzymes, and reagents may be initially free of thiols, the latter can arise by base catalysed degradation of disulfides. Under any reasonable conditions very little thiol will be produced, but only catalytic amounts are needed to cause extensive disulfide rearrangement during long incubations.

2.1.1 Enzymatic digestion

Native proteins with disulfide bonds tend to resist proteolysis, and natural protease inhibitors are commonly 'knotty'. Despite the presence of appropriate cleavage sequences, the peptide chain may not adopt a conformation suitable for binding, or the susceptible sequence may be buried. Digestion must then be carried out for prolonged times at high enzyme/substrate ratios. To minimize exchange under these circumstances one should:

- exclude thiols, not only in solution but as vapours (e.g. mercaptoethanol)
- work at acidic pHs that discourage thiol generation from base attack on disulfides
- include a thiol trapping agent such as 0.1 mM *N*-ethylmaleimide if digestion is at pH > 7

To check whether exchange is occurring during digestion:

- carry out control incubations of peptide without enzyme under similar conditions
- examine more than one time point and enzyme/substrate ratio

The most useful proteases have been trypsin and chymotrypsin (used below their optimum pH), endoproteinase Glu-C, and thermolysin. Use purified 'sequencing grade' enzymes, because their low content of extraneous proteins and other proteases leads to cleaner chromatograms. Examine the amino acid sequence of your protein and choose an enzyme or combination of enzymes that 'should' result in chain cleavage between each successive pair of Cys residues (see Chapter 5). Specificity guidelines for preferred (not absolute) sites of cleavage are as follows:

(a) Trypsin: after Lys and Arg, not before Pro.

(b) Chymotrypsin: after Trp, Tyr, Phe, Leu, not before Pro.

(c) Glu-C: after Glu, sometimes Asp.

(d) Thermolysin: before hydrophobics.

Protocol 1. Enzymatic digestion of proteins and peptides

Reagents

- Enzymes: trypsin, chymotrypsin, endopro-teinase Glu-C, thermolysin (sequencing grade from Boehringer, Promega, Sigma, or similar sources)
- Buffer A (trypsin, chymotrypsin): 50 mM NH$_4$OAc pH 6.8, containing 0.002% (w/v) NaN$_3$ (6)
- Buffer B (Glu-C): 25 mM NH$_4$OAc pH 4.0 (6)
- Buffer C (thermolysin): 0.1 M Mes pH 6.0, containing 0.1 mM CaCl$_2$ and 0.002% (w/v) NaN$_3$ (4)

Method

1. Dissolve the enzyme in a minimal volume of a solvent suitable for storage.[a]

2. Dissolve the protein in the appropriate digestion buffer, at a concentration of 10–500 μM.

3. In a microvial combine aliquots of the enzyme and protein solutions to achieve the desired enzyme/substrate ratio. For initial experiments try 1/10 by weight for trypsin, chymotrypsin, and endoproteinase Glu-C, and 1/50 for thermolysin. These ratios may have to be increased to as high as 1/1 if digestion is too limited.[b]

4. Cap the vial tightly, and incubate it at 37 °C; thermolysin can be used up to 55 °C.

5. At time intervals of 6, 12, 24 h, remove samples for HPLC evaluation (Section 2.2).

6. Set-up and evaluate a matched 'enzyme control' vial in which the enzyme is incubated under exactly the same conditions, but in the absence of substrate.[c]

[a] Follow the manufacturer's suggestion. Use small volumes to obtain a maximum enzyme concentration.
[b] Test separately so that the combination of solvents does not change the pH. If it does, then readjust the digest.
[c] This is particularly important when using large amounts of protease and long digestion times, so that one can distinguish autolysis fragments from substrate digestion products.

2.1.2 Chemical fragmentation

Specific chemical methods can be used for splitting the peptide chain at residues such as Met and Trp residues (7) (see Chapter 5). The rarity of these amino acids makes this approach useful only as a supplement to enzymatic digestion. Partial hydrolysis by acid is much more general, but generates extremely complex digests. Unless one is dealing with a relatively small molecule, the problem is almost intractable without HPLC MS: remember that one needs to check every fragment to be sure that one is not missing evidence of disulfide exchange. If acid hydrolysis is attempted, use oxalic acid (8) or TFA rather than HCl, which catalyses exchange by a different chemical process (1).

2.2 Peptide purification

2.2.1 Separation of peptides

The best general system is reversed-phase HPLC, although ion exchange may be better for retaining small peptides. Within available instrumentation, use the following guidelines:

(a) Column packing material: C-18, end-capped.
(b) Column size: analytical scale (4.6 mm i.d.) or microbore (1–2 mm i.d.) for sensitivity.
(c) Elution buffers: acetonitrile gradient in a volatile buffer such as 0.1% (v/v) TFA, or 20–50 mM NH_4OAc, or 20–50 mM triethylammonium acetate.
(d) Work at pH 6 or below to avoid exchange—0.1% TFA (pH 2) is ideal.
(e) Start at 0% acetonitrile to help retain small peptides.
(f) Monitor by UV absorbance at 210–220 nm so that all peptides can be detected.
(g) Collect fractions manually to make best use of column resolution.
(h) Evaluate all chromatographic peaks; absorbance varies widely among peptides, being dependent on length, and on content of aromatic amino acids.

Some laboratories will have access to more sophisticated instruments such as on-line evaluation by MS: this is beyond the scope of the present article.

2.2.2 General evaluation of digest

Analyse the 'enzyme control' digest by HPLC. Identify which fractions may be from protease or reagents, but do not assume they contain nothing from your protein.

Carry out trial digests of minimal amounts of protein sample. A serious concern is that proteins with mispaired disulfides are often much more susceptible to digestion than is the native form. The rate-limiting step in digestion can then be the exchange reaction—resulting in a mixture of peptides that is mostly from rearranged protein! If digestion is slow (incomplete at 24–48 h), increase the enzyme/substrate ratio rather than extending the time: this will tip the balance towards digestion and away from exchange. A clean digestion is usually indicated by rapid disappearance of starting protein, with production of a stable pattern of fragment peaks.

2.3 Identification of cystine-containing peptides

Most peptides in a digest are irrelevant. Which contain cystine? Analysis must be done in a way that allows us be sure that the other peptides do not contain cystine, and hence that disulfide exchange is not significant. The most important approaches are:

- direct analysis/detection of cystine
- mobility-shift assays
- MS

2.3.1 Direct analysis for cystine

Sensitivity and simplicity are major issues here, since many fractions must be dealt with.

1. Amino acid analysis. Historically, every peptide fraction was hydrolysed and subjected to quantitative analysis. Although time-consuming, this approach provides so much information that it should be used whenever possible: a precise amino acid composition may identify where the peptide fits in the protein chain. It also gives a good measure of peptide recovery, which lessens the chance of being misled by minor products. This approach works well with proteins of moderate size, but gets more difficult with larger and scarcer proteins—there are many more fragments to analyse, and quantification of small amounts of cystine is less reliable.

2. Electrochemical detection of cystine-containing peptides in HPLC effluent has been reported (9). If a suitable on-line detector is available, this approach should be considered, because of its sensitivity and immediacy. It should be used only as a supplemental detector, with a UV monitor to assess the overall progress of digestion.

2.3.2 Mobility-shift analysis

The archetype of this approach is the well known 'diagonal method'. Breaking a disulfide bond almost always changes a peptide's chromatographic behaviour, and often generates two peptides from a single original. Brown and Hartley (10) used performic acid vapour to convert cystine-containing peptides into pairs of cysteic acid-containing peptides: the resulting change of electrophoretic mobility on paper allowed them to deduce that two peptides were originally connected, and to purify and identify them. Although we cannot pump an HPLC column at right angles, in the way that one can electrophorese sheets of paper at right angles, we can compare HPLC analyses of a digest before and after reduction. If the mixture is not too complex, the major disulfide-containing peptides can be identified by their being absent after reduction, and further purified from the main HPLC fractions.

Protocol 2. Mobility shift assay on whole digest[a]

Reagents
- Reduction buffer A: 0.1% (v/v) β-mercaptoethanol in 0.2 M Tris pH 8.5

Method

1. Digest the protein with enzyme under conditions that were established as optimal (*Protocol 1*).

2. Take two equal aliquots of the digest, as little as will give a reliable chromatogram. Reserve one of these as the control digest.

3. Mix the other aliquot with an equal volume of reduction buffer A, and incubate the mixture at 37 °C for 2–5 h to obtain the reduced digest.

4. Compare the chromatographic profiles of the two digests by HPLC.

[a] Adapted from ref. (4). Use caution to avoid unwanted exposure of other digests to mercaptoethanol vapour. Other reducing agents such as DTT or TCEP may be used.

This is convenient, but may not locate every cystine peptide. A more rigorous method employs a mobility shift assay on separated fractions. This is much more work, but both arms of the bridged peptide usually shift their elution times, while contaminants that lack cystine do not. Like the original diagonal method, it thus serves both to identify and to purify the wanted peptides. In either protocol, cystine-containing peptides are most likely present in fractions that show a major reduction in absorbance at the original elution time (allowing for any change in gradient), or that give rise to multiple components after reduction (*Protocol 3*). Remember that contaminating peptides will show up as residual absorbance at the original position after reduction.

Protocol 3. Mobility shift assay on single fractions

Reagents
- TCEP3: 20 mM TCEP (Pierce Chemical) in 0.2 M sodium citrate pH 3.0

Method

1. Digest the protein with enzyme under suitable conditions, and separate the digest by HPLC (see text). Collect all significant fractions.

Protocol 3. *Continued*

2. From each fraction, remove a sample sufficient to give a reliable chromatogram. Place it in a microvial, and add an equal volume of TCEP3. Cap the vial tightly, and incubate it at 65°C for 10–15 min.[a]

3. Examine each reduced fraction by HPLC. Minimally, use a short gradient spanning 5–10% (v/v) acetonitrile on either side of the expected elution time. More elaborately, use the full gradient (e.g. 0–60% (v/v) acetonitrile) to locate both halves of a cystine-containing peptide.

[a] This temperature is more than adequate in most cases, but very hydrophobic peptides can be relatively slow to reduce. The reduction can be carried out batchwise, and the samples frozen until they can be analysed.

2.3.3 Mass-shifting analysis

Modern mass spectrometers allow relatively quick and sensitive measurement of intact masses for peptides in the size range of interest (see Chapter 2). They make it reasonable to compare small aliquots of HPLC fractions before and after reduction with either TCEP (11) or dithiothreitol (12). Opening a disulfide changes the peptide mass:

(a) If the bridge formed a loop within a single peptide, the mass increases by 2 amu.

(b) If it joined two peptides, the subfragments appear at lower masses whose total equals the original plus 2 amu.

Both can be used with the typical matrices used in MALDI MS and FAB/LSIMS.

2.4 Identification of subfragments

The procedures given in Section 2.3 should locate peptide fractions that are likely to contain cystine. Analyse all such fractions to establish where the halves of that cystine originate in the protein. Ideally, every peptide will fit with only one site, and all the data will point to only one disulfide bridging pattern. Sequence analysis is the most rigorous approach, followed by MS; the methods complement each other nicely. In comparison, amino acid analysis requires purer samples, and high precision is needed to make a secure assignment.

Sequence analysis can be done on the intact cystine-containing peptide, or on the separated reduction products. A few cycles usually suffice to pin-point the origin of a single peptide, while the more complex data from bridged peptides may require extra cycles to identify the two halves uniquely. Because the protein's sequence is known, all possible origins can be tested visually. Mass analysis of the intact peptide and/or the reduction fragments provides strong back-up, and knowing the amino terminus of each fragment allows one to focus the search for matching masses.

MS (Chapter 2) can also be used as a first-line method, depending on available facilities and expertise. More sophisticated instruments use fragment ions to establish partial sequences: these are then located in the whole by visual inspection. Intact mass analysis is more widely available, employing soft ionization methods (ES, FAB/LSIMS, MALDI) and simpler instruments. Precision varies widely according to the post-ionization analyser: double-focusing instruments are likely to give $\pm$ 0.1 amu, while TOF analysis following laser desorption may be as poor as $\pm$ 2 amu.

It is important that the peptide chemist and the mass spectroscopist work closely to obtain a reliable interpretation of intact mass data. Points for the peptide chemist to be conscious of:

(a) What is the precision of the measured mass?

(b) What does the given mass number refer to?
   i.   Is it the molecular mass (M), or that of the protonated molecule (M + H)?
   ii. Could it be a cation adduct such as (M + Na)?
   iii. Is it the average mass, or the monoisotopic mass?

(c) If a computer-based search is used to identify all possible fragment peptides that match the observed mass within stated precision (as in point a):
   i.   Again, be clear whether the program searches by M or (M + H).
   ii. Include non-cystine peptide fragments in the search.
   iii. Consider modifications that may have occurred in handling—methionine oxidation (+16 amu), loss of amide (+1 amu), or glutamine cyclization (−17 amu).

Given these cautions, the search is likely to provide one or at most a few possible matches to the observed mass for a single peptide. Knowing the selectivity of the cleavage methods used to generate the fragments will usually eliminate all but one. Bridged peptides require an additional search for matches among all possible pairwise combinations of Cys residues. Further support should be obtained by way of mass-shifting after reduction or Edman degradation (12).

2.5 Specific examples

The reader will profit greatly by consulting original papers that use the above methods. Two are recommended for giving good procedural descriptions, with attention to disulfide exchange.

2.5.1 *Euplotes* pheromones (4)

These 40 residue peptides contain three disulfides, with the cystines being well dispersed. The authors used thermolysin digestion with 8 nmol of natural peptide. Major Cys-containing fragments were identified by mobility-shifting after reduction. Amino acid analysis was done on all HPLC fractions to confirm the absence of rearranged products. Peptide identification was by sequencer degradation, with MS confirmation (TOF analysis, laser desorption, or ES ionization).

2.5.2 *Agelenopsis* toxins (6)

Two 48 residue toxins had the same amino acid sequence and the same mass. The initial presumption was that they differed in the connections of their four disulfide bridges, but careful analysis showed that the difference lay in epimerization of a single Ser residue. Vigorous digestion with trypsin (1/1, w/w, 20 h) gave a cross-linked core peptide that contained six of the eight Cys residues. Further digestion with endoproteinase Glu-C, and partial reduction with TCEP, gave enough fragments to define the whole pattern. Peptide identification was by sequence degradation and MS (quadrupole analyser, ES ionization). The work used approximately 200 nmol natural toxin.

3. The TCEP strategy

Peptide toxins, growth factors, and hormones often have cysteine residues so close together in the sequence that definitive enzymatic fragments are not obtained. Disulfide analysis of such molecules was very difficult until the introduction of TCEP (3,5,13). This water soluble phosphine remains active down to pH 3, so it can reduce disulfides under conditions where exchange is minimized. One can then separate the partially reduced peptides, alkylate the free thiols, and hence identify the linkages. We have since used this approach with numerous peptides. It is a dynamic method in which one must remain flexible in dealing with competing reactions. The description that follows is based closely on the original work, plus insights gained with new peptides. Again, the emphasis is on conveying practical insights rather than rigid directions.

Before committing a valuable peptide, it is highly recommended to practice reduction and alkylation on a well-understood system such as insulin, conotoxin GI, or endothelin. These peptides are relatively cheap to purchase; their disulfide reduction intermediates are all characterized, spanning the range between very labile and remarkably stable (3).

3.1 Partial reduction

One's aim is to optimize the yield of peptides having at least one bridge opened, while at least one remains closed. It is best to use a large excess of TCEP and control the extent of reduction by varying the time and temperature.

3.1.1 Reaction conditions

TCEP functions down to pH 3, but slows markedly below that pH. This provides a good working compromise: reduction times are relatively short (in the range of 2–10 min), while disulfide exchange is not much of a problem. Another convenience is that at pH 3 TCEP is very stable towards air oxidation, and solutions can be kept for many weeks at room temperature.

Protocol 4.   Partial reduction by TCEP

Equipment and reagents

- Temperature controlled heating block
- Microvials, 0.5 and 1.5 ml, virgin polypropylene (e.g. Sarstedt 72.690.051)[a]
- TCEP3: 20 mM TCEP in 0.2 M sodium citrate pH 3.0

- Peptide solution as eluted from HPLC, or in a similar solvent (0.1% (v/v) TFA plus acetonitrile)[b]

Method

1. If the incubation is to be at an elevated temperature, pre-equilibrate both the TCEP3 and peptide to that temperature for 5–10 min before mixing.[c]

2. Place the peptide solution in a microvial, add an equal volume of TCEP3, and cap the tube tightly.

3. Incubate the tube for an appropriate time at the desired temperature (Section 3.1.2).

4. If necessary, dilute the sample with 0.1% TFA to lower the acetonitrile concentration. Inject the reaction mixture directly on to the HPLC column.

[a] Avoid products that contain plasticizers, which leach out with acetonitrile and confuse chromatograms.
[b] Peptides may become less soluble upon reduction. If this is a problem, add more acetonitrile.
[c] Heat transfer through polypropylene is slow, and the initial reduction can be very sensitive to temperature.

3.1.2 Trial experiments to optimize reaction temperature and time

Individual disulfides vary widely in reaction rates, depending on their accessibility and degree of strain—factors that may change as a multicyclic peptide is opened by reduction. Finding reaction conditions that give suitable intermediates involves a few trial experiments with small amounts of peptide (5–200 pmol, according to instrumental sensitivity). Most of the sample is easily recovered for reuse. Use *Protocol 4*, and carry out reductions of greater or less severity:

(a) Reagent blank. Check TCEP3 solution for potential contaminating absorbance peaks.

(b) Complete reduction. Reduce a minimal aliquot for 15–20 min at 65°C to obtain fully reduced peptide (R). Compare HPLC elution (Section 3.2.2) with that of unreduced peptide (N).

(c) Mild reduction. Treat a larger sample for 2–5 min at room temperature. Analyse by HPLC. Partially reduced peptides (PR) often, though not always, elute between N and R.

(d) Evaluate and adapt. Use these guidelines for deciding how to change the conditions:
   i. If there is some PR but very little R: increase the reaction time.
   ii. If no PR or R is produced: increase the reaction temperature by 5–10 degrees.
   iii. If there is more R than PR: decrease the TCEP concentration if working at room temperature, or decrease the reaction time, or lower the temperature by 2–5 degrees.

It usually takes only a few trials to find conditions in which useful amounts of PR are obtained without wasteful conversion to R. A good compromise is reached when the amounts of R and PR are comparable. Most peptides reduce satisfactorily at room temperature in 2–15 min using 10 mM TCEP. A minority appear very resistant until heated above some threshold, at which point they become sensitive to small changes in temperature. In such cases it is important to maintain close control over incubation conditions (see *Protocol 4*).

3.1.3 Kinetic factors in reduction

Peptides vary a lot in their overall patterns of reduction. A general understanding of the kinetic possibilities is helpful in making the most of any particular situation.

For example, consider a tricyclic peptide (ABC) with bridges A, B, and C. In the absence of disulfide exchange the reduction proceeds by way of three bicyclic (aBC, AbC, ABc) and three monocyclic (abC, aBc, Abc) intermediates to fully reduced peptide (abc). The system dynamics are described by 12 pseudo-first order rate constants, because we are using an irreversible reductant in very large excess. The overall pathway, and the ease of obtaining intermediates is set by the relative reaction rate constants. A few important patterns are shown in *Figure 1*.

These patterns have all been encountered. They may be described as:

(a) Selective *(Figure 1a)*: one bridge reacts rapidly while the others are resistant. A single partially reduced product is obtained in high yield. The most effective way of handling this situation is to isolate the stable intermediate, label its thiols, and reduce it again under more vigorous conditions. Several tricyclic peptides have given selective reduction patterns, including BPTI. The four-bridged echistatin (5) was a dramatic example.

(b) Random *(Figure 1b)*: all rates are comparable. Most intermediates are

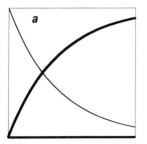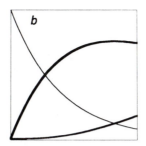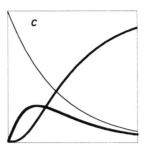

Figure 1. Kinetics of reduction of a tricyclic peptide under pseudo-first order conditions (large excess of TCEP, pH 3). Normalized to show about four half-lives of the original peptide (monotonically decreasing light lines), with the production of one- and two-disulfide intermediates (heavy lines), and their conversion into fully reduced peptide (medium lines). (a) Rate constants for reduction of bridge A is fixed, while those for bridges B and C are zero in all forms of the peptide. (b) All rate constants are set equal. (c) As for (b) except that the rate constants for reduction of bridges in 'native' tricyclic peptide are set ten times lower than those in partially reduced peptides.

produced in fair yield, but some are easier to purify than others. They may also vary greatly in their rate of disulfide exchange when it comes to alkylation. Begin by analysing the most easily purified peptides, and proceed to others as needed. Many bicyclic peptides such as α-conotoxins and endothelin show an essentially random pattern. Insulin is a tricyclic peptide for which all intermediates can be isolated, though some are quite labile.

(c) Accelerating *(Figure 1c)*: the initial step is slow, but once the peptide is partially reduced, the remaining steps are faster. This pattern is quite common with tricyclics (14). It is the hardest to deal with because the concentration of intermediates is never high. One must work in the narrow time range where PR is higher than R, and recycle the unreduced peptide.

3.2 Preparation of incompletely reduced peptides (full-length)

3.2.1 Scale-up of reduction

Having found suitable reduction conditions, prepare usable amounts of the intermediates. Under the pseudo-first order reaction system, with excess TCEP, no changes are needed for varying quantities of peptide. When reduction goes by the random or selective pathways, a single reduction and HPLC run is enough. Two options are available for the accelerating pattern *(Figure 1c)*:

(a) If the peptide is relatively easy to obtain, as with synthetic material, commit 50–100 nmol to a single short reduction. Most will be recovered as unchanged N, while giving adequate PR.

(b) If the peptide is scarce (2–10 nmol), use a short reaction time and recycle unreacted N. Do this while the HPLC is re-equilibrating at the end of a run: simply mix the eluted N with an equal volume of TCEP3, making sure that both solutions have reached the proper temperature. Do a series of recyclings, storing the intermediates at −70°C.

3.2.2 Purification and storage of intermediates

To minimize exchange reactions, we separate by HPLC at pH 2, using acetonitrile gradients in 0.1% (v/v) TFA. The range of products is narrower than that obtained in a typical enzymatic digest:

(a) Relatively few intermediates are expected (no more than two for a bicyclic, six for a tricyclic).

(b) All are full-length, so they adsorb well to the column, and their molar extinction is comparable.

A suitable acetonitrile gradient should be worked out while optimizing the reduction procedure (see Section 3.1.1). Reduced and partially reduced peptides usually elute later than N, but a complete reversal may occur with peptides that contain many hydrophobic residues (14). Be sure to allow for this possibility when choosing the initial acetonitrile concentration.

Collect peak fractions manually, to make the best use of column resolution. Store the intermediates at −10°C or −70°C until they are needed for further analysis. If several fractions are to be combined, as after recycling, pool them and repurify using a steep gradient: this ensures a high final peptide concentration that makes labelling easier (see Section 3.3).

If the peaks are unduly crowded, try one of the following:

(a) Slow the gradient. A disadvantage is that peak fractions tend to become more dilute and may need to be reconcentrated.

(b) Isolate one intermediate and use it as the starting point for further partial reduction.

(c) Enzymatically digest N (see Section 2.1.1) and work with one or more fragments.

It is strongly recommended that labelling be done as soon as possible after isolating intermediates. Although most are stable, some undergo internal disulfide exchange at a significant rate even at pH 2. Stored peptides should always be checked before use.

3.3 Labelling the intermediates

Scrupulous attention to detail is needed to achieve good results. Partly reduced peptides retain some folded structure, so that thiols may be poised close to each other, and close to unopened disulfides. This produces a twin risk of reoxidation and of thiol–disulfide exchange:

(a) Reoxidation.

    i. At acid pH dissolved oxygen is of little consequence, but becomes significant when the pH is raised for alkylation. Include EDTA as a chelating agent to prevent heavy metal catalysis.

    ii. A more serious problem is iodine, which is active even at acid pH: it is produced from iodoacetamide by light and oxygen. Include ascorbic acid as a scavenger.

(b) Exchange. Almost all the usable alkylating agents react only with the thiolate ion RS⁻, not with RSH. Because thiolate ion attacks disulfides, there is inevitable competition between exchange and labelling. A few thiolate peptides 'scramble' on a millisecond time-scale, so it is critical to avoid raising the pH without there being a large excess of alkylating agent present.

3.3.1 Choice of alkylating agent

To compete effectively with exchange, use a high concentration of reactive alkylating agent, with close attention to technique. Iodoacetamide (2.2 M) has been most successful (*Protocol 5*). Provided it is absolutely colourless, indicating absence of iodine, it may be used as purchased. Iodoacetamide's main drawback is that the PTH-Cys(cam) obtained during sequence analysis is difficult to separate from PTH-Glu. One promising alternative is *N*-methyl-iodoacetamide which gives a better chromatographic separation (15).

Other reagents have been tested with less satisfactory results.

(a) Iodoacetic acid is not as reactive as the amide.

(b) 4-Vinylpyridine cannot be used at high concentration because UV absorbing impurities interfere with HPLC detection.

(c) *N*-Ethylmaleimide is prochiral and gives rise to multiple epimeric forms.

(d) Benzamidomethanol was used successfully to analyse the bridges of a spider toxin (6), taking advantage of its ability to react with thiols in strong acid (100% TFA). However, it also reacts with tyrosine and tryptophan, restricting its generality. When applied to a BPTI folding intermediate this reagent gave a very heterogeneous mixture of products (G. Bulaj, personal communication).

3.3.2 Primary labelling protocol

The recommended method for labelling a partially reduced peptide is given in *Protocol 5*. Not all intermediates require such vigour, but this is not known ahead of time—it is better to work on the conservative assumption that exchange is a risk. Several points are worth emphasizing:

(a) The order of doing things is important. Add peptide to the iodoacetamide solution, and not vice versa. Otherwise the pH rises faster than the iodoacetamide concentration, and exchange is favoured.

(b) Make the alkylating solution fresh for every individual reaction, and use it within a few minutes. Do not prolong the heating unnecessarily. Tris reacts with iodoacetamide, generating acid, and the resultant lowering of pH can prevent alkylation.

(c) Use caution when vortexing the iodoacetamide solution. Check that the mixer is set to a speed that swirls the solution thoroughly without splashing it out of the tube.

(d) A fine syringe needle helps maintain a fast liquid stream for adding peptide to reagent.

Using these precautions, a short reaction time, and the acid quench, we have had no significant problems with alkylation of other amino acids. It is strongly recommended that a practice run be carried out, and that the reagent mixture be analysed by HPLC to test for interfering impurities: there is more than a million-fold excess of iodoacetamide over peptide!

Protocol 5. Rapid labelling of thiols

Equipment and reagents
- Heating block (65°C), plastic microvials (see *Protocol 4*)
- 500 μl syringe fitted with fine needle (25 gauge or finer)
- Acid quench solution: 0.5 M citric acid
- Tris8 buffer: 0.5 M Tris–acetate pH 8, 2 mM Na_2-EDTA, purged 5–10 min with N_2
- Ascorbic acid solution: 100 mM in Tris8 buffer

Method

1. Thaw the peptide solution and allow it to reach room temperature as needed.

2. Weigh 100 mg iodoacetamide into a 1.5 ml microvial. Add 200 μl Tris8 buffer, and cap the tube. After vortexing the tube briefly, place it in the heating block and cover the tube to exclude light. Repeat the mixing periodically until the solid is completely dissolved.

3. Add 10 μl ascorbic acid solution to scavenge traces of iodine. Allow the iodoacetamide solution to cool for a few minutes, but avoid letting crystals appear.

4. While the iodoacetamide solution is cooling, load 300 μl peptide solution into the fine-tipped syringe. Load 400 μl of the acid quench solution into a pipette.

5. Uncap the iodoacetamide solution, and swirl it at moderate speed in a Vortex mixer. While it is swirling, rapidly add the peptide solution as a fine jet.

6. After 20–30 sec add the acid quench solution, and immediately inject the whole mixture on to the HPLC column.

3.3.3 Isolating the alkylated products, and monitoring for disulfide exchange

Once the reaction mixture is injected on to the HPLC column, wash the column with starting buffer until the absorbance of the effluent drops below 0.05 at 220 nm. The baseline absorbance remains above zero due to concentrated reagents trapped in minute 'dead spots' in the liquid flow path. After the run is complete, and while the acetonitrile concentration is still high, flush the injection port to remove these residues.

Peptides that have been alkylated with iodoacetamide generally elute close to the unmodified peptides, so a similar gradient can be used. Once alkylated and purified, the peptides need no special handling. Chromatograms typically show one of two patterns in the region of interest:

(a) Single major peak, with or without minor satellites. This is the commonest situation, and the major peak almost always represents the correctly alkylated peptide: sequence analysis of a set of such peptides provides a uniquely consistent disulfide pattern. Very occasionally the major peak has been due to reoxidized peptide.

(b) Two or more significant peaks, often with other satellites. This occurs in perhaps 10–20% of cases. Only rarely has it been due to heterogeneity of the starting intermediate, or to incomplete alkylation. The latter only occurred when the iodoacetamide solution was heated too long, and the pH had dropped (see Section 3.3.2). It is almost always a sign that thiol–disulfide exchange has partially converted the peptide to other disulfide isomers.

If you obtain multiple peaks do not assume that the largest one represents the correct product, though it usually does. To identify the correct product positively, carry out a second experiment with a small amount of starting peptide:

(a) Repeat the alkylation according to *Protocol 5*, but using only 10–20 mg iodoacetamide.

(b) Evaluate by HPLC. The peak corresponding to the correct product will be diminished because the lower reagent concentration allows exchange to occur faster than alkylation.

3.3.4 Secondary labelling of remaining Cys residues with 4-vinylpyridine

It can be helpful to obtain positive identification of every Cys residue in the peptide, rather than using absence of Cys(cam) to indicate non-reduced disulfides (Section 3.4). This is best achieved by alkylating with 4-vinylpyridine, generating *S*-pyridylethylcysteine, which is the preferred derivative for sequence analysis. A suitable method is as follows:

(a) Mix the eluted peptide (see Section 3.3.3) with an equal volume of TCEP3, and heat the mixture for 15 min at 65°C. Purify the reduced peptide by HPLC.

(b) For each 500 µl of peptide solution add 3 µl 4-vinylpyridine, followed by 3 µl pyridine. Mix well. Cap the tube, and incubate it in a dark place for 20 min at room temperature.

(c) Purify the resulting peptide by HPLC. Because pyridylethyl groups are polar, these peptides usually elute earlier than the parent molecules. Allow for this by:

  i. diluting the reaction mixture well with 0.1% TFA
  ii. starting the gradient at a lower acetonitrile concentration
  iii. washing the column until the effluent absorbance falls below 0.05.

3.4 Identifying the labelled cysteines

Sequence analysis is the most comprehensive method for locating alkylated cysteines. This becomes expensive for long peptides, and researchers may prefer to use some kind of peptide mapping. Such an approach is also useful when the peptide has a blocked amino terminus.

3.4.1 Sequence analysis

Unequivocal analysis is obtained with 10–100 pmol of labelled product, depending on the length of the peptide and the facilities available. Analysis must be carried through to the last Cys in the peptide, not merely until a consistent set of alkylated residues has been identified. The extent to which label is limited to a small set (usually two or four) of positions is a good measure of how well disulfide exchange was controlled. Such a judgement is best made by plotting the yield of derivative versus residue position along the peptide. Numerous plots of this type are given in the various publications from this laboratory (3,5,14), and their interpretation is simple.

In low level analyses the repetitive yield is 85–95%, leading to an exponentially decreasing recovery. Because of the many-fold decrease by the end of a long peptide, it may be hard to decide whether absence of Cys(cam) is due to lack of labelling, or to low recoveries of all residues. Using a secondary label of those residues that did not label with iodoacetamide is then recommended. Pyridylethylation is the method of choice for this step (Section 3.3.4).

3.4.2 Peptide mapping analysis

In some situations it may be simpler to identify alkylation sites by digesting the labelled peptide with proteases, and characterizing the fragments as discussed in Section 2. This happens with peptides that are long, or that are blocked—in such cases, the original sequence analysis probably included some mapping steps. It becomes especially attractive when doing repeated analyses of several related molecules, folding intermediates, or disulfide isomers. Details of such an analysis will vary among proteins, but the following principles are widely applicable:

(a) Reduce any remaining bridges, and label the newly exposed thiols with a reagent other than iodoacetamide. Since disulfide exchange is no problem, there are many options.

(b) Ideally, use proteases that generate a unique cleavage product for each Cys residue. Fully alkylated, linear peptides digest more readily than the original, and there is no risk of disulfide exchange during digestion.

(c) Separate the fragments by HPLC, using the guidelines discussed in Section 2.2.1.

(d) Identify the origins of the Cys-containing peptides by MS or sequencing.

If several related structures are to be analysed, IAEDANS (Molecular Probes) is an excellent choice as the second alkylating agent (16). Its derivatives are relatively hydrophobic, so smaller peptide fragments are better retained on the HPLC column. It also absorbs well at 340 nm, allowing one to monitor only the Cys-containing peptides. In this approach, one should first reduce a peptide sample fully and alkylate it with IAEDANS, then digest it and identify all the Cys-containing fragments. Any peptide that was partially alkylated with iodoacetamide will generate a diagnostic subset of the IAEDANS labelled fragments. When digestion does not give unique fragments for every Cys, sequencing may be needed to resolve ambiguities.

3.5 Comparing disulfide fingerprints with TCEP

It is sometimes important to test whether two peptides have the same disulfide pattern, even before knowing what that pattern is. We make frequent use of a simple 'disulfide fingerprint' technique (*Protocol 6*) for comparing natural and synthetic peptides in this way. In several cases it has played a crucial role with peptides that have been too scarce for direct analysis. This situation arises when only a few nanomoles of peptide are isolated from a venom, but elicit a promising biological response—how then should one proceed?

(a) Use < 1 nmol to obtain a complete sequence analysis and accurate molecular mass. This defines the linear structure, including C terminal amidation.

(b) Synthesize the corresponding linear peptide, and fold it under oxidizing conditions. Purify the product that elutes at the same position as natural peptide.

(c) Compare disulfide fingerprints of 10–200 pmol of natural and synthetic products.

(d) If they match, analyse the disulfide bridges of the synthetic peptide by partial reduction, and reserve the natural peptide for biological comparison with synthetic material.

The fingerprint test relies on the subtlety of TCEPs reaction kinetics, which is affected by the accessibility and degree of strain in the various disulfides (Section 3.1.2). Two peptides having the same amino acid sequence but different bridging patterns may not separate on HPLC, but it is extremely unlikely that their partially reduced products would all elute at equivalent positions, and be produced in the same ratios.

Protocol 6.   Disulfide comparison by TCEP fingerprinting

Equipment and reagents
- Heating block, microvials, TCEP3 (see *Protocol 4*)

Method
1. Use small aliquots (10–100 pmol) of synthetic peptide to find partial reduction conditions that give a distinctive pattern of intermediates (see Section 3.1.2 and *Protocol 4*).
2. When suitable conditions have been identified, set-up three microvials: RB, reagent blank; SP, synthetic peptide; and NP, natural peptide. SP and NP should contain equal amounts (e.g. 50 pmol) of the corresponding peptides, dissolved in equal volumes of the same solvent (e.g. HPLC elution buffer). RB should contain the same volume of solvent without peptide.
3. In succession, carry out partial reduction on the three samples. Pay close attention to factors such as pre-equilibrating the reactants to temperature, reaction times, and the time taken to load and inject on to the HPLC. Carry each sample through the entire experiment before starting the next. If natural material is scarce, recover any unreduced peptide.
4. Compare the elution profiles with respect to retention times and relative sizes of peaks.

A typical comparison of a synthetic peptide with an authentic standard by this method is shown in *Figure 2*. Note the close correspondences between the two chromatograms.

4. Concluding comment

Many technical advances in peptide purification and analysis have combined to make disulfide assignment accessible for a wider range of peptides and proteins. The analysis can still represent a big commitment in effort and material, however. It is hoped that the methods presented here will serve as a

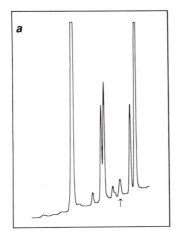
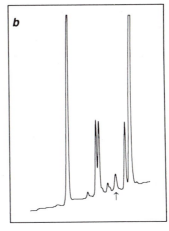

Figure 2. Comparison of HPLC profiles of partial reduction products of ω-conotoxin MVIIA. Peptide samples (200 pmol) were reduced by treatment with 10 mM TCEP at pH 3.0, for 4 min at 65 °C. (a) Authentic material. (b) Synthetic peptide. The *arrow* indicates a reagent impurity.

practical guide for those who are faced with deciding how best to meet the challenge—and that many more structures will enrich our understanding of protein biology.

References

1. Ryle, A. P. and Sanger, F. (1955). *Biochem. J.*, **60**, 535.
2. Ryle, A. P., Sanger, F., Smith, L. F., and Kitai, R. (1955). *Biochem. J.*, **60**, 541.
3. Gray, W. R. (1993). *Protein Sci.*, **2**, 1732.
4. Stewart, A. E., Raffioni, S., Chaudhary, T., Chait, B. T., Luporini, P., and Bradshaw, R. A. (1992). *Protein Sci.*, **1**, 777.
5. Gray, W. R. (1993). *Protein Sci.*, **2**, 1749.
6. Heck, S. D., Kelbaugh, P. R., Kelly, M. E., Thadeio, P. F., Saccomano, N. A., Stroh, J. G., *et al.* (1994). *J. Am. Chem. Soc.*, **116**, 10426.
7. Savige, W. E. and Fontana, A. (1977). In *Methods in enzymology* (ed. C. H. W. Hirs), Vol. 47, p. 459. Academic Press, New York.
8. Zhou, Z. and Smith, D. L. (1990). *J. Protein Chem.*, **9**, 523.
9. Lazure, C., Rochemont, J., Seidah, N. G., and Chretien, M. (1985). *J. Chromatogr.*, **326**, 339.
10. Brown, J. R. and Hartley, B. S. (1966). *Biochem. J.*, **101**, 214.
11. Fischer, W. H., Rivier, J. E., and Craig, A. G. (1993). *Rapid Comm. Mass Spectrom.*, **7**, 225.
12. Zhou, Z. and Smith, D. L. (1990). *Biomed. Environ. Mass Spectrom.*, **18**, 782.
13. Burns, J. A., Butler, J. C., Moran, J., and Whitesides, G. M. (1991). *J. Org. Chem.*, **56**, 2648.

14. Shon, K. J., Hasson, A., Spira, M. E., Cruz, L. J., Gray, W. R., and Olivera, B. M. (1994). *Biochemistry*, **33**, 11420.
15. Hunter, M. J. and Komives, E. A. (1995). *Anal. Biochem.*, **228**, 173.
16. Weissman, J. S. and Kim, P. S. (1991). *Science*, **253**, 1386.

Analysis of protein conformation by gel electrophoresis

DAVID P. GOLDENBERG

1. Introduction and basic principles

Electrophoresis through polyacrylamide gels (PAGE) is one of the simplest and most commonly used methods for characterizing protein molecules (1,2). As described in Chapter 1, gel electrophoresis in the presence of the detergent sodium dodecyl sulfate (SDS) is widely used to determine the molecular weights of polypeptides. Electrophoretic methods can also be used to determine the isoelectric point of a protein and to count the number of reactive groups (Chapter 6). This chapter describes the use of non-denaturing, or 'native', gel electrophoresis to analyse the conformations of proteins as well as a method for studying protein folding–unfolding reactions using gels containing transverse gradients of denaturants.

Electrophoretic methods for studying protein conformation offer a number of advantages. The apparatus required is relatively simple and inexpensive and is available in almost any biochemistry or molecular biology laboratory. The method requires only small amounts of protein, less than 1 μg if the gel is stained with Coomassie blue, and even less if more sensitive staining methods are used. Also, the protein sample may not need to be pure, since electrophoresis can often separate different molecules in a sample. The use of selective visualization techniques such as autoradiography of radiolabelled molecules, antibody staining, or enzymatic assays may also aid in characterizing complex mixtures of proteins. On the other hand, gel electrophoresis is primarily a qualitative method. As discussed below, electrophoretic mobilities are determined by several interacting factors, and it is usually possible to interpret only changes in mobility arising from covalent modifications or conformational changes. More quantitative or detailed information is best obtained using other hydrodynamic or spectroscopic methods (Chapters 9, 11, and 12).

The velocity of a macromolecule moving through a gel in an electric field depends upon both the net charge of the molecule and its size and shape. The greater the net charge, the greater will be the electrostatic force generated by interaction of the macromolecule with an external field, leading to a greater mobility. Molecules that differ by only a single unit charge frequently have readily distinguishable mobilities. This sensitivity to charge can be used to 'count' reactive residues in a polypeptide, as described in Chapter 6. Quantitative treatments of the interactions between a charged molecule and an external field are very complex, however, and must take into account the cloud of counterions surrounding the macromolecule.

The relationships between molecular size and shape and electrophoretic mobility are equally complex. In the absence of a gel support, the electrophoretic mobility of a macromolecule is expected to be inversely related to its frictional coefficient. The frictional coefficient, however, depends upon not only the size and shape of a macromolecule, but also its interactions with the solvent. The situation is further complicated by the presence of a gel matrix, which generates a sieving effect. Although the physical basis of gel sieving is not fully understood (3,4), it can be measured by determining electrophoretic mobilities in gels of varying composition. Increasing the total concentration of acrylamide in a gel ($\%T$) (while maintaining a constant ratio of bisacrylamide cross-linker to acrylamide) has been found to lead to reduced mobilities (R_f) according to the empirical relationship:

$$\log R_f = \log Y_o - K_r(\%T), \qquad [1]$$

where Y_o is the so-called free mobility corresponding to the mobility in the absence of a gel, and K_r is a retardation coefficient (3–6).

The extrapolated free mobility is expected to depend on the molecule's charge and frictional coefficient, as discussed above. The retardation coefficient is expected to be qualitatively related to the hydrodynamic volume of the molecule and has been observed to be correlated with molecular weight (4,6–8). However, the apparent gel sieving effect may also depend upon specific interactions between the molecule and the gel, and retardation coefficients must be interpreted with caution.

In spite of these limitations in interpreting electrophoretic mobilities, it is easy to demonstrate the ability of gel electrophoresis to separate proteins of evenly slightly different hydrodynamic volumes, as illustrated in *Figure 1* for different forms of bovine pancreatic trypsin inhibitor (BPTI), all with the same net charge. The native conformation of BPTI can be disrupted simply by reducing the protein's three disulfide bonds, leading to an extended unfolded conformation with reduced electrophoretic mobility, as shown in the first two lanes of *Figure 1* (9). The third lane of *Figure 1* contains the reduced form of a BPTI variant in which four neutral amino acid residues are inserted. Although the insertion is expected to increase the radius of gyration of the reduced protein by only 2–3%, it leads to a clearly detectable decrease in mobility (10). Cross-linking the BPTI termini, to generate a circular polypeptide, causes a much larger change in the mobility of the reduced protein, as shown in the fourth lane.

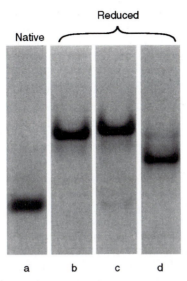

Figure 1. Non-denaturing gel electrophoresis of different forms of bovine pancreatic trypsin inhibitor (BPTI), illustrating the sensitivity of gel electrophoresis to conformation. (a) Native wild-type BPTI. (b) Reduced and unfolded wild-type BPTI. (c) Reduced and unfolded BPTI variant with an insertion of four neutral amino acids between residues 12 and 13 of the wild-type sequence. (d) Reduced and unfolded circular BPTI, in which the termini were cross-linked using a water soluble carbodiimide. Reduced samples were prepared by incubating protein for 30 min at 25 °C in the presence of 8 M urea, 0.1 M Tris–HCl pH 8.7, and 10 mM DTT. The samples were electrophoresed through a 15% polyacrylamide gel towards the cathode at pH 6.1 and 21–24 °C. The gel was stained with Coomassie blue R-250. Adapted, with permission, from ref. 10.

A protein in solution may exist as a mixture of interconverting conformations. If the electrophoretic separation is rapid compared to the rate of the conformational interconversion, it may be possible to resolve the isomerizing species. Quantitative treatments of the electrophoretic behaviour of isomerizing molecules have been developed (11), and can be used to predict the patterns expected for various rates of electrophoresis and isomerization, as illustrated in *Figure 2*. If the half-time for interconversion between conformations is less than about one-tenth the duration of the electrophoresis, there will be no separation, and a single band will be observed with a mobility that is the weighted average of the mobilities of the two forms. At the other extreme, if the half-time of the interconversion is more than about ten times the electrophoresis time, the different conformations persist and may be cleanly separated. At intermediate rates of isomerization, only a fraction of the molecules will isomerize during the separation, leading to a smear of protein between the positions of the two forms. If there is only one species present at the beginning of the electrophoresis, only a single band is expected under any circumstances, although there may be an additional diffuse distribution of

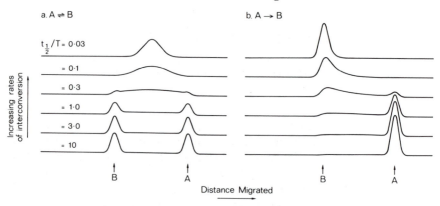

Figure 2. Calculated electrophoretic profiles of interconverting molecules. The expected distributions of protein after electrophoresis through a gel were calculated using the probabalistic model of Mitchell (11). The profiles on the left are those expected if an equilibrium mixture composed of equal amounts of the fast (A) and slow (B) migrating forms of protein were applied to the gel and reversible interconversion between the two forms took place during electrophoresis, at the indicated relative rates. The profiles on the right are those expected if the fast migrating form (A) was applied to the gel and was irreversibly converted to the slow migrating form (B) during electrophoresis. Such a profile would be generated, for instance, if a native protein was subjected to electrophoresis in a gel containing urea where unfolding took place and refolding was not significant. For all of the calculated profiles, the two forms of the protein, A and B, were taken to have electrophoretic mobilities (indicated by the *arrows*) of 0.2 and 0.1 mm/min, respectively, and both to have diffusion coefficients of 3.3×10^{-7} cm³/sec. The time of electrophoresis (T) in all cases was 200 min. The rates of interconversion are indicated as the half-time ($t_{1/2}$) relative to the time of electrophoresis. Reprinted, with permission, from ref. 1.

protein due to slow isomerization. By comparing observed electrophoretic patterns with those predicted for various rates of isomerization, it is possible to estimate the rate constant for an interconversion.

The most dramatic conformational change that proteins undergo is the unfolding transition in which the compact and relatively fixed native state is converted to a distribution of many rapidly interconverting unfolded conformations, most of which are much more extended than the native conformation. A relatively simple gel electrophoresis method has been developed to study unfolding transitions induced by urea or other non-ionic denaturants (12). In this method, illustrated in *Figures 4* and *5*, proteins are applied across the top of a slab gel containing a gradient of urea concentration perpendicular to the direction of electrophoresis. Thus, the mobility can be determined at continuously increasing urea concentrations. The cooperative unfolding transition is detected as a relatively sharp decrease in mobility. This method can be used to study both the thermodynamics and kinetics of protein folding–unfolding reactions.

2. Non-denaturing gel electrophoresis of proteins

The following sections discuss some of the practical aspects of using non-denaturing gel electrophoresis for the study of protein conformation.

2.1 Apparatus

Many different designs of apparatus for polyacrylamide gel electrophoresis have been described, and several are available commercially. Although cylindrical gels can be used, for most applications slab gels are more convenient because several samples can be analysed on the same gel and the mobilities of proteins in different samples can be easily compared.

For studies of protein conformation, it is often important to control the temperature during electrophoresis. For this purpose, a device in which most or all of the gel is surrounded by buffer is advantageous. Some commercial apparatuses also contain cooling coils that can be connected to a circulating water-bath to facilitate temperature control.

For different applications, it may be advantageous to be able to prepare slab gels of different thicknesses. Thinner gels generate less heat at a given voltage, but thicker gels appear to work better with some buffer systems (such as the low pH discontinuous buffer system described in *Protocol 2*). It is often useful to be able to prepare several gels in advance of their use. For this purpose, a casting box to hold several sets of glass plates and spacers is helpful. The use of a casting box also eliminates the need to seal the gel 'sandwiches' individually.

Several manufacturers (including Bio-Rad Laboratories, Hoefer Scientific Instruments, and Pharmacia LKB) have introduced devices for preparing and running 'mini-gels' about 10 cm long and 10 cm wide, or smaller. The resolution on mini-gels often seems to be as good as that on larger gels, and the small gels require less material to prepare, and less time to run.

Most procedures for gel electrophoresis of proteins do not require particularly large power supplies. A supply capable of providing 100 mA at 500 V should be adequate for running two or more slab gels simultaneously.

2.2 Buffer systems

2.2.1 Continuous buffer systems

The simplest buffer systems for gel electrophoresis are those in which the compositions of the buffer in the gel and electrode reservoirs are identical. *Table 1* lists several buffer systems devised by McLellan (13) for electrophoresis of proteins at different pH values. These buffers have been designed so that both the anionic and the cationic component provide some buffering capacity. As a consequence, the ionic strength of a solution with a given buffering capacity is minimized, as is the heat generated during electrophoresis.

Table 1. Buffers for non-denaturing gel electrophoresis of proteins[a]

pH	Basic component[b]	Acidic component[b]
3.8	30 mM β-Alanine	20 mM Lactic acid
4.4	80 mM β-Alanine	40 mM Acetic acid
4.8	80 mM GABA	20 mM Acetic acid
6.1	30 mM Histidine	30 mM Mes
6.6	25 mM Histidine	30 mM Mops
7.4	43 mM Imidazole	35 mM Hepes
8.1	32 mM Tris	30 mM Epps
8.7	50 mM Tris	25 mM Boric acid
9.4	60 mM Tris	40 mM Caps
10.2	37 mM Ammonia	20 mM Caps

[a] Reprinted, with permission, from ref. 13.
[b] Abbreviations: Caps, 3-(cyclohexylamino)-1-propanesulfonic acid; Epps, 4-(2-hydroxyethyl)-1-piperazinepropanesulfonic acid; GABA, γ-amino-*n*-butyric acid; Hepes, *N*-2-hydroxyethylpiperazine-*N*'-2-ethanesulfonic acid; Mes, 2-(*N*-morpholino)ethanesulfonic acid; Mops, 3-(*N*-morpholino)-propanesulfonic acid; Tris; tris(hydroxymethyl)aminomethane.

Although continuous buffer gels lack a 'stacking gel' (such as described in the following section) to concentrate the protein sample at the onset of electrophoresis, they are capable of good separation if the volume and ionic strength of the sample are low. As the protein molecules enter the gel, they become concentrated simply because the first molecules to enter are immediately retarded. This effect may be enhanced if the ionic strength of the sample is less than that of the gel, so that the protein molecules that enter the gel first experience a lower electric field. For best results, however, the band of sample applied to a continuous buffer gel should be no more than 2–3 mm thick. If the sample is so dilute that larger volumes must be used, or if the ionic strength of the sample is significantly greater than that of the gel buffer, better results will probably be obtained using a discontinuous buffer system, as described below.

2.2.2 Discontinuous buffer systems

Discontinuous buffer systems, first introduced by Davis and Ornstein (14,15) and by Reisfeld *et al.* (16), utilize two gel components, a 'stacking gel' where the protein sample is concentrated, and a 'separating gel' where the electrophoretic separation occurs. The theory of these systems is complex and has been described in detail elsewhere (15,17, and Chapter 1).

In brief, the stacking gel concentrates the protein sample by generating a sharp voltage gradient between two zones of different buffer composition. The zone ahead of the protein sample contains an ion with a mobility greater than that of the proteins, while the trailing zone contains an ion with a much lower mobility. Because the leading and trailing ions carry most of the electric current in their respective zones, the difference in their mobilities generates a region of high voltage behind the protein sample and a low voltage region ahead of it. The sample is concentrated because those protein molecules ahead of the voltage gradient are exposed to the lower voltage and slow down, while those behind the gradient are accelerated by the increased voltage of the trailing zone. The stacking gel is prepared with very large pores, so that there is no significant gel sieving at this stage.

In the separating gel, the mobility of the trailing ion is increased due to a change in pH, so that it now migrates faster than the protein molecules. The proteins are no longer trapped in the voltage gradient and migrate with varying mobilities. For instance, in the system for cationic proteins described in *Protocol 2*, β-alanine is used as the low mobility ion in the stacking gel. The pH of the stacking gel is about 5, so that β-alanine (with pK_as of 3.6 and 10.2) carries a small partial positive charge. However, in the separating gel the pH is about 3.5, so that the positive charge and the mobility of β-alanine increases greatly. (The pH values during electrophoresis differ from those of the buffers used to prepare the gels because the ionic compositions change during electrophoresis.) The separating gel is designed with pores that will differentially sieve the molecules of interest.

A detailed theoretical treatment of discontinuous buffer systems has been developed and used to design several hundred different systems using different components at a variety of pH values (17,18). However, only a few systems have been used extensively. One of the most commonly used is the original anionic system described by Davis and Ornstein (14,15). This system works very well for proteins that are negatively charged and stable at pH 8.3 to 9.5 and is described in *Protocol 1*.

Protocol 1. Non-denaturing gels for acidic proteins

Reagents

- Acrylamide/bisacrylamide stock for separating gel: 30% (w/v) acrylamide, 0.8% (w/v) *N,N'*-methylene -*bis*-acrylamide
- Acrylamide/bisacrylamide stock for stacking gel: 10% (w/v) acrylamide, 2.5% (w/v) *N,N'*-methylene-*bis*-acrylamide
- 4 × buffer for separating gel: 18.2 g Tris base dissolved in about 40 ml water, 0.23 ml TEMED,1 M HCl to pH 8.9, water to 100 ml
- Riboflavin stock solution: 0.004% (w/v) in water
- 8 × buffer for stacking gel: 5.7 g Tris base dissolved in about 40 ml water, 0.46 ml TEMED,1 M H_3PO_4 to pH 6.9, water to 100 ml
- Ammonium persulfate stock solution: 10% (w/v) in water (prepared fresh daily)
- Overlay solution: 20% (v/v) ethanol
- 10 × electrode buffer: 0.6% (w/v) Tris base, 2.88% (w/v) glycine
- 5 × glycerol–tracking dye solution: 50% (w/v) glycerol, 0.01% (w/v) bromphenol blue

Method

1. Clean and dry the glass plates and assemble the gel sandwiches. For most purposes, scrubbing the plates with a liquid detergent and allowing them to air dry is adequate preparation, but an additional rinse with ethanol may be helpful.

193

Protocol 1. *Continued*

2. Mix the separating gel solution in a vacuum flask:
   - 6 ml water
   - 3 ml 30% acrylamide stock
   - 3 ml pH 8.9 buffer stock

   Caution: acrylamide, bisacrylamide, and TEMED are all toxic. Plastic gloves should be worn when handling both gel solutions and polymerized gels, and pipetting by mouth should be strictly avoided.

3. Stopper the vacuum flask and connect it to a water aspirator vacuum pump for about 5 min. Molecular oxygen inhibits acrylamide polymerization, and must be largely eliminated from the gel solution. Degassing can be accelerated by warming the solution to about 40°C.

4. Immediately before casting the gel, add to the solution 84 μl of the 10% ammonium persulfate solution.

5. Fill the gel sandwich to a level that will be 1 cm below the bottom of the well-former, when it is later inserted. Carefully overlay the gel solution with a solution of 20% ethanol. A syringe with a needle sufficiently fine to fit between the glass plates and long enough to reach just above the gel solution is helpful in overlaying the gel.

6. Allow the gel to polymerize for 30–60 min.

7. Mix the stacking gel solution in a vacuum flask:
   - 4 ml water
   - 2 ml 10% acrylamide stock
   - 1 ml pH 6.9 buffer stock
   - 1 ml riboflavin stock

8. Stopper the vacuum flask and connect it to a water aspirator vacuum pump for about 5 min.

9. Rinse the 20% ethanol solution from the top of the separating gel, first with water, and then with a few millilitres of the stacking gel solution.

10. Fill the remaining volume with the stacking gel solution and insert the well-former.

11. Place the gel in front of a fluorescent lamp or in sunlight and allow the gel to polymerize for 30–60 min.

12. Remove the well-former and rinse the wells.

13. Mount the gel in the electrophoresis chamber and fill the buffer reservoirs with electrode buffer.

14. Prepare samples for electrophoresis by mixing 4 vol. of sample with 1 vol. of the glycerol–bromphenol blue solution. If the gel is to be

stained with Coomassie blue, the volume to be applied to the gel (typically 10–50 μl) should ideally contain about 1 μg of each protein species to generate strong bands in a gel lane about 0.5 cm wide.

15. Apply the samples to the wells in the gel using an air-displacement micropipettor with a special gel-loading tip.

16. Connect the gel apparatus to the power supply with the lower electrode connected to the positive power supply lead.

17. Electrophorese at a constant current of 5 mA per gel (for a gel 1 mm thick and 10 cm wide) until the bromphenol blue has entered the separating gel. Then increase the current to 10–20 mA per gel.

18. When the bromphenol blue has migrated to within 1–2 cm of the gel bottom, turn off the power supply, disassemble the gel apparatus, and stain the gel.

For proteins that are positively charged at pH 3.5 to 5, the buffer system described in *Protocol 2* (18), which is similar to that described by Reisfeld *et al.* (16), is recommended. Unfortunately, this system does not work quite as well as the anionic system. The most common problem is distortion of the protein bands into a 'dumbbell' shape, with most or all of the protein migrating at the edges of the lane. This effect seems to be minimized in gels that are 1 mm or more thick and is made less objectionable if the lanes of the gel are well separated. (It is sometimes helpful to use alternating wells when loading the samples.)

Protocol 2. Non-denaturing gels for basic proteins

Reagents

- Acrylamide/bisacrylamide stock for separating gel: 30% (w/v) acrylamide, 0.8% (w/v) *N,N'*-methylene-*bis*-acrylamide
- Acrylamide/bisacrylamide stock for stacking gel: 10% (w/v) acrylamide, 2.5% (w/v) *N,N'*-methylene-*bis*-acrylamide
- 8 × buffer for separating gel: 12.8 ml glacial acetic acid diluted in about 30 ml water, 1.0 ml TEMED, 1 M KOH to pH 4.0 (about 35 ml), water to 100 ml
- Overlay solution: 20% (v/v) ethanol
- Riboflavin stock solution: 0.004% (w/v) in water
- 8 × buffer for stacking gel: 4.3 ml glacial acetic acid diluted in about 30 ml water, 0.46 ml TEMED,1 M KOH to pH 5.0 (about 50 ml), water to 100 ml
- 4 × electrode buffer: 14.2 g β-alanine, dissolved in about 800 ml water, acetic acid to pH 4.0, water to 1 litre
- 5 × glycerol–tracking dye solution: 50% (w/v) glycerol, 0.2% (w/v) methyl green

Method

1. Clean and dry the glass plates and assemble the gel sandwiches.

2. Mix the separating gel solution in a vacuum flask:
   - 3 ml water
   - 6 ml 30% acrylamide stock

Protocol 2. *Continued*

- 1.5 ml pH 4 buffer stock
- 1.5 ml riboflavin stock

Caution: acrylamide, bisacrylamide, and TEMED are all toxic. Plastic gloves should be worn when handling both gel solutions and polymerized gels, and pipetting by mouth should be strictly avoided.

3. Stopper the vacuum flask and connect it to a water aspirator vacuum pump for about 5 min.

4. Fill the gel sandwich to a level that will be 1 cm below the bottom of the well-former, when it is later inserted. Carefully overlay the gel solution with a solution of 20% ethanol.

5. Place the gel in front of a fluorescent lamp or in sunlight and allow the gel to polymerize for 30–60 min.

6. Mix the stacking gel solution in a vacuum flask:
   - 4 ml water
   - 2 ml 10% acrylamide stock
   - 1 ml pH 5 buffer stock
   - 1 ml riboflavin stock

7. Stopper the vacuum flask and connect it to a water aspirator vacuum pump for about 5 min.

8. Rinse the 20% ethanol solution from the top of the separating gel, first with water and then with a few millilitres of the stacking gel solution.

9. Fill the remaining volume with the stacking gel solution and insert the well-former.

10. Place the gel in front of a fluorescent lamp or in sunlight and allow the gel to polymerize for 30–60 min.

11. Remove the well-former and rinse the wells.

12. Mount the gel in the electrophoresis chamber and fill the buffer reservoirs with electrode buffer.

13. Prepare samples for electrophoresis by mixing 4 vol. of sample with 1 vol. of the glycerol–methyl green solution. If the gel is to be stained with Coomassie blue, the volume to be applied to the gel (typically 10–50 ml) should ideally contain 1–2 μg of each protein species to generate strong bands in a gel lane about 0.5 cm wide.

14. Apply the samples to the wells in the gel using an air-displacement micropipettor with a special gel-loading tip.

15. Connect the gel apparatus to the power supply with the lower electrode connected to the negative power supply lead.

16. Electrophorese at a constant current of 5 mA per gel (for a gel 1 mm thick and 10 cm wide) until the methyl green has entered the separating gel. Then increase the current to 10–20 mA per gel.

17. When the methyl green has migrated to within 1–2 cm of the gel bottom, turn off the power supply, disassemble the gel apparatus, and stain the gel.

2.3 Gel composition

The pore size of a gel, and therefore its ability to separate different protein molecules, is determined by the concentrations of acrylamide and cross-linking agent (usually *N,N'*-methylene-*bis*-acrylamide) (see Chapter 1). Higher concentrations of acrylamide lead to smaller pores, but the effects of the cross-linker concentration are more complicated. At a given acrylamide concentration the minimum pore size is obtained when the cross-linker is about 5% of the total monomer concentration (6). At higher cross-linker concentrations, bundles of polyacrylamide form, leading to larger pores. High cross-linker concentrations are often used in stacking gels to generate large pores that will not sieve the protein molecules.

Typically, separating gels contain 7.5% to 15% acrylamide and a ratio of acrylamide to bisacrylamide in the range of 30:1 to 150:1. Gels of these compositions should sieve proteins of molecular weights between 5000 and 100 000. Gels with either higher or lower acrylamide concentrations are considerably more difficult to handle. Stacking gels typically contain 2.5% acrylamide and 0.625% bisacrylamide.

2.4 Polymerization catalysts

Polymerization of acrylamide requires a catalyst to generate free radicals. The most commonly used method (e.g. *Protocol 1*) uses ammonium persulfate and *N,N,N',N'*-tetramethylethylenediamine (TEMED) as catalysts. In another method (e.g. *Protocol 2*) ammonium persulfate is replaced by riboflavin, which generates free radicals upon exposure to visible light. A third method, using methylene blue to initiate photoactivated polymerization, has recently been introduced (19), and this method appears to have significant advantages over the two more traditional methods.

The use of photoactivated catalysts allows the timing of polymerization to be easily controlled. The gel solutions can be prepared and poured into the gel mould under subdued lighting (absolute darkness is not required) and polymerization initiated by exposing the gel to daylight or a fluorescent lamp; one or two 15 W cool-white fluorescent tubes are usually adequate. Photopolymerization may also leave fewer reactive compounds in the gels after polymerization. The major disadvantage of using riboflavin as the catalyst is that polymerization may not be as complete as with ammonium persulfate, particularly at neutral or alkaline pH values. Methylene blue, in conjunction with a redox couple (toluenesulfinate and diphenyliodonium), has been shown to be a much more reliable and efficient photoactivated catalyst, and use of this method is highly recommended. A protocol for using these catalysts is provided below.

Protocol 3.   Photopolymerization with methylene blue

Reagents[a]

- 2 mM methylene blue
- 20 mM sodium toluenesulfinate
- 1 mM diphenyliodonium chloride

Method

1. For 10 ml of gel solution, mix acrylamide,[b] buffer components, and water to yield a total of 8.5 ml.

2. Before degassing, add the following to the solution:
   - 0.5 ml sodium toluenesulfinate solution
   - 0.5 ml diphenyliodonium chloride solution

3. Stopper the flask and connect to a water aspirator for 5 min.

4. Add to the gel solution 0.5 ml of the methylene blue solution.

5. Pour the solution into the gel form and either overlay with 20% ethanol, for separating gels, or insert the well-former.

6. Place the gels in front of one or two 15 W fluorescent lamps, or in direct sunlight.

7. Allow the gels to polymerize until the blue colour of the dye is no longer visible, typically less than 20 min.

[a] These compounds are available from Aldrich Chemicals. Store all of the solutions in the dark at 4°C, and prepare them fresh weekly, except the methylene blue, which has been reported to retain its reactivity for up to one month.
[b] Because methylene blue polymerization is very efficient, leading to small pore sizes, it may be necessary to use lower acrylamide concentrations than with other catalysts, particularly riboflavin.

It is not always appreciated that TEMED is a strong base that can significantly affect the pH of a gel buffer. If the pH is to be carefully controlled, it is best to include the TEMED in the buffer when the pH is adjusted. There does not appear to be any significant loss of effectiveness of the TEMED in buffers stored for several months.

2.5 Sample preparation, application, and electrophoresis

For gels that will be stained with Coomassie blue, the sample should ideally contain 1–2 μg of each protein species to generate strong bands in a gel lane about 0.5 cm wide. For continuous buffer gels, the protein concentration should be about 1 mg/ml, so the protein can be applied in a few microlitres. For discontinuous buffer gels, samples of up to 50 µl can be applied, depending on the thickness of the gel. Thus, samples as dilute as 0.02 mg/ml can be used. If possible, the salt concentration of the samples should be less than 0.1 M. Although discontinuous buffer systems will tolerate higher salt concentrations, some distortion of the bands is likely.

In order to be applied to the gel, the samples must be more dense than the electrode buffer, as by adding glycerol to a final concentration of 10% (w/v). The glycerol also minimizes convection at the beginning of electrophoresis, as the samples enter the gel and are warmed by Joule heat.

A marker dye should also be added to the sample to allow the progress of the electrophoresis to be monitored. For electrophoresis towards the anode, about 0.5 µg bromphenol blue generates a readily visible band during electrophoresis. For electrophoresis towards the cathode, about 10 µg methyl green is suggested.

Samples can be applied in the wells of the gel (after filling the upper reservoir with buffer!) using a syringe (such as a Hamilton microsyringe) or micropipette. An air-displacement micropipette (e.g. Gilson or Eppendorf) is particularly convenient if used with a special narrow tip designed for this purpose and available from many suppliers.

Electrophoresis is usually carried out at a current of 5–20 mA per gel (for a gel 1 mm thick and 10 cm wide). Better results may be obtained if the current is lower when electrophoresis is begun and then increased after the proteins have entered the separating gel. With continuous buffer gels, it is advisable to pre-electrophorese the gel for about one hour (at the same current to be used for electrophoresis) before applying the sample. This helps remove reactive compounds in the gel that may cause artefacts. Unfortunately, discontinuous gels cannot be pre-electrophoresed.

2.6 Staining gels

The most common method for visualizing protein bands in gels is staining with Coomassie blue R-250 (this dye is distinct from Coomassie blue G-250, which is used in some other staining solutions). This procedure is simple and provides reasonable sensitivity. Coomassie blue appears to bind similarly to most proteins, approximately equally on a mass basis, so that band intensity can be used as a reasonable measure of relative concentration. The mechanisms of dye binding to proteins are not well understood, however, and probably vary somewhat from protein to protein. Thus, any quantification based on gel band intensity should be calibrated for the protein of interest.

In order to be permanently stained, proteins in a gel must be fixed by precipitation. Many proteins are efficiently fixed if the gel is stained in a solution containing 0.1% Coomassie blue R-250, 50% methanol, and 7.5% acetic acid, a solution commonly used for staining SDS–polyacrylamide gels (Chapter 1, *Protocol 6*). However, some small proteins, such as BPTI, are not fixed by this treatment, particularly if they have not been unfolded before electrophoresis. For such proteins, and for gels containing urea, the stain described in *Protocol 4* is recommended. The high acid concentration in this solution makes it much more effective in fixing small proteins, and, since the dye is only slightly soluble under these conditions, background stain in the gel is low.

Protocol 4. Coomassie blue staining

Reagents

- Stain solution: 0.5 g Coomassie blue R-250 dissolved in about 400 ml water (stir well for 1 h before adding acids), 50 g TCA, 50 g 5-sulfosalicylic acid, water to 500 ml (caution: these acids are very caustic—wear gloves when handling either the solids or the stain solution)

- Rinse solution: 50% (v/v) methanol, 7.5% (v/v) glacial acetic acid
- Destain solution: 5% (v/v) methanol, 7.5% (v/v) glacial acetic acid

Method

1. After electrophoresis, place the gel in a plastic box containing sufficient stain solution to cover the gel.
2. Gently rock the gel in the plastic box for at least 2 h.
3. Transfer the gel to a clean plastic box containing the rinse solution. Do not leave the gel in this solution for more than 5 min.
4. Transfer the gel to a clean box containing the destain solution.
5. Allow the gel to destain with gentle agitation. Two changes of destain solution over a 24 h period are usually adequate to obtain a nearly clear background. Extended treatment may lead to the loss of stain from the protein bands.

Recently, stains based on the precipitation of silver have gained popularity (20). Silver staining is at least 10–50-fold more sensitive than Coomassie blue, and is valuable for analysing samples available in only small amounts or in dilute solutions. However, this procedure, which is described in Chapter 1, is considerably more expensive and complicated than Coomassie blue staining.

Protein bands in a gel can also be visualized by autoradiography (21), enzymatic staining, or antibody staining, either directly or after the proteins are transferred to a membrane (22) (see Chapters 1, 3, and 5).

2.7 Drying gels

Polyacrylamide gels can be dried on to a support to provide a permanent record of an electrophoresis experiment. In addition, it is usually necessary to dry

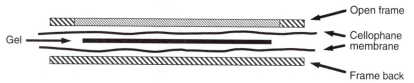

Figure 3. A simple method for drying polyacrylamide gels between two sheets of cellophane membrane, as described in *Protocol 5*. The drying frame is constructed from two pieces of Plexiglass, 13 cm × 16 cm × 0.4 cm thick, one of which has a 10 cm × 13 cm opening cut into it. The assembly is held together with four 2 inch wide binder clips.

a gel before autoradiography, particularly for the relatively weak β particles emitted by ^{14}C, ^{35}S, and ^{3}H.

Gel drying devices that heat the gel under vacuum are available from several manufacturers including Bio-Rad Laboratories and Hoefer Scientific Instruments. The gel can be dried on to either chromatography paper (e.g. Whatman 3MM) or cellophane membrane. In order to avoid cracking of the gel, it is important that the vacuum not be broken before the gel is completely dried.

An alternative method is the use of a simple plastic frame to dry the gel between two sheets of cellophane membrane without vacuum, as illustrated in *Figure 3* and detailed in *Protocol 5*. This procedure preserves the gel in a transparent form that can be readily photographed or scanned with a densitometer, although it is usually best to photograph or analyse the gel before drying because there is usually some distortion of the gel and a significant risk of cracking during drying. For gels to be autoradiographed, the lower sheet of cellophane should be replaced with plastic wrap. After drying, the plastic wrap is removed, to allow the gel to be placed in direct contact with the autoradiography film.

Protocol 5.   Drying slab gels

Equipment and reagents

- Plexiglass gel drying frame, as illustrated in *Figure 3*
- Two sheets of cellophane membrane (Bio-Rad Laboratories) cut to a size slightly larger than the frame

- Six binder clips, 2 inch wide
- Glycerol destain solution: 3% (w/v) glycerol, 5% (v/v) methanol, 7.5% (v/v) glacial acetic acid

Method

1. Soak the gel in the glycerol destain solution for 1 h or longer.
2. Soak the cellophane membrane in water for approx. 15 min.
3. Place one sheet of the cellophane on the solid drying frame back. Carefully smooth out any bubbles between the cellophane and the Plexiglass.

Protocol 5.   *Continued*

4. Place the gel on the cellophane, again removing any bubbles that may be trapped.

5. Spread a few millilitres of the glycerol destain solution over the top of the gel.

6. Place the second cellophane sheet over the gel. Remove any bubbles.

7. Place the open drying frame over the gel. Clamp the assembly together with the six binder clips.

8. Allow to dry for at least 24 h at room temperature. For faster drying, place a 40 W incandescent lamp approximately 10 inches from the open side of the drying frame.

2.8 Documentation and quantification of electrophoresis patterns

Although a dried gel can last for several years and can, for some purposes, serve as an adequate record of an electrophoresis experiment, it is often important to create a more permanent record or a figure for publication. Coomassie blue stained gels can be photographed effectively using a panchromatic black and white film and a dark orange filter (e.g. Wratten 15G) for optimum contrast. Gel images can also be recorded digitally using a video camera or scanner, as discussed below. When printed on a high quality printer capable of true grey-scale reproduction, such as a dye-sublimation printer, the quality of digitized images can be nearly indistinguishable from that of conventional photographs.

For many purposes, it is important to quantify the relative intensities of gel bands, and several scanning micro-densitometers suitable for analysing gels are available commercially (e.g. from Bio-Rad Laboratories, Hoefer Scientific Instruments, and Pharmacia LKB). Alternatively, a digital image of the gel can be made using a video camera or flat-bed scanner and then analysed with a computer. Video cameras offer the advantages of relatively low cost and very rapid image capture, and the resolution of conventional video cameras is adequate for most purposes. For some applications, such as large gels or autoradiographs, a flat-bed scanner may offer higher resolution than a video camera, but it is more difficult to use such a scanner with wet gels. Digitizing video images requires a 'frame-grabber' board for the computer and appropriate software. An excellent public domain image analysis program for the Apple Macintosh, NIH Image, has been written by Dr Wayne Rasband and is available via anonymous FTP from `zippy.nimh.nih.gov`, as well as from many other internet sites and computer bulletin boards. The program is distributed with a set of macros specifically designed for quantifying gel bands, and the documentation provides general guidelines for many aspects of video densitometry, including the selection of an appropriate video camera and digitizing board. Other programs for a variety of different computers are available commercially.

First-time users of densitometers or imaging systems are often disappointed when they find that the densitometer trace appears to have less resolution than the original gel. This is often not a fault of the instrument, but reflects the ability of the human mind to distinguish easily bands that are in fact not fully resolved. Thus, densitometry does not usually provide a means of extracting additional resolution from a gel (unless the results are analysed with sophisticated image processing techniques). In addition, a densitometer cannot usually detect the faintest bands that are visible to the eye. However, densitometry does provide a means of quantifying objectively the relative concentrations of different species in a sample.

In order to provide reliable estimates of concentration, some form of calibration is required. Scanning densitometers are generally designed to provide an output that is proportional to absorbance, or optical density, which according to the Beer–Lambert law is proportional to concentration. Digital scanners or video frame-grabbers, however, usually assign to each pixel in the image an integral value that is linearly related to the light intensity detected by the scanner or camera. In order to use the digitized image to quantify concentrations, this value must be converted to a quantity linearly related to absorbance, which is a logarithmic function of light intensity. This conversion can be made computationally, for instance by using the 'Uncalibrated OD' calibration function in the NIH Image program. Alternatively, the image can be calibrated using optical filters of known absorbance or gel samples containing known quantities of protein.

Because the absorbance of light by a gel band depends upon the amount of stain bound, which is very sensitive to the exact conditions of staining and destaining, it is usually possible only to compare the relative amounts of protein in different bands on the same gel. To determine the absolute concentration of protein in a band, standards containing known amounts of protein must be electrophoresed and stained on the same gel. Ideally, the standards should contain the same protein as the bands of interest.

The major difficulties in quantifying band intensities usually arise in determining the baseline for a trace and resolving adjacent peaks. Unfortunately, these processes are often quite subjective, and the best that can be done is to treat the different traces from an experiment systematically and to check that the final quantitative results are consistent with a qualitative visual impression of the gel. Some of the software available for analysing densitometer data or digitized images offers a great deal of flexibility in manipulating the data, including baseline subtraction. Such manipulations can generate very impressive looking traces, but must be used with great caution.

3. Transverse gradient gel electrophoresis

Gel electrophoresis is primarily a comparative technique, in which only changes in the mobility of a protein can be readily interpreted. For studies of protein conformation, it is often particularly useful to compare the electrophoretic mobility under continuously varying conditions. This may be accomplished by incorporating a transverse gradient within the gel, perpendicular to the direction of electrophoresis, so that a band of protein applied across the top of the gel will migrate under the varying conditions.

Many types of gradients can be incorporated, provided only that they do not markedly interfere with the electrophoretic process. Only a few, however, have been investigated. For example, the relationship between mobility and acrylamide concentration (Equation 1) can be determined using a single gel with a linear transverse gradient of acrylamide concentration (1). Establishing a pH gradient across a gel before electrophoresis can produce a graphic illustration of a protein's pH titration curve. Unfolding and refolding of the native conformation of a protein can be studied using gels containing transverse urea gradients (12) or by generating a temperature gradient across the gel (23). The use of transverse gradient electrophoresis will be illustrated here with urea gradient gels, but many of the considerations are applicable to other types of gradients.

In a transverse urea gradient gel, unfolding of the protein is detected as a decrease in electrophoretic mobility because the unfolded protein has a greater hydrodynamic volume. This technique has been tested with several well characterized proteins, and has yielded results that are generally consistent with those obtained with more conventional biophysical methods used for studying unfolding transitions. The electrophoretic method is relatively simple, requires only small amounts of protein, and can be used with mixtures of proteins, since the various species can often be separated by electrophoresis and analysed individually. Additional selectivity for individual components in a complex mixture can be attained using antibodies or other specific ligands to stain the gel (24,25). Urea gradient gels are also particularly well suited to studies of mutant proteins, where it is desirable to screen many different samples rapidly (25–27).

Urea gradient gel electrophoresis can provide information about both the thermodynamics and kinetics of the unfolding transition. The thermodynamic stability of a native protein is defined relative to the unfolded state, and is determined by measuring the equilibrium between the two states under conditions where both states are detectable. In urea gradient gels, this equilibrium determines the average mobility of a population of interconverting molecules. As with most other methods for measuring protein stability, data from the transition zone must be extrapolated to zero denaturant concentration. (See Chapter 12 for a detailed discussion of the measurement of protein stability.) Urea gradient gels can also indicate the number of independent

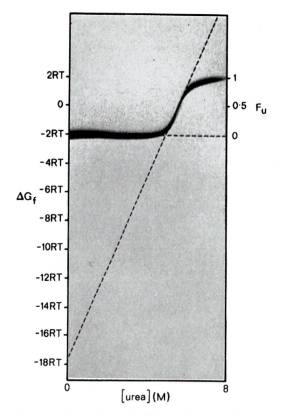

Figure 4. Urea gradient gel electrophoresis of horse ferricytochrome *c*, illustrating a method for estimating the net stability of the native protein by extrapolation. Protein was applied across the top of the gel containing a transverse gradient of urea and subjected to electrophoresis toward the cathode at pH 4 at 15°C. The mobility of the protein at each urea concentration indicates the fraction of protein unfolded, F_U, as indicated on the right-hand scale. Near the midpoint of the transition, where $F_U = 0.5$ and the net stability of the folded state, ΔG_f, equals 0, F_U is approximately a linear function of urea concentration. The net stability in the absence of denaturant is estimated by extrapolating ΔG_f from the transition region to the position of zero urea concentration, assuming a linear relationship between ΔG_f and urea concentration (29). The plateau positions of the U and N states correspond to the positions on the left-hand scale where $\Delta G_f = 2RT$ and $-2RT$ (R is the gas constant and T is the absolute temperature), as derived in ref. 30. Reprinted, with permission, from ref. 1.

domains a protein is made up of, since separate domains may undergo distinct unfolding transitions, as shown for calmodulin in *Figure 5*. The kinetics of the transition are reflected in the continuity of the band of protein in a urea gradient gel; if the folding equilibrium is rapid compared to the time of electrophoresis a sharp continuous band is generated through the transition zone, but if the transition is slow the band will be smeared out or discontinuous.

Calmodulin

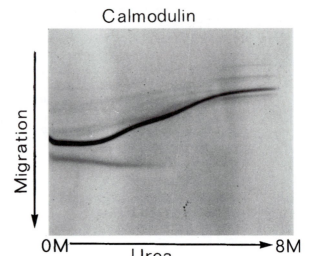

Figure 5. Urea gradient gel electrophoresis of calmodulin. Calmodulin was subjected to electrophoresis in the absence of either added Ca^{2+} or chelating agents at pH 8.4 toward the anode. The two separate unfolding transitions are probably due to the sequential unfolding of different domains. A number of minor contaminants also demonstrate unfolding transitions. Reprinted, with permission, from ref. 1.

Interpretation of the patterns generated by urea gradient gels is discussed further in Section 3.4.

3.1 Apparatus for transverse gradient gel electrophoresis

Transverse gradient gel electrophoresis requires some minor modifications of most devices designed for more conventional electrophoretic methods. The major requirement is some arrangement for casting the gradient gels. Generally, the gels are cast with the glass plates turned so that the gradient is vertical, and, after polymerization, the gels are turned 90 degrees for electrophoresis. This requires spacers that fit along the edges of the glass plates that will ultimately correspond to the top and bottom of the gel, as well as an arrangement for holding the gel sandwich perpendicular to the orientation used for electrophoresis. Since preparing the gradients is somewhat more time-consuming than preparing simpler gels, a casting box for making several gels at once is advantageous. A set of spacers and a casting box for the preparation of gels to be used with a Bio-Rad Mini-Protean™ II apparatus are illustrated in *Figure* 6 and are available commercially from Aquebogue Machine and Repair Shop, Inc.

Some arrangement is also needed to mix the solutions to generate a linear concentration gradient. A variety of small gradient formers are available commercially, or a three-channel peristaltic pump can be used to generate a linear gradient, as illustrated in *Figure 7* and described in *Protocol 6*. The use

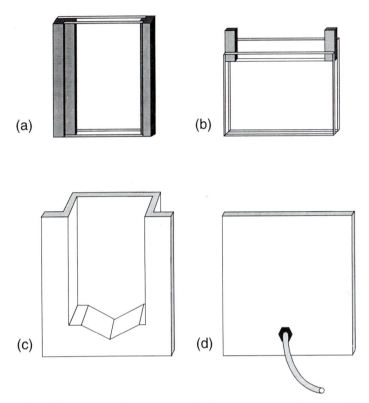

Figure 6. Special spacers and casting box for the preparation of urea gradient gels for use in a Bio-Rad Mini-Protean™ II gel apparatus. (a) The spacers are designed to fit along the sides of the glass plates as the gel is oriented for casting and produce gels 1 mm thick. The standard glass plates for the Mini-Protean™ II apparatus are used. (b) After the gels are cast, the spacers are removed and the gel rotated 90° for electrophoresis. Short spacers are then placed at the two sides of the top of the gel. These spacers form the edges of the space where the sample will be applied and form a seal with the electrophoresis chamber. (c) The casting box shown here is constructed of clear acrylic plastic and holds the glass plates and spacers for five 1 mm thick gels oriented as shown in (a). After assembling the plates and spacers and placing them in the box, the cover (d) is clamped on with binder clips. The inlet for the gel solutions is on the cover and directs the solution to the bottom of a 'V'-shaped chamber below the glass plates. This arrangement allows the solution to enter the five sets of gel forms evenly with a minimum of mixing. Adapted, with permission, from ref. 34.

of a peristaltic pump has the advantage that the flow of gel solution into the casting chamber is very reproducible and easily controlled.

3.2 Solutions for urea gradient gel electrophoresis
As with non-denaturing gel electrophoresis, the choice of buffer depends upon the particular protein and experiment. Since some proteins do not

unfold in urea at neutral pH, it may be necessary to carry out the electrophoresis at acidic or alkaline pH to observe an unfolding transition. Although discontinuous buffer systems with a stacking gel can be used (24), continuous buffer systems (see *Table 1*) are much simpler and have been found to provide good results provided that the protein sample is sufficiently concentrated. Buffers that have been used successfully in urea gradient gels include:

- 0.05 M Tris–acetate pH 4 (0.05 M acetic acid adjusted to pH 4 with Tris base)
- 0.05 M imidazole–Mops (both components at 0.05 M, pH ~ 7.0)
- 0.05 M Tris–acetate pH 8 (0.05 M Tris base adjusted to pH 8 with acetic acid)
- 0.05 M Tris–bicine (both components at 0.05 M, pH ~ 8.4)

In addition, the buffers listed in *Table 1* should work well. It is convenient to prepare 10 × stocks of the buffers.

Urea is an unstable compound, yielding cyanates that can react with the amino groups of proteins. For this reason, the highest available grade of urea should be used (e.g. Schwarz-Mann 'Ultra-Pure'), and urea solutions should be prepared immediately prior to use.

It has been found that the presence of increasing concentrations of urea causes a continuous decrease in the electrophoretic mobility of a protein in the absence of any unfolding transition (12). Presumably this effect is due to the increased viscosity of the solution or a change in the sieving properties of the gel. This effect can be largely compensated for by superimposing on the urea gradient an inverse gradient of acrylamide concentration (e.g. 15–11% or 11–7%).

Because of the time required for preparing the gels, photopolymerization is particularly convenient for gradient gels. Although many of the published experiments have utilized riboflavin to polymerize urea gradient gels, methylene blue appears to produce much more efficient and reproducible photopolymerization. Urea gradient gels polymerized with methylene blue and containing gradients of 11–7% acrylamide appear to have sieving properties similar to those of riboflavin-polymerized gels containing 15–11% acrylamide (27).

3.3 Sample preparation and electrophoresis

The sample for a 10 cm wide urea gradient gel to be stained with Coomassie blue should contain about 50 μg of protein in a volume of about 75 μl: a smaller volume is difficult to apply evenly across the top of the gel and a larger volume will lead to broader protein bands. If possible, the ionic strength of the sample should be lower than that of the electrophoresis buffer, to lead to concentration of the protein as it enters the gel. The sample should also contain glycerol and a marker dye.

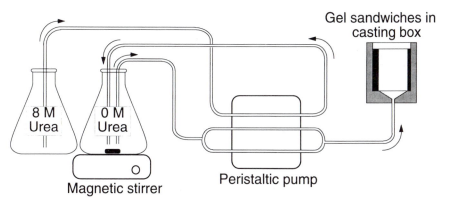

Figure 7. Preparation of urea gradient gels using a three-channel peristaltic pump. The glass plates are held in a clear plastic casting box, as illustrated in *Figure 6*. The gradient is formed by pumping from the low urea concentration solution in the mixing vessel into the gel forms, using two channels of the pump, while simultaneously mixing the high urea concentration solution, pumped through a single channel of the pump, with the low urea concentration solution. The solution in the mixing vessel is vigorously stirred with a magnetic stirrer. The volume of the gradient is chosen to leave room for regions containing 0 M and 8 M urea at the top and bottom of the gels (as the gels are oriented for casting), respectively. Reprinted, with permission, from ref. 34.

As discussed in Section 3.4, it is often useful to compare the patterns generated by urea gradient electrophoresis when the protein is applied to the gel in the native and unfolded states. Small samples of unfolded proteins can be prepared by adding solid urea to a sample, as described in *Protocol 6*. The time required for unfolding must be determined experimentally for each protein, but is likely to be less than 1 h.

Urea gradient gels are typically electrophoresed at 10–20 mA per gel. Better results may be obtained if a lower current is applied as the proteins enter the gel. The optimum time for electrophoresis must be determined empirically for a particular protein and buffer, but 3–4 h should be a reasonable starting point. If the electrophoresis is too brief, the difference in mobility between the native and unfolded forms may not be large enough to clearly define the unfolding transition. On the other hand, extended electrophoresis may not yield better results because diffusion of the protein bands may counteract any increase in separation. Because the pattern generated depends on the rate of the unfolding transition relative to the time of electrophoresis, comparing patterns generated by electrophoresis at different voltages, and correspondingly different total electrophoresis times, can provide useful information about the rate of the transition (26). Significant separations have been obtained after only eight minutes of electrophoresis (28).

Protocol 6. Transverse urea gradient gel electrophoresis

Equipment and reagents

- Glass plates, spacers, and casting box: spacers and a casting box for preparing gels to be used with the Bio-Rad Mini-Protean™ gel apparatus, as illustrated in *Figure 6*, are available from Aquebogue Machine and Repair Shop, Inc.
- Three-channel peristaltic pump (e.g. Isco Tris™)
- Magnetic stirrer and a small stir bar
- Acrylamide/bisacrylamide stock: 30% (w/v) acrylamide, 0.8% (w/v) *N,N'*-methylene-*bis*-acrylamide

- Solid urea (highest grade available, e.g. Schwarz-Mann Ultra-Pure)
- 10 × buffer solution, with composition and pH chosen for the particular experiment
- Solutions for methylene blue photopolymerization, as detailed in *Protocol 3*
- Overlay solution: 20% (v/v) ethanol
- 5 × glycerol–tracking dye solution, as detailed in *Protocols 1* and *2* for acidic and basic proteins, respectively

Method

These instructions are for preparing urea gradient gels 10 cm wide (as oriented for electrophoresis) in a casting box for which the total volume of the gels is 40 ml; the volumes indicated should be adjusted proportionally for gels of other dimensions. The gels are prepared so that the top and bottom 2 cm (as the gels are cast) contain 0 M and 8 M urea, respectively. The flow rate of the peristaltic pump should be about 1 ml per min per channel, so that the gel solutions are pumped into the casting box in about 20 min.

1. Assemble the gel sandwiches and the casting box and connect the tubing as in *Figure 7*.

2. Degas approx. 50 ml of the ethanol overlay solution for 5 min with a water vacuum aspirator.

3. Add the ethanol solution to the casting box so that the solution fills the bottom inch or two of the gel sandwiches. Using a syringe, draw the solution into the tubing between the mixing vessel and the casting box. Make certain that there are no bubbles in the tubing or casting box and that the solution is still above the level of the bottoms of the glass plates. Clamp the tubing in the peristaltic pump.

4. Prepare the following solution, containing 0 M urea and 11% acrylamide, in a vacuum flask:
   - 11 ml acrylamide stock solution
   - 3 ml 10 × buffer solution
   - 1.5 ml toluenesulfinate stock
   - 1.5 ml diphenyliodonium chloride stock
   - 11.5 ml water

   Degas this solution for 5 min.

5. Prepare the following solution, containing 8 M urea and 7% acrylamide, in a vacuum flask:
   - 14.4 g solid urea
   - 7 ml acrylamide stock solution
   - 3 ml 10 × buffer solution
   - 1.5 ml toluenesulfinate stock
   - 1.5 ml diphenyliodonium chloride stock
   - 4.4 ml water

   After the urea is completely dissolved, degas this solution for 5 min.

6. Immediately before casting the gels, add 1.5 ml of the methylene blue stock to each gel solution.

7. Place 12 ml of the 8 M urea solution in its reservoir. Use a syringe to fill the tubing from this vessel to the mixing vessel with the 8 M urea solution. Clamp the tubing in the peristaltic pump and temporarily place the outlet of this tubing in the 8 M urea reservoir.

8. Place 8 ml of the 0 M urea solution in the mixing vessel. Turn on the peristaltic pump to pump this solution into the casting chamber. (While the 0 M urea solution is being pumped into the gel forms, the 8 M urea solution will simply be pumped back into its reservoir.) When the last of this solution has just entered the tubing (and before any air has entered) stop the pump.

9. Place 12 ml of the 0 M urea solution in the mixing vessel and position the outlet of the tube from the 8 M urea solution in the mixing vessel. Turn on the magnetic stirrer and the peristaltic pump. The concentration of urea in the mixing vessel will now increase linearly as it is pumped into the gel forms and the 8 M urea solution is simultaneously mixed in. When the last of the solution from the mixing vessel has entered the tubing, immediately stop the pump.

10. Place about 20 ml of the 8 M urea solution in the mixing chamber and turn on the peristaltic pump. Pump the 8 M urea solution into the casting chamber until the boundary between the ethanol and the 0 M urea solution reaches the top of the glass plates. Stop the pump, close the inlet of the casting box, and disconnect it from the pump.

11. Place the gels in front of a light source to initiate polymerization.

12. Promptly rinse out the tubing with water to prevent polymerization in the tubing.

13. After the gels have polymerized, remove them from the casting box and remove the spacers. Place two small spacers on the sides of the sample well, as shown in *Figure 6b*, and mount the gels in the electrophoresis chamber. The gels can be stored one or two days at 4 °C without significant dissipation of the urea gradient.

Protocol 6. *Continued*

14. Fill the electrode buffer chambers and connect the apparatus to the power supply.

15. Pre-electrophorese the gel at 10–20 mA for 1 h to remove charged reactive species that may have arisen from decomposition of the urea or the polymerization process.

16. Prepare the sample for electrophoresis. For a 10 cm wide gel to be stained with Coomassie blue, the sample should contain about 50 μg of protein in a volume of 75 μl. For a native protein, the sample can be prepared conveniently by mixing:

    • 50 μg of protein
    • water and buffer to a total volume of 60 μl
    • 15 μl of the appropriate glycerol–dye solution (bromphenol blue for acidic proteins, methyl green for basic ones)

    To prepare a 100 μl sample of a urea unfolded protein, mix:

    • 65 μg of protein
    • water and buffer to a total volume of 63 μl (because the urea solution will be much more dense than the electrophoresis buffer, it is not necessary to add glycerol, though it may be convenient to add a tracking dye)
    • 48 mg of solid urea

    The time required for urea unfolding must be determined experimentally for each protein, but is likely to be less than 1 h.

17. After pre-electrophoresing the gel, use a Pasteur pipette to rinse away any urea that may have diffused from the top of the gel.

18. Apply the sample evenly across the top of the gel using an air-displacement micropipettor with a gel-loading tip.

19. Electrophorese the protein at a current of 10–20 mA per gel. The optimum time for electrophoresis must be determined empirically for a particular protein and buffer, but 3–4 h should be a reasonable starting point.

20. After electrophoresis, disassemble the gel sandwiches and place the gels in stain solution (*Protocol 4*), or prepare for other visualization techniques.

3.4 Interpreting urea gradient gel patterns

The pattern generated by urea gradient electrophoresis depends upon the thermodynamics and kinetics of the urea-induced unfolding transition. The number of possible patterns is large, particularly if there are multiple species present that do not interconvert rapidly. Some of the possible patterns are

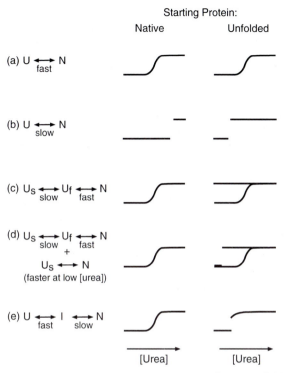

Figure 8. Urea gradient electrophoresis patterns expected for some folding transitions. See Section 3.4 of the text for discussion of these patterns and their corresponding folding mechanisms.

described below and illustrated schematically in *Figure 8*.

3.4.1 Transitions in rapid equilibrium

The simplest patterns are those generated by proteins that undergo cooperative two-state unfolding transitions with the native and unfolded forms rapidly interconverting on the time-scale of the electrophoresis. At low urea concentrations the native form predominates, while at higher urea concentrations the unfolded form, with a reduced electrophoretic mobility, predominates. At intermediate urea concentrations, both forms are present at significant levels and the apparent electrophoretic mobility is the average of the mobilities of the two forms, weighted by their relative abundance. A rapid two-state transition is indicated by a continuous band of protein with a single inflexion point in the transition zone, as illustrated in *Figure 8a*. In addition, identical patterns should be generated when either native or unfolded protein is applied to the gel.

If a protein undergoes a rapid two-state unfolding transition, urea gradient electrophoresis can be used to estimate the stability of the native conforma-

213

tion relative to the unfolded state. The average mobility (M) of the protein is linearly related to the fraction of protein unfolded (f_u) according to:

$$f_u = \frac{[U]}{[U] + [N]} = \frac{(M - M_N)}{(M_u - M_n)},$$ [2]

where M_u *and* M_n are the mobilities of the unfolded and native forms, respectively. Thus, the electrophoretic pattern can be thought of as a graphical plot of f_u as a function of urea concentration. The equilibrium constant for unfolding, K_u, is given by:

$$K_u = \frac{[U]}{[N]} = \frac{f_u}{1 - f_u},$$ [3]

and the free energy change for unfolding, ΔG_u, is given by:

$$\Delta G_u = -RT\ln(K_u)$$ [4]

where R is the gas constant and T the absolute temperature. The value of this equilibrium constant can be determined directly only in the transition region; to estimate the stability of the native state in the absence of urea, K_u must be extrapolated. Unfortunately, there are great uncertainties and inaccuracies in such extrapolations, largely because of our current lack of understanding of the mechanisms by which denaturants act.

Several methods of extrapolating unfolding data have been proposed, as discussed by Pace and Scholtz in Chapter 12. Schellman (29) has argued that ΔG_u should be a linear function of denaturant concentration, and this behaviour has been observed experimentally for many different proteins. This relationship can be expressed by the following equation:

$$\Delta G_u = \Delta G_u^\circ + m \cdot C = m(C - C_m),$$ [5]

where ΔG_u° is the value of ΔG_u in the absence of urea, C is the urea concentration, C_m is the urea concentration at the transition midpoint, and m is an empirical parameter describing the steepness of the transition. As shown by Hollecker and Creighton (30), the derivative of f_u with respect to C evaluated at the midpoint is proportional to m:

$$\frac{df_u(C_m)}{dC} = -\frac{m}{4RT}.$$ [6]

The derivative can be evaluated from a urea gradient gel pattern by drawing a line tangent to the gel band at the midpoint:

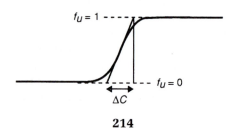

$$\frac{df_u(C_m)}{dC} = \frac{\Delta f_u}{\Delta C} = \frac{1}{\Delta C}. \qquad [7]$$

and m can be calculated from Equation 7 (26). From the values of m and C_m, the value of ΔG_u can be calculated for any other urea concentration, including 0 M, where $\Delta G_u = -m \cdot C_m$. This analysis requires knowing the urea concentration at different points in the gel, which must be estimated from the dimensions of the gel and the total volumes of the linear gradient and the regions of 0 M and 8 M urea on each side of the gel.

Using the same principle, the stability of a protein in the absence of urea can also be estimated graphically directly from the gel, as illustrated in *Figure 4*. The mobilities of the native and unfolded forms define the free energy of folding ($\Delta G_f = -\Delta G_u$) values of $-2RT$ and $+2RT$, respectively. (See ref. 30 for a derivation of this simple method of extrapolation.) *Figure 4* provides graphic illustration of the difficulties inherent in measuring protein stability; the large extrapolation greatly amplifies any errors in the estimates of f_u in the transition zone, and small deviations from linearity in the relationship between ΔG_f and urea concentration may result in quite large errors in the extrapolated value of ΔG_f. None the less, when several methods, including urea gradient gel electrophoresis, have been used to estimate the stability of the same protein, surprisingly consistent results have usually been obtained (30) (see Chapter 12).

Some proteins undergo more complex unfolding transitions in which partially folded intermediates are present at significant levels at equilibrium. Many of the proteins that have been observed to undergo multistate unfolding transitions appear to be composed of more than one domain that unfold independently. If they occur at sufficiently different urea concentrations, distinct transitions can be detected in urea gradient electrophoresis, as illustrated in *Figure 5* for calmodulin. Unless there is a clear plateau between the transitions, extrapolating the stabilities of the native protein and intermediates is not as simple as for a two-state transition. In such a case, it is necessary to use a mathematical curve-fitting procedure to analyse the data.

3.4.2 Kinetic analysis of folding transitions by urea gradient electrophoresis

If the native and unfolded forms of the protein do not interconvert rapidly during electrophoresis, the band of protein will not be continuous through the transition zone, and the patterns obtained when electrophoresis is initiated with native or unfolded protein will not be identical. In such a case, the resulting pattern cannot be used to estimate the relative free energies of the native and unfolded states, but can provide information about the kinetics and mechanism of folding and unfolding. The ability to kinetically resolve interconverting conformations can be enhanced by carrying out the electrophoresis at lower temperatures (where most folding–unfolding transitions are slower) and minimizing the time of electrophoresis by using higher voltages (28).

In general, rates of unfolding increase and rates of folding decrease with increasing urea concentration (31,32). As a consequence, both folding and unfolding are often relatively slow in the transition zone; at low urea concentrations folding is rapid and at high urea concentrations unfolding is rapid. *Figure 8b* illustrates the patterns expected for a situation where folding and unfolding in the transition region are slow compared to the time of the electrophoresis. When native protein is applied to the gel, a significant fraction of the protein remains native at urea concentrations within or above the transition zone, simply because the time of electrophoresis is short compared to the half-time for unfolding at these urea concentrations. Conversely, when unfolded protein is applied to the gel, a fraction of the molecules remain unfolded even at urea concentrations where the native form would be significantly populated at equilibrium. A smear of protein between the native and unfolded bands is generated by molecules that undergo the transition only once or a few times during the electrophoresis. Thus, at urea concentrations corresponding to the position of this smear, the rate constant for unfolding, if native protein has been applied to the gel (or folding, if unfolded protein has been applied) is of the order of $1/t$, where t is the time of electrophoresis. More quantitative analyses of the kinetics of urea gradient gel electrophoresis have been presented elsewhere (12).

Many proteins have been observed to refold with complex multiphasic kinetics (32). Much of this kinetic complexity is due to the presence of multiple unfolded forms that interconvert slowly and refold with different rates. There is now considerable evidence that the unfolded forms differ by the isomeric state of peptide bonds preceding proline residues, but there has also been much uncertainty about the relationships between particular slow folding phases and individual proline isomerization events. The simplest model predicts that folding is completely blocked by any incorrect proline isomers (that is, isomers different from those found in the native conformation). Since proline isomerization in the unfolded state is expected to be independent of urea concentration, the rate of folding of slow folding forms should be independent of urea concentration.

Figure 8c illustrates the urea gradient gel patterns expected if interconversion of the unfolded forms is slow compared to the time of electrophoresis (at 4°C the half-time for proline isomerization is about 20 min). When native protein is applied to the gel, a smooth continuous band of protein is expected because there is no significant formation of the slowly refolding unfolded form during the electrophoresis. However, when unfolded protein (in which the unfolded forms have equilibrated) is applied to the gel, only the rapidly refolding form folds during the electrophoresis. In the extreme case illustrated here, none of the slowly folding form refolds and a continuous straight band is seen.

Patterns indicating the existence of multiple unfolded forms have been observed with ribonuclease A, chymotrypsinogen, and α-chymotrypsin (28). However, the observed patterns are more complex than those illustrated schematically in *Figure 8c*. In particular, the rate of folding of the slowly folding form often does increase at low urea concentration, so that most or all of the protein is folded on the low urea side of the gel, as illustrated schematically in *Figure 8d*. In addition, the compactness of the slowly folding form often increases at low urea concentrations. These results are consistent with those obtained by more conventional kinetic analyses and indicate that proline isomerization in the slowly folding molecules does not completely block folding.

Kinetic intermediates in folding or unfolding may be detectable if one of the steps in the mechanism is slow compared to the time of electrophoresis. If the unfolded form rapidly interconverts with an intermediate, I, that is only slowly converted to or from the native state:

$$U \xrightleftharpoons{\text{fast}} I \xrightleftharpoons{\text{slow}} N \qquad [8]$$

then the intermediate may be detected during refolding, but not unfolding, as illustrated in *Figure 8e*. The intermediate in equilibrium with the unfolded form will give rise to an increase in average mobility at lower urea concentrations. However, the intermediate will not be observed if the native protein is applied to the gel, because it is rapidly converted to the unfolded form at the urea concentrations where the native protein is unfolded at a detectable rate. Such a situation has been observed with β-lactamase (33).

In summary, urea gradient electrophoresis can quickly provide considerable information about unfolding–refolding reactions over a range of denaturant concentrations, using quite small amounts of protein. The estimates of rate constants are not as precise as those obtained from spectroscopic methods (Chapter 10), but the electrophoretic method does provide complementary information about the hydrodynamic volumes of the native, unfolded, and intermediate states.

References

1. Goldenberg, D. P. and Creighton, T. E. (1984). *Anal. Biochem.*, **138**, 1.
2. Hames, B. D. and Rickwood, D. (ed.) (1990). *Gel electrophoresis of proteins: a practical approach* (2nd edn). IRL Press, Ltd., Oxford.
3. Rodbard, D. and Chrambach, A. (1970). *Proc. Natl. Acad. Sci. USA*, **65**, 970.
4. Tietz, D. (1988). In *Advances in electrophoresis* (ed. A. Chrambach, M. J. Dunn, and B. J. Radola), Vol. 2, p. 111. VCH, Weinheim.
5. Ferguson, A. O. (1964). *Metabolism*, **13**, 985.
6. Morris, C. J. O. and Morris, P. (1971). *Biochem. J.*, **124**, 517.
7. Rodbard, D. and Chrambach, A. (1971). *Anal. Biochem.*, **40**, 95.

8. Gonenne, A. and Lebowitz, J. (1975). *Anal. Biochem.*, **64**, 414.
9. Creighton, T. E. (1974). *J. Mol. Biol.*, **87**, 579.
10. Goldenberg, D. P. and Zhang, J. X. (1993). *Proteins: Struct. Funct. Genet.*, **15**, 322.
11. Mitchell, R. M. (1976). *Biopolymers*, **15**, 1717.
12. Creighton, T. E. (1979). *J. Mol. Biol.*, **129**, 235.
13. McLellan, T. (1982). *Anal. Biochem.*, **126**, 94.
14. Davis, B. J. (1964). *Ann. N.Y. Acad. Sci.*, **121**, 404.
15. Ornstein, L. (1964). *Ann. N.Y. Acad. Sci.*, **121**, 321.
16. Reisfeld, R. A., Lewis, U. J., and Williams, D. E. (1962). *Nature (London)*, **195**, 281.
17. Jovin, T. M. (1973). *Biochemistry*, **12**, 871, 879, 890.
18. Jovin, T. M. (1973). *Ann. N.Y. Acad. Sci.*, **209**, 477.
19. Lyubimova, T., Caglio, S., Gelfi, C., Righetti, P. G., and Rabilloud, T. (1993). *Electrophoresis*, **14**, 40.
20. Merril, C. R., Joy, J. E., and Creed, G. J. (1994). In *Cell biology: a laboratory handbook* (ed. J. E. Celis), Vol. 3, p. 281, Academic Press, San Diego.
21. Laskey, R. A. (1980). In *Methods in enzymology* (ed. C. Grossman and K. Moldave), Vol. 68, p. 363. Academic Press, London.
22. Gershoni, J. M. and Palade, G. E. (1983). *Anal. Biochem.*, **131**, 1.
23. Riesner, D., Henco, K., and Steger, G. (1991). In *Advances in electrophoresis* (ed. A. Chrambach, M. J. Dunn, and B. J. Radola), Vol. 4, p. 170. VCH, Weinheim.
24. Attanasio, R., Stunz, G. W., and Kennedy, R. C. (1994). *J. Biol. Chem.*, **269**, 1834.
25. Yu, H., Lee, K. N., and Kim, J. (1995). *Nature Struct. Biol.*, **2**, 363.
26. Klemm, J. D., Wozniak, J. A., Alber, T., and Goldenberg, D. P. (1991). *Biochemistry*, **30**, 589.
27. Creighton, T. E. and Shortle, D. (1994). *J. Mol. Biol.*, **242**, 670.
28. Creighton, T. E. (1980). *J. Mol. Biol.*, **137**, 61.
29. Schellman, J. A. (1978). *Biopolymers*, **17**, 1305.
30. Hollecker, M. and Creighton, T. E. (1982). *Biochim. Biophys. Acta*, **701**, 395.
31. Tanford, C. (1968). *Adv. Protein Chem.*, **23**, 121.
32. Schmid, F. X. (1992). In *Protein folding* (ed. T. E. Creighton), p. 197. Freeman, New York.
33. Creighton, T. E. and Pain, R. H. (1980). *J. Mol. Biol.*, **137**, 431.
34. Goldenberg, D. P. (1996). In *Current protocols in protein science* (ed. J. E. Coligan, B. Dunn, H. L. Ploegh, D. W. Speicher, D. W. and P.T. Wingfield), p. 7. 4. 1. Wiley, New York.

9

Hydrodynamic properties of proteins

STEPHEN E. HARDING

1. Renaissance of hydrodynamic techniques

Since the first edition of this book, there has been something of a renaissance of hydrodynamic methods for the determination of the mass, quaternary structure, gross conformation, and interaction properties of proteins and other macromolecules in solution. By 'hydrodynamic' (Greek for 'water-movement') techniques, we mean any technique involving motion of a macromolecule with or relative to the aqueous solvent in which it is dissolved or suspended. This therefore includes not only gel filtration, viscometry, sedimentation (velocity and equilibrium), and rotational diffusion probes (fluorescence anisotropy depolarization and electric–optical methods), but also 'classical' and 'dynamic' light scattering, which both derive from the relative motions of the macromolecular solute in relation to the solvent. This definition also includes electrophoretic methods (considered in Chapter 8 and not covered here), which are powerful tools for separation, purification, and identification of proteins, but also, with 'SDS' methodology, provide an estimate of polypeptide molecular weight (see Chapter 1). The present chapter therefore considers the hydrodynamic determination of 'molar mass' or molecular weight and quaternary structure (subunit composition and arrangement, self-association phenomena, and polydispersity). It will also consider the measurement of protein conformation in dilute solution with particular reference to the use of the analytical ultracentrifuge, a technique although of considerable antiquity (70th birthday in 1993) that has been the centre of the revival of hydrodynamic methodology.

After a brief description of the methodology in each case, practical tips and advice about the measurement and analysis will be provided—largely of the type not to be found in the manuals of commercial manufacturers. The interested reader can then find any other information needed from the latter and from the key references given.

2. Mass and quaternary structure measurement

It is worth stressing here that, unlike a polypeptide mass from sequence analysis or a protein structure from crystallography or NMR, a hydrodynamic mass or a conformation is a 'soft' quantity as opposed to a 'hard' one. That is to say, it will always come with a '±' and often with assumptions (about thermodynamic ideality, hydration, etc.). Although the molecular weight of an unglycosylated polypeptide can be determined to an accuracy of ± 1 Da from sequence information or from mass spectrometry (see Chapter 2), a similar precision cannot be obtained for glycosylated proteins because of polydispersity deriving from the variability of a cell's glycosylation process. Many proteins contain more than one non-covalently linked protein chain, particularly at higher concentrations. This can be uncovered by carrying out analyses under both native and denaturing conditions. An important role of hydrodynamic methods for mass analysis in protein chemistry is to give the molecular weight of the 'intact' or 'quaternary' structure and also to provide an idea of the strength of binding of these non-covalent entities through measurement of association constants.

2.1 Gel filtration and size exclusion chromatography

The simplest method of measuring molar mass is gel filtration (1), commonly referred to as 'gel permeation chromatography' or now 'size exclusion chromatography' (SEC), since the chemical inertness of the separation medium is assumed. Originally this was conceived as a method for the separation and purification of macromolecules, but has developed over the years in its 'calibrated' form as a very popular method for measuring protein molar masses both in native and dissociative conditions.

The separation medium is a cross-linked gel, traditionally cross-linked polysaccharide or polyacrylamide beads equilibrated with the appropriate buffer. The degree of cross-linking dictates the separation range of the gel: looser gels separate bigger molecules (see Chapter 1). Proper packing of columns requires some skill, and the user manuals as supplied by the commercial manufacturers are usually very comprehensive. The availability of HPLC versions makes the measurement particularly attractive for protein chemists.

Gel filtration or SEC depends on the principle that some of the space inside the gel particle is available to smaller molecules, but unavailable to larger molecules, which are excluded. Thus, when a solution is applied to a properly packed gel column (*Figure 1a*) only the dead space—between gel particles—is available to the excluded molecules, which therefore come off first when elution is commenced. The excluded molecules—the larger molecules—will thus have a smaller elution volume, V_e, and will elute first from the column (*Figure 1b*). Smaller macromolecules, having progressively

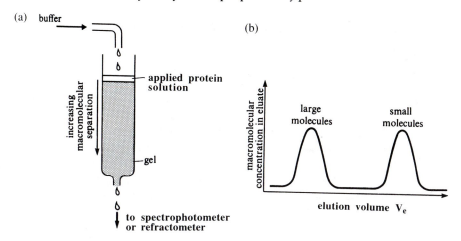

Figure 1. Principle of gel filtration and size exclusion chromatography. (a) Experimental set-up. (b) Example of an elution profile. Adapted from ref. 2.

more and more space available to them as molecular weight decreases, are accordingly eluted only at higher values of V_e. The separation is sometimes given in terms of the partition coefficient, K_{av}, defined by:

$$V_e = V_o + K_{av} (V_t - V_o) \qquad [1]$$

where V_o and V_t are the 'void volume' and 'total volume' of the column, respectively. They are determined from separate elutions using solute species having partition coefficients of zero (totally excluded) and one (non-excluded), respectively. Elution of proteins as they emerge from the column is usually monitored spectrophotometrically. If the buffer contains absorbing reagents, like ATP, azide, etc., highly sensitive differential refractometers are now available, which are arguably preferable now as the detection method of choice.

All other things being equal, M_r and V_e are related empirically by the expression (1):

$$V_e = A - B \log_{10} M_r \qquad [2]$$

where parameters A and B are properties of the column. This equation is valid only over the fractionation range of the gel; it also does not hold if other separation mechanisms are operating (4). To obtain M_r of a protein molecule or mixture of molecules, the column is first calibrated by the use of standard proteins of known size. Linear regression analysis is then used to evaluate A and B; hence M_r of the unknown protein can be found from its measured value of V_e. The calibration can only be applied within the fractionation range of the gel which depends on the pore size (*Figure 2*). Fractionation ability can be enhanced by running differing gel columns in

221

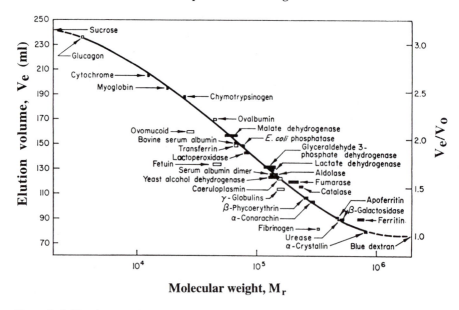

Figure 2. Calibration plot for proteins eluting from a Sephadex G200 column. From ref. 3.

series, a practice common with HPLC systems because of the much shorter elution times.

Equation 2 is valid only for molecules of similar shape and conformation. Thus calibration using globular protein standards would be inappropriate for fibrinogen and asymmetric muscle proteins like myosin and titin and for heavily glycosylated proteins. These calibration problems can be avoided by coupling an absolute molar mass detector (e.g. a light scattering photometer) downstream from the column (see Section 2.5).

The theory behind Equation 2 is not rigorous, but, at least for globular proteins, it seems to represent the data very well. For linear macromolecules of limited stiffness, there appears to be growing acceptance that the separation is more a logarithmic function of the hydrodynamic volume of a macromolecule ($\approx M_r[\eta]$ where $[\eta]$ is the intrinsic viscosity of a molecule) (see Section 3) and its corresponding hydrodynamic or 'effective' radius, r_H. This has culminated in a proposal for a 'universal calibration' (5). This may be more appropriate for proteins with disulfide bonds broken and in denaturing solvents, e.g. 6 M GdmCl. For such unfolded proteins, wider pore gels (such as Sepharose) are the most appropriate separation medium.

The procedure for gel filtration/ SEC analysis is given in *Protocol 1*.

2.2 Dynamic light scattering (DLS)

The appearance of simple to use fixed-angle (90°) dynamic light scattering photometers has made DLS an increasingly popular tool amongst protein chemists. After certain assumptions and approximations, largely involving an assumed spherical shape, remarkably reliable estimates for the mol. wt of globular proteins have been obtained (6). When used in isolation, this method is, like gel filtration, a relative one, requiring calibration using standard proteins of known mol. wt. For asymmetric proteins like fibrinogen and myosin, the single angle approximation fails, but extraction of mol. wt and related parameters is still possible if a multi-angle instrument is used. Also, the primary parameter that comes from DLS measurements is the translational diffusion coefficient, D (in units of cm^2/sec), and it can be combined with results from sedimentation analysis in the analytical ultracentrifuge to determine M_r more accurately (see Equation 9 in Section 2.3.2).

Protocol 1. Estimating the size of a protein by low-pressure gel filtration/SEC

Equipment and reagents

- Column, with optional reservoir to assist packing
- Gel filtration matrix (e.g. Sephadex of the appropriate grade) equilibrated with buffer at temperature at which analysis is to be made

- Peristaltic pump
- Mol. wt calibration standards of similar shape and other conformational properties to the protein to be characterized

Method

1. Mount the empty column vertically, with the aid of a plumb-line. Attach outlet tubing and fill the column with buffer, removing all dead space. Close outlet.

2. Pack the column with the matrix; pour in a thick gel slurry (preferably degassed) in a single operation, avoiding air bubbles and keeping the temperature approximately constant.

3. Close off the the column without trapping any air; repeat with any additional columns that are to be used in series.

4. Attach peristaltic pump to the first column, and run through at least three column volumes of buffer to ensure equilibration (check tubing joints for leaks!). The maximum flow rate will depend upon the matrix (see manufacturers' specifications); typically, it is in the range 0.2–6 ml/min.

5. Attach a UV recorder downstream from the last column. To monitor most proteins, set the monitoring wavelength to 278 nm or, if the buffer is sufficiently transparent, to 210–230 nm, which will give greater sensitivity.

6. Measure the absorbance baseline of the buffer.

7. Inject samples of the mol. wt standards on to the column and measure their V_e.

Protocol 1. *Continued*

8. Calibrate the column by plotting V_e versus mol. wt for the standards.
9. Inject the test protein and measure its V_e under the same conditions.
10. Estimate the mol. wt of the test protein from its V_e and the calibration curve for the column.
11. Wash the column with three volumes of buffer.
12. If the column is to be used again at a later time, keep buffer flowing slowly through it; or include an anti-microbial agent, such as sodium azide, and store it in the cold.

2.2.1 Principle

The principle of DLS experiments is very simple (*Figure 3a*) and is based on the high intensity, monochromaticity, collimation, and coherence of laser light. Laser light is directed on to a thermostatted protein solution, and the intensity is recorded at either a single or multiple angles using a photomultiplier/photodetector. The intensities recorded will fluctuate with time caused by Brownian diffusive motions of the macromolecules; this movement causes a 'Doppler' type of wavelength broadening of the otherwise monochromatic light incident on the protein molecules. Interference between light at these wavelengths causes a 'beating' or fluctuation in intensity in much the same as a listener perceives a radio station with superposition of other radio stations at nearby frequencies. How rapid the intensity fluctuates (nsec to μsec time intervals) depends on the mobility or diffusivity of the protein molecules. A purpose-built computer, known as an autocorrelator, 'correlates' or interprets these fluctuations. It does this by evaluating a 'normalized intensity autocorrelation function'($g^{(2)}$) as a function of the 'delay time', τ (in the range of milli- to microseconds). The decay of the correlation, $g^{(2)}(\tau)$ as a function of τ, averaged over longer time intervals (usually minutes) can then be used, by an interfaced PC or equivalent, to obtain the value of D. Larger and/or asymmetric particles that move more sluggishly will have slower intensity fluctuations, slower decay of $g^{(2)}(\tau)$ with τ, and hence smaller D values compared to smaller and/or more globular particles. The delay time τ itself is the product of the 'channel number' b (taking on all integral values between 1 and 64, or up to 128 or 256 depending on how expensive the correlator) and a user-set 'sample time', τ_s; its value is typically ~ 100 nsec for a rapidly diffusing protein of low mol. wt (e.g. about 20 000) and increasing up to milliseconds for microbes. In the past, τ_s was selected by trial and error, but now modern data acquisition software usually does this automatically.

For spherical particles, a single term exponential describes the decay of Γ with τ:

$$g^{(2)}(\tau) - 1 = e^{-Dk^2\tau} \qquad [3]$$

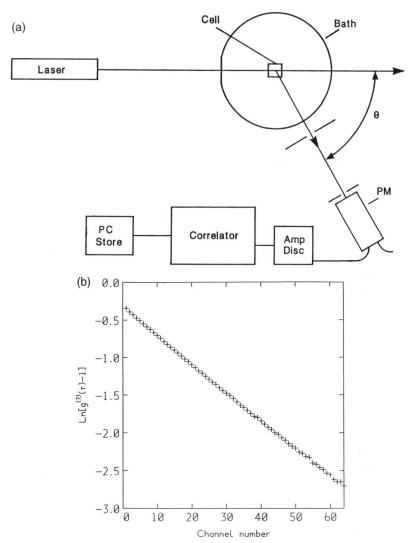

Figure 3. Principle of dynamic light scattering. (a) Experimental set-up. (b) Normalized autocorrelation decay plot for the protein assembly dynein (in 40 mM NaCl). $D^0_{20,w}$ = 1.1 × 10^{-7} cm^2/sec; M_r (from Equation 9) = 2.5 × 10^6. From ref. 7.

where k is the Bragg wave vector whose magnitude is defined by:

$$k = \{4\pi n/\lambda\} \sin (\theta/2) \qquad [4]$$

n is the refractive index of the medium, θ the scattering angle, and λ the wavelength of the incident light. Equation 3 can be reasonably applied to quasi-spherical particles like globular proteins or spheroidal protein assemblies (*Figure 3b*).

2.2.2 Fixed-angle (90°) DLS photometer

For globular proteins and spheroidal assemblies, application of Equation 3 at only a single fixed-angle is usually sufficient. Low angles are usually avoided because they magnify problems due to any contamination with dust or other supramolecular particles: an angle of 90° is normally used. For a given laser power at a given protein concentration, the smaller the protein the lower the intensity of scattered light, and hence the longer the averaging required to give a sufficient signal. A commercial instrument is available based on this single fixed-angle principle (6) (*Figure 4a*). Its operation is described in *Protocol 2*.

To obtain mol. wt information from the value of D, a calibration curve of log D versus log M_r is produced, based on globular protein standards and known as an 'MHKS' (Mark–Houwink–Kuhn–Sakurada) scaling relation (8) (*Figure 4b*). It is assumed that the same relation holds for the unknown protein.

Protocol 2. Measuring the diffusion coefficient and approximate molecular weight of a globular protein by fixed-angle DLS

Equipment
- Fixed-angle DLS photometer, such as the Protein Solutions 801 instrument
- Sterile syringe with appropriate filter, 0.1–0.45 μm, depending upon the size of the protein
- Deionized, distilled water
- Sample of protein in an appropriate buffer and close to the optimal concentration (for the Protein Solutions 801 instrument, 2 mg/ml for a 30 kDa protein, proportionally less for larger proteins)

Method
1. Inject water or buffer, via a 0.1 μm filter, into the warmed-up DLS photometer to obtain the clean water count rate.
2. Inject the sample in the same way, using the appropriate filter, and measure the count rate.
3. If the count rate is below the manufacturer's threshold, check the instrument alignment or increase the protein concentration.
4. Use the instrument's software to obtain the diffusion coefficient and, where appropriate, the in-built calibration to obtain directly the approximate mol. wt.
5. Rinse and dry the flow cell of the photometer.

Figure 4. Single-angle dynamic light scattering. (a) Photometer DynaPro 801 (courtesy of Protein Solutions Ltd.) incorporates a 20 mW infrared (780 nm) semiconductor laser. Photons scattered at an angle of 90° are collected by a lens and conducted to an avalanche photodiode via an optical fibre; this produces a single electrical pulse for each photon received and these are stored and correlated by an integral computer. The optical bench

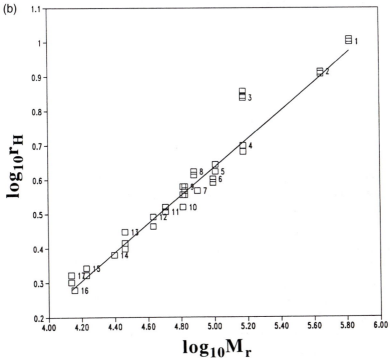

measures only 25 × 5 × 5 cm (6). (b) Double logarithmic calibration plot of r_H versus M_r for: 1, thyroglobulin; 2, apoferritin; 3, IgG; 4, yeast alcohol dehydrogenase; 5, hexokinase; 6, amyloglucosidase; 7, horse alcohol dehydrogenase; 8, transferrin; 9, bovine serum albumin; 10, haemoglobin; 11, hexokinase subunit; 12, ovalbumin; 13, carbonic anhydrase; 14, chymotrypsinogen; 15, myoglobin; 16, lysozyme; 17, ribonuclease A. From ref. 6.

Other approximations and practical requirements with the operation of this type of fixed-angle instrument have to be made:

(a) Solutions must be as free as possible from dust and supramolecular aggregates. This requirement is met by injection of the sample into the (scrupulously clean) scattering cell via a Millipore filter(s) of appropriate size (0.1–0.45 μm).

(b) The diffusion coefficient is a sensitive function of temperature and the viscosity of the solvent. The log D versus log M_r calibration must be made under the same temperature (kept constant during the measurement) and solvent viscosity conditions.

(c) The diffusion coefficient measured at a single concentration is an apparent one, D_{app}, because of non-ideality effects (finite volume and charge). These effects become vanishingly small as the concentration approaches zero. The approximation is made—usually reasonably for proteins—that $D_{app} \approx D$, or that any non-ideality effects are the same as for the calibration standards.

Despite these approximations, the values of diffusion coefficients and M_r obtained in this way have been remarkably reliable. For non-globular proteins, however, the log D versus log M_r calibration becomes invalid and Equation 3 no longer applies; resort has then to be made to an instrument with a multi-angle facility.

2.2.3 Multi-angle instruments

Measurements using multi-angle equipment (*Figure 5a*) are more time-consuming, and the instrumentation larger and more expensive. Data analysis is also more complicated. Equation 3 no longer applies, largely because of the added complication of rotational diffusion effects. These effects vanish, however, as the scattering angle θ approaches zero. It is therefore possible to use Equation 3 in terms of an apparent diffusion coefficient D_{app}, with contributions from both concentration and rotational diffusion effects. D_{app} is measured at several angles and extrapolated back to zero angle to give D if concentration effects are negligible. If, however, concentration dependence effects are suspected, a double extrapolation can be performed on the same plot (called a 'Dynamic Zimm plot') of D_{app} to zero angle and to zero concentration (10). The common intercept gives the 'ideal' (in a thermodynamic sense) diffusion coefficient, D^0. Because this quantity is not only an intrinsic property of the protein but also of the viscosity, η, and the temperature of the buffer, it has to be corrected to standard conditions (viscosity of pure water at 20°C, $\eta_{20,w}$), either before or after the extrapolation (11):

$$D^0_{20,w} = D^0 \, (\eta/\eta_{20,w}) \, (T/293.15). \qquad [5]$$

(a)

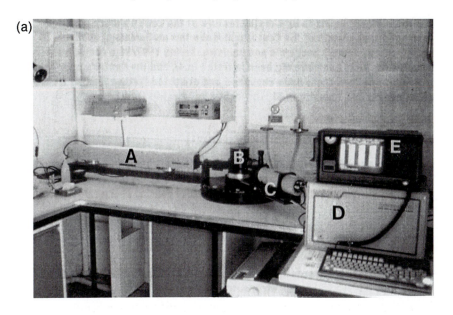

(b)

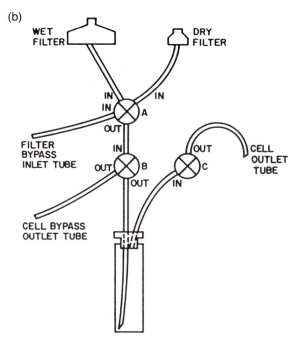

Figure 5. Multi-angle DLS. (a) Photometer Malvern Instruments 4700 system in our laboratory. A, 15 mW He-Ne laser; B, water-bath, goniometer; C, photomultipliers/amplifier discriminator; D, autocorrelator; E, PC. (b) Schematic of specially constructed cuvette designed to minimize the dust problem. From ref. 9.

The size of a protein, as represented by its equivalent hydrodynamic radius, r_H, is related to $D^0_{20,w}$ by the Stokes equation:

$$r_H = k_B T / (6\pi\eta_{20,w} D^0_{20,w}) \qquad [6]$$

where k_B is Boltzmann's constant. To obtain an absolute measure of M_r of a protein from $D^0_{20,w}$, without assumptions concerning the shape of the protein, requires combination with the sedimentation coefficient from the analytical ultracentrifuge, as described in Section 2.3. Some modern software attempts to evaluate M_r directly from the diffusion coefficient; this should be treated with some caution.

2.2.4 Further notes

(a) For multi-angle measurements, preferences vary in terms of the type of cuvettes used. Square cuvettes are optically more reliable, but cell corners are obviously prohibited. Cylindrical cuvettes, if used, should be of the wide diameter type (> 2 cm) to avoid internal and stray reflections.

(b) Scrupulous attention to sample and cuvette clarity is mandatory, particularly for macromolecules of $M_r < 10^5$, which give low scattering signals, and if low angles are employed, where the effects of supramolecular contaminants are at their maximum. Special cuvette filling arrangements are used for clarification purposes (*Figure 5b*).

(c) The angular extrapolation of D_{app} can provide an estimate for the rotational diffusion coefficient, albeit to a lower precision than conventional methods (fluorescence depolarization, electric birefringence).

(d) If the protein is polydisperse or self-associating, the logarithmic plot of the type shown in *Figure 3b* will tend to be curved, and the corresponding diffusion coefficient will be a *z*-average (12). The spread of diffusion coefficients is indicated by a parameter known as the 'polydispersity factor' (12) which most software packages evaluate.

(e) Various computer packages are available from the commercial manufacturer for data acquisition and evaluation. In our laboratory, we prefer to acquire the data in ASCII format using the data capture software of the commercial manufacturer and then use our own in-house routine 'PROTEPS' (S. E. Harding, J. C. Horton, and P. Johnson, unpublished data) for the evaluation of diffusion coefficients and polydispersity factors.

(f) More advanced routines are available, such as 'CONTIN', designed for the study of heterogeneous systems by going beyond the use of polydispersity factors and inverting the autocorrelation data directly to give distributions of particle size. These methods have been recently reviewed (13).

(g) DLS is particularly valuable for the investigation of changes in macromolecular systems when the time-scale of changes is minutes or hours, and not seconds or shorter (14).

(h) For charged macromolecular systems, DLS provides a useful tool for monitoring electrophoretic mobilities (15), and commercial instrumentation is available for this purpose.

2.3 Sedimentation velocity in the analytical ultracentrifuge

Combination of the sedimentation coefficient, s, from sedimentation velocity with the diffusion coefficient, D, from DLS gives an absolute value for the mol. wt of a protein, without assumptions about conformation. This method for mol. wt measurement was given by T. Svedberg (16), the founder of the analytical ultracentrifuge: a technique which is now undergoing something of a renaissance with the launch of a new commercial instrument (*Figure 6a*) (17).

The basic principle of the technique is as follows: a solution of the protein is placed in a specially designed cell with sector-shaped channel and transparent end windows (*Figure 6b*). This in turn is placed in an appropriately balanced rotor and run in high vacuum at the appropriate speed (typically 50000–60000 r.p.m. for a protein of M_r 10^4 to 10^5, lower speeds for larger molecules). A light source positioned below the rotor transmits light via a monochromator or filter through the solution and a variety of optical components. The moving boundary is recorded at appropriate time intervals, either on photographic film, on chart paper, or as digital output fed directly into a PC. Measurement of the rate of the movement of the boundary (per unit centrifugal field) enables evaluation of the sedimentation coefficient. For an introduction, see ref. 11; for the state of the art, see two recent books (18,19).

2.3.1 Optical systems

Three principle optical systems can be employed:

- absorbance (in the range 200–700 nm)
- 'Schlieren' (refractive index gradient)
- Rayleigh interference

The simplest system is the absorbance system, and it is used in the Optima XL-A analytical ultracentrifuge available commercially, so it will be described here. Use of the other optical systems requires more specialist knowledge, and the interested protein chemist needs really to consult an expert.

Examples of sedimenting boundaries recorded using absorption optics are shown in *Figure 7*, using a highly purified preparation of an enzyme (*Figure 7a*) and a heterogeneous preparation of a DNA binding protein (Pf1) with a macromolecular component and a fast moving aggregate (*Figure 7b*). The procedure for obtaining such data is described in *Protocol 3*.

(a)

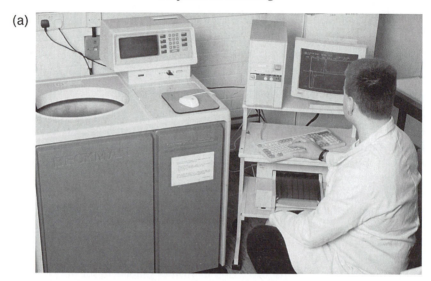

(b)

Figure 6. Modern analytical ultracentrifugation. (a) Beckman Optima XL-A in our labora-tory, equipped with scanning absorption optics, with full on-line data capture and analy-sis. The rotor is stable down to $\approx$ 1000 rev. min, permitting the analysis of large macromolecular assemblies. (b) Components of an analytical cell (12 mm optical path length).

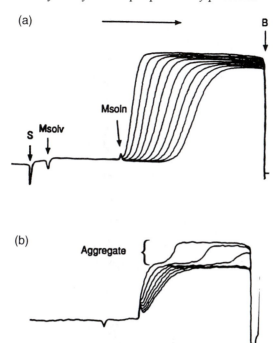

Figure 7. Sedimentation velocity diagrams obtained using scanning absorption optics. (a) Methylmalonyl mutase, 0.7 mg/ml. Monochromator wavelength 295 nm; scan interval 9 min; rotor speed 44 000 r.p.m.; temperature 20 °C; measured s_{20} = (7.14 ± 0.04)S. (b) Gene 5 DNA-binding protein, 0.7 mg/ml. Monochromator wavelength, 278 nm; scan interval, 8 min; rotor speed, 40 000 r.p.m., temperature, 20 °C; measured $s^0_{20,w}$ = (35.5 ± 1.4)S (faster boundary) and (2.6 ± 0.1)S (slower boundary).

Protocol 3. Sedimentation velocity measured with an analytical ultracentrifuge with scanning absorption optics detection system

Equipment
• Beckman Optima XL-A ultracentrifuge

Method

1. Concentration requirements for the protein. This depends on the extinction coefficient of the protein (see Chapter 10). The lower the protein concentration the better, since it minimizes problems of thermodynamic non-ideality. For proteins of average absorbance at 280 nm ($\approx$ 500 ml/g/cm), concentrations as low as 0.2 mg/ml are possible with 12 mm optical path length cells. This can be made even

Protocol 3. *Continued*

lower if the buffer is transparent and the peptide bond wavelength can be used (210–230 nm). For absorbance values > 3, shorter path length cells need to be employed (the minimum is about 3 mm; below this, cell window problems become significant), 'off-maxima' wavelengths used (with caution), or, more desirably, a different optical system used (interference or Schlieren).

2. Choose the appropriate buffer/solvent. If possible, work with an aqueous solvent of sufficiently high ionic strength (> 0.05 M) to provide adequate suppression of non-ideality phenomena deriving from macromolecular charge effects (see below). If denaturing/dissociating solvents are used, appropriate centre-pieces need to be used (e.g. of the 'Kel-F' type; Beckman Instruments).

3. Load the sample into the cell. Double sector cells are used with the protein solution (0.2–0.4 ml) in one sector and the reference buffer or solvent in the other; the latter is filled to a slightly higher level to avoid complications caused by the signal coming from the solvent meniscus: the scanning system subtracts the absorbance of the reference buffer from that of the sample. Electronic multiplexing allows multiple hole rotors to be used, so that several samples can be run at a time.

4. Choose the appropriate temperature. The standard temperature at which sedimentation coefficients are quoted is now 20°C (sometimes 25°C). If the protein is thermally unstable (a sedimentation velocity run can take between one and a few hours), temperatures down to about 4°C can be used without difficulty.

5. Choose the appropriate speed. For a small globular protein of sedimentation coefficient ~ 2 Svedbergs (S, where $1S = 10^{-13}$ sec), a rotor speed of 50 000 r.p.m. will give a measurable set of optical records after some hours. For larger protein systems (e.g. 12S globulins or 30S ribosomes), speeds of < 30 000 r.p.m. can be employed.

6. Measure the sedimentation coefficient, *s*. The sedimentation coefficient, *s*, is defined by the rate of movement of the boundary per unit centrifugal field: $s = (dr/dt)/\omega^2 r$, where *r* is the radial position of the boundary at time *t*, and ω is the angular velocity in radians/sec (ω = r.p.m. $\times 2\pi/60$). Commercial software is available for identifying the centre of the sedimenting boundary (strictly the '2nd moment' of the boundary is more appropriate; practically there is no real difference). Personal choices vary, but we find the most satisfactory method—if requiring a little more effort—is:

   (a) To plot out the boundaries (recorded at appropriate time intervals) using a high resolution printer or plotter and graphically draw a line through the user-identified boundary centres.

(b) Then use a graphics tablet to recapture the central boundary positions as a function of radial position.

Computer routines such as XLA-VEL (H. Cölfen and S. E. Harding, unpublished data) yield the sedimentation coefficient and a correction to the loading concentration for average radial dilution during the run (caused by the sector shape of the cell channels).

7. Correct the results to standard conditions. For each protein concentration used, correct the sedimentation coefficient, s, to standard conditions of buffer/solvent density and viscosity (water at 20°C, $\rho_{20,w}$ and $\eta_{20,w}$ respectively):

$$s_{20,w} = s\,(\eta/\eta_{20,w})\,\{(1 - \bar{v}\,\rho_{20,w})/(1 - \bar{v}\,\rho)\} \tag{7}$$

where ρ is the density of the solvent. Knowledge of a parameter known as the 'partial specific volume', $\bar{v}$ (essentially the reciprocal of the anhydrous macromolecular density) is needed; this can usually be obtained for proteins from amino acid composition data, or measured with a precision density meter (20). Typically for proteins, $\bar{v}$ is close to 0.73 ml/g.

8. Extrapolate to zero protein concentration. Plot $s_{20,w}$ versus concentration (corrected for radial dilution) and extrapolate (usually linearly) to zero concentration (*Figure 8*) to give a parameter, $s^0_{20,w}$ which can be directly related to the frictional properties of the macromolecule (the so-called 'frictional ratio') and from which size and shape information can be inferred. If the protein is very asymmetric or solvated, plotting $1/s_{20,w}$ versus concentration generally gives a more useful extrapolation. The downward slope of a plot of $s_{20,w}$ versus concentration is a result of non-ideality behaviour and is characterized by the parameter k_s in the equation:

$$s_{20,w} = s^0_{20,w}\,(1 - k_s\,c). \tag{8}$$

The value of k_s, which reflects non-ideality effects of the system, will depend on the size, shape, and charge of the protein. If the solvent used is of a sufficient ionic strength, charge effects can be suppressed.

2.3.2 Evaluation of molecular weight

The molecular weight, M_r, can be found by combination of $s^0_{20,w}$ with $D^0_{20,w}$ using the Svedberg equation:

$$M_r = (s^0_{20,w}/D^0_{20,w})\,\{R\,T/(1 - \bar{v}\,\rho_{20,w})\}. \tag{9}$$

An accurate estimate for $\bar{v}$ as described above is normally required, since errors are tripled for proteins; e.g. an error of $\pm\,1\%$ in $\bar{v}$ results in an error of $\pm\,3\%$ in M_r. This means that care has to be made if the protein is glycosylated, since the $\bar{v}$ of carbohydrate is typically 0.6 ml/g.

For a heterogeneous system, $s^0_{20,w}$ will be a weight average and $D^0_{20,w}$ a

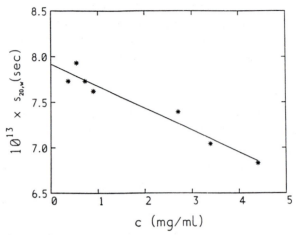

Figure 8. Sedimentation coefficient $s_{20,w}$ as a function of concentration for a rat IgE antibody. Measured $s^0{}_{20,w} = (7.92 \pm 0.06)$S.

z-average: the M_r calculated will also be a weight average (12), thus distinguishing it from the M_r obtained by osmometry (21), which is a number average.

A further approximate estimate can be obtained simply by combining $s^0{}_{20,w}$ with k_s (22):

$$M_r = (6\pi \, \eta_{20,w} s^0{}_{20,w})^{1.5} \, \{(3\bar{v})/4\pi).[(k_s/2\bar{v})-(\bar{v}_s/\bar{v})]\}^{0.5} \qquad [10]$$

where $\bar{v}_s$ is a specific volume allowing for hydration of the protein; since $(\bar{v}_s/\bar{v})$ in Equation 10 is usually small in comparison with $(k_s/2\bar{v})$, only an approximate estimate is needed. This method has given reliable estimates for standard protein molecules of known mol. wt. The parameter k_s is itself valuable for shape measurement. The form of the concentration-dependence can also be used as an assay for self-associating systems (23), although sedimentation equilibrium methods are usually superior (see Section 2.4).

2.3.3 Limitations

Sedimentation velocity is not so convenient for evaluating the molecular weights of proteins in denaturing/dissociating solvents, since their sedimentation coefficients are much smaller, due to greater frictional forces: *s* values of < 1S are difficult to measure with any precision because of the upper limit of rotor speed (60000 r.p.m.). If these solvents are used, care has to be expressed concerning inertness of the cells used.

2.4 Sedimentation equilibrium

The 'sedimentation-diffusion' method for giving the mol. wt, although an absolute method, is rather inconvenient in that it requires two sets of measurements. A simpler method is to use one measurement by sedimentation equilibrium, and it is probably the method of choice for mol. wt determination of intact protein assemblies, and for the investigation of interacting systems of proteins (24). The same instrument and optical system(s) for sedimentation velocity are used, the principal differences being:

- the much lower rotor speeds employed
- the longer run times
- the shorter solution (and buffer) columns in the ultracentrifuge cell; hence the smaller amount of material required

Sedimentation equilibrium, unlike sedimentation velocity, gel filtration, and dynamic light scattering, is not a transport method. In a sedimentation equilibrium experiment, the rotor speed is chosen to be sufficiently low so that the forces of sedimentation and diffusion on the macromolecular solute become comparable and an equilibrium distribution of solute is attained. This equilibrium can be established after a period of 2 to 96 hours, depending on the macromolecule, the solvent, and the run conditions. Since there is no net transport of solute at equilibrium, the recording and analysis of the final equilibrium distribution (*Figure 9*) will give an absolute estimate of the

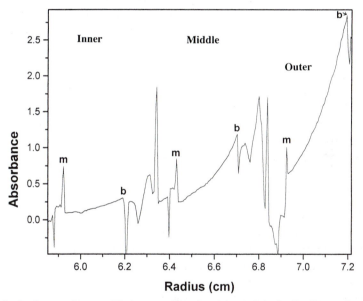

Figure 9. Sedimentation equilibrium profiles for β-lactoglobulin B. Absorption optics, with wavelength 280 nm. Rotor speed, 15 000 r.p.m.; temperature, 20 °C. A multichannel cell (12 mm optical path length) was used allowing three solution/solvent pairs, with 0.12 ml in the solvent channels and 0.10 ml in the sample channels. The initial protein concentrations were 0.1 mg/ml (inner profile); 0.2 mg/ml (middle); 0.3 mg/ml (outer). Only absorbances < 1.5 could be used with the outer channel; this difficulty could have been overcome by using a longer wavelength. With the inner channel, the signal could have been increased by using a far-UV wavelength (210–230 nm).

protein mass and associated parameters, since frictional (i.e. shape) effects are not involved.

Protocol 4 refers only to the absorption system—because of its simplicity and availability—for recording the equilibrium distribution of solute in the ultracentrifuge cell. The most accurate method is, in fact, the interference system, but it requires considerable more expertise to operate correctly (11,18,19).

Protocol 4. Measuring the sedimentation/equilibrium profile of a protein

Equipment
- As in *Protocol 3*

Method

1. Choose the appropriate conditions. These are similar to those applying to sedimentation velocity (see *Protocol 3*). As with sedimentation velocity, a temperature of 4°C can be used without difficulty. Sample volume requirements are lower than for sedimentation velocity: 0.1–0.2 ml gives a column length of about 0.1–0.2 mm with 12 mm cells. The longer the column, the greater the precision and the more information that can be extracted. The shorter the column, the quicker equilibrium will be reached (27), which may be important if many samples need to be run and/or the protein is relatively unstable.

2. Load the sample in the cell as in *Protocol 3*. As with sedimentation velocity, multiple cells can be run simultaneously in multihole rotors and electronically multiplexed. Further, because of the shorter columns needed for sedimentation equilibrium, special multichannel cells containing three sample/solvent pairs can be used (*Figure 9*). So, for a four-hole ultracentrifuge rotor (with one hole needed for the counterpoise with reference slits for calibrating radial positions in the cell), nine solutions can be run simultaneously. Eight-hole rotors are now available.

3. Choose the appropriate rotor speed.

4. Run the rotor until equilibrium is reached, when scans separated by sufficient time are identical. Smaller molecules get to equilibrium faster than larger ones. Less than 24 h are required for molecules of $M_r < 10^4$; large, slower diffusing molecules take 48–72 h. The time to equilibrium can be decreased by initial 'overspeeding', i.e. running at higher speed for a few hours before setting to the final equilibrium speed. It may, in some applications, be desirable to use shorter columns (as short as 0.5 mm); although the accuracy of the measure-

ments will be lower, this 'short column' method offers the advantage of reaching equilibrium in a few hours.

5. Record the equilibium profile. The parameter measured is the absorbance of the protein, A, as a function of the radial distance from the centre of the rotor, r. If scanning absorption optics are used, equilibrium patterns such as *Figure 9* can be read directly into an attached PC.

6. Measure the absorbance baseline. If the proteins are not too small, after the final equilibrium pattern has been recorded, the rotor is run for a short time at a higher speed (up to 60 000 r.p.m. or the upper limit for the particular centre-piece) to deplete the solution—or at least the meniscus region—of solute: the residual absorbance gives the baseline absorbance of the solvent. With small proteins, careful dialysis of the protein solution versus the reference solvent before the run may be necessary.

7. Calculate the molecular weight. The average slope of a plot of ln A versus r^2, (*Figure 10a*) will yield M_r:

$$M_r = (\mathrm{dln}\ A/\mathrm{d}r^2) \times 2RT/(1-\bar{v}\rho)\ \omega^2. \qquad [11]$$

As with Equation 9, an accurate estimate for the partial specific volume $\bar{v}$ is required; ρ is the density of the solvent.

8. Analyse for heterogeneity. For a non-associating, monodisperse system, the plot of ln A versus r^2 will be linear (*Figure 10b*); for a heterogeneous protein (containing interacting or non-interacting species of different molar mass), it will be curved upwards. This situation occurs with self-associating systems (see below) and with mixed solute or heavily glycosylated protein systems such as mucus glycoproteins. In this case the data can be treated in one of two ways:

   (a) An average slope is measured. This yields, as with Equation 9, the weight average mol. wt, M_w. For strongly curving plots or for systems where the cell baseline is not clearly defined, a procedure that uses a function known as M^* (25,26) is useful.

   (b) Local slopes (using a sliding strip procedure) (28) along the ln A versus r^2 curve can be obtained to give what is called 'point' weight average mol. wts, $M_w(r)$, as a function of either radial position (or the equivalent local concentration or absorbance). This procedure is particularly useful for the investigation of self-association phenomena and other types of heterogeneity; it also provides a method for extracting the z-average mol. wt, M_z:

$$M_z = \frac{\{M_w(r = \text{cell base}) - M_w(r = \text{meniscus})\}}{\{M_w(r = \text{cell base})A(\text{cell base}) - M_w(r = \text{meniscus})A(\text{meniscus})\}} \qquad [12]$$

The ratio M_z/M_w can be used as an index of the heterogeneity of the sample and, for non-interacting systems, is a measure of the

Protocol 4. *Continued*

inherent polydispersity of a system; this is particularly relevant to the study of heavily glycosylated systems.

9. Examine whether any apparent heterogeneity in mol. wt is due to association of the protein. If the system is self-associating or involved in 'heterologous' association (i.e. complex formation), either the ln A versus r^2 plot or the $M_w(r)$ versus A plot can be used to measure the stoichiometry and strength of the interaction. There are several commercial software packages available,[a] and a recent article has reviewed three using the dimerizing β-lactoglobulin as a model system for self-associations (29). Methods are also available for distinguishing between self-associating and non-interacting mixtures.

10. Consider non-ideality. For larger macromolecules ($M_r \geq 10^5$), such as protein assemblies and heavily glycosylated systems and/or for more concentrated solutions, non-ideality (through macromolecular exclusion and any unsuppressed charge effects) may become significant, which will tend to cause downward curvature in the ln A versus r^2 plots. This can obscure heterogeneity phenomena, and the two effects (non-ideality and heterogeneity) can occasionally cancel to give a linear plot that can be misleading; this can be avoided by running at more than one initial protein concentration. If the sample is not significantly heterogeneous, a simple extrapolation from a single experiment of point (apparent) mol. wt to zero concentration (absorbance) can be made, to give the infinite dilution 'ideal' value (in general, reciprocals are usually plotted as in *Figure 11*) (30). Alternatively, several sedimentation equilibrium experiments performed at different initial concentrations and extrapolation of 'whole cell' molecular weights, $M_{w,app}$, to zero concentration are necessary.

[a] Software currently available from the commercial manufacturer tends to require an assumed model prior to the analysis (ideal monomer, self-association, non-ideal self-association, etc.). We find two other general packages, not requiring assumed models, of use. These are:
(a) MSTAR, written in-house (26) and now available for PC (H. Cölfen and S. E. Harding, unpublished data), which evaluates $M_{w,app}$ (using the M^* function), $M_{z,app}$, or $M_{w,app}$ (and also $M_{z,app}$, if the data are of sufficiently high quality) versus r or A.
(b) XLASe, which evaluates $M_{w,app}$ and $M_{z,app}$ (M. D. Lechner, Universität Osnabrück, unpublished data).
After these model-independent analyses have been performed, resort is then made to the more specialist packages (self-association, polydispersity, etc.). There exists now a highly useful e-mail system called RASMB for the exchange of software and other matters concerning analytical ultracentrifugation (RASMB database; W. F. Stafford, stafford@edu.harvard.eri.bbri).

2.5 Classical light scattering

This is another powerful absolute method for the determination of mol. wts of intact macromolecules, and it is particularly suited to the study of large

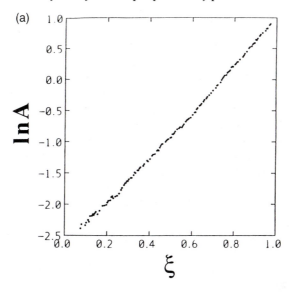

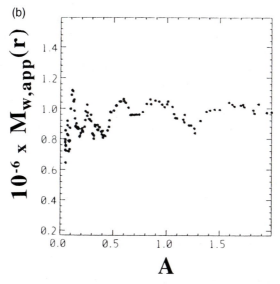

Figure 10. Sedimentation equilibrium data analysis for human IgM₁. Phosphate buffer pH 6.8; ionic strength, 0.1 M; protein concentration, ≈ 0.6 mg/ml. Scanning wavelength, 278 nm; rotor speed, 5000 r.p.m.; temperature, 20°C. (a) Log absorbance versus radial displacement squared plot. $\xi = (r^2 - a^2)(b^2 - a^2)$ where r is the radial displacement at a given point in the solute distribution and a and b the corresponding positions at the meniscus and cell base, respectively. From M^* analysis (25,26) of this data, $M_r = (1.00 \pm 0.02) \times 10^6$. (b) Plot of point average (apparent) M_r versus local concentration (expressed in absorbance units) in the solute distribution. Apart from noise near the meniscus there is no trend in the data, confirming a monodisperse, nearly ideal system. Adapted from ref. 26.

241

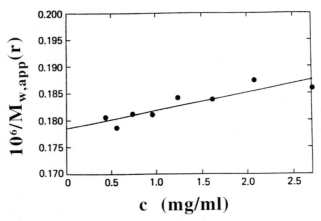

Figure 11. Plot of the reciprocal point (apparent) average molecular weight as a function of local concentration for turnip yellow mosaic virus. The measured M_r (from extrapolation to zero concentration) is $(5.8 \pm 0.2) \times 10^6$. Adapted from ref. 30.

macromolecular assemblies, up to a maximum of 50×10^6 M_r; beyond this the simple theory (known as the 'Rayleigh–Gans–Debye' approximation) breaks down. By 'classical' light scattering (as opposed to DLS) we mean the total or time-integrated intensity of light scattered by a macromolecular solution compared with the incident intensity for a range of concentrations and/or angles. Although a more rapid and, in principle, more convenient alternative to either the sedimentation-diffusion method or sedimentation equilibrium, the application of classical light scattering has until relatively recently suffered greatly from the 'dust problem', namely all solutions/scattering cells having to be scrupulously clear of dust and supramolecular particles, particularly for the analysis of proteins of mol. wt < 50 000; unlike for DLS, except for small proteins, measurements at low angles (where dust problems are their greatest) are mandatory. This has resulted in many cases in the requirement for unacceptably large amounts of purified material: experiments on incompletely purified solutions have been of little value.

Two developments have made the technique now worthy of serious consideration by protein scientists (31):

(a) The use of laser light sources, providing high collimation, intensity, and monochromaticity.
(b) The coupling of SEC-HPLC systems on-line to a light scattering photometer via the incorporation of a flow cell.

These facilitate considerably the analysis of mixtures of proteins and, more significantly, provide a very effective on-line 'clarification' system from dust and supramolecular contaminants. An example of such a set-up is given in *Figure 12*.

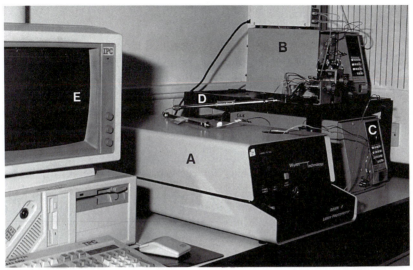

(b)

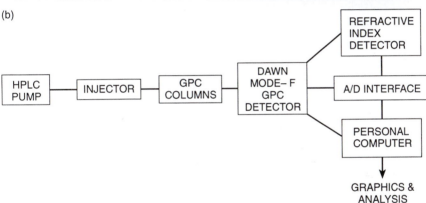

Figure 12. Multi-angle laser light scattering coupled to size exclusion chromatography (SEC-MALLS). (a) Experimental set-up in our laboratory. A, Dawn-F (Wyatt Technology); B, HPLC pump; C, refractive index detector; D, two SEC columns in series; E, interfaced PC system. (b) Schematic (courtesy of Wyatt Technology).

2.5.1 Principle

The intensity of light scattered by a protein solution is measured as a function of angle with a light scattering photometer (*Figure 12a*). For solutions of macromolecules or macromolecular assemblies, the basic equation for the angular dependence of light scattering is the Debye–Zimm relation:

$$Kc/R_\theta \approx \{1 + (16\pi^2 R_g^2/3\lambda^2) \sin^2 [(\theta/2)]\}[(1/M_r) + 2Bc] \qquad [13]$$

where it is assumed that the second virial coefficient B (in units of ml mol g^{-2}) is sufficient to represent non-ideality (i.e. third and higher order terms are assumed to be negligible); c is the protein concentration. R_θ is the Rayleigh excess ratio—the ratio of the intensity of excess light scattered (compared to pure solvent) at an angle θ to that of the incident light intensity (a $\cos\theta$ correction term is necessary if unpolarized light is used). K is an experimental constant dependent on the square of the buffer or solvent refractive index, the square of the refractive index increment (dn/dc in ml/g, analogous to the partial specific volume for proteins, and with a value of about 0.19 ml/g for proteins), and the inverse fourth power of the incident wavelength, λ. The parameter R_g is usually referred to as the 'radius of gyration' of the macromolecule, and is useful for conformation studies (see Section 3). If the macromolecular solute is heterogeneous, M_r will, as with sedimentation-diffusion and sedimentation equilibrium, be a weight average, M_w. Equation 13 is valid for particles of maximum dimension $< \lambda$ (i.e. $M_r < 50 \times 10^6$).

Normally, a double extrapolation to zero scattering angle and to zero protein concentration is necessary, using a procedure known as a Zimm plot (32). However:

(a) For particles of dimensions $< \lambda/20$ (i.e. $M_r < 50\,000$), the angular term in Equation 13 is small (i.e. $\sin^2(\theta/2) \approx 0$) and no angular dependence measurements are in principle necessary to obtain M_r (although this comes at a price: R_g cannot be measured if a conventional light source is used, although it can be measured using electromagnetic radiation of a lower wavelength—namely X-ray and neutron scattering (33)).

(b) More significantly, if the concentration is small enough (< 0.5 mg/ml for proteins and protein assemblies), the concentration term in Equation 13 is small (i.e. $Bc \approx 0$), and only an angular extrapolation is necessary. This is usually the situation with modern photometers designed with a flow cell for coupling on-line to an SEC system (31): after dilution through the column, the effective scattering concentration is usually ≤ 0.5 mg/ml. In these cases, Equation 13 becomes:

$$Kc/R_\theta \approx (1/M_r)\{1 + (16\pi^2 R_g^2/3\lambda^2) \sin^2 [(\theta/2)]\} \qquad [14]$$

For the special case of globular proteins of $M_r < 50\,000$, the term Kc/R_θ is approximately equal to $1/M_r$ and no angular extrapolation is necessary. In this case, a large scattering angle (90°) is normally chosen, since at lower angles the greater noise/signal ratio is much more serious compared with the case for larger scatterers. To a further approximation:

$$R_\theta/Kc \approx M_r\{1-(16\pi^2 R_g^2/3\lambda^2) \sin^2 [(\theta/2)]\}. \qquad [15]$$

2.5.2 SEC-MALLS

An example of a multi-angle laser light scattering photometer (MALLS) coupled to SEC is illustrated in *Figure 12*. The photometer is the DAWN-F system (Wyatt Technology). The angular scattering envelope is measured simultaneously by an array of photodiodes, unlike the moving photomultiplier system used by multi-angle dynamic light scattering photometers (*Figure 5a*). Equations 14 or 15 are used, or Equation 13 if the term Bc is significant and is known. From Equations 13–15, it is clear that it is necessary to have also a concentration detector, as well as the MALLS detector; this is normally a highly sensitive differential refractometer, also equipped with a flow cell (see corner of *Figure 12*).

For proteins, the principle value of this method is that it allows on-line clarification of the material from supramolecular aggregates. The method is, however, most valuable for the analysis of mixtures or for polydisperse heavily glycosylated protein systems such as mucus glycoproteins, since it provides weight-average masses and mass distributions without recourse to calibration standards required by SEC (Section 2.1). *Protocol 5* and *Figure 13* describe the various stages of analysis.

Protocol 5. SEC-MALLS analysis

Equipment

- SEC chromatography apparatus (see *Protocol 1*).[a] A pulse-free HPLC pump is essential. A guard filter upstream is desirable, as is pre-filtering solutions through an appropriate Millipore filter (e.g. 0.22 μm). For the Dawn-F system, a ≈ 100 μl microinjection loop is desirable. A column by-pass option can be installed if fractionation is not required (namely the Zimm plot or full application of Equation 13 is desired for a range of loading concentrations).

- Light scattering photometer, which must be calibrated (but not in a protein standards sense), usually with a strong Rayleigh (i.e. maximum dimension < λ/20) scatterer such as toluene, whose scattering properties are known (31). Calibration is necessary because the ratio of the intensities of the scattered and incident beams is usually very small (≈ 10^{-6}).[b]

Method

1. Determine accurately the delay in eluant volume or time between the light scattering photometer and the concentration (refractive index or UV absorbance) detector, so that the Kc/R_θ term in Equations 13–15 can be synchronized.

2. Record the SEC elution profile using the concentration detector (refractometer) and the light scattering. Only the 90° light scattering signal is shown in *Figure 13*.

3. Subject each elution volume V_e as it passes through the detectors to measurement of Kc/R_θ over the range of the angular scattering envelope. The resulting 'Debye plot' (Equation 15) yields the molar mass of each volume element.

4. Calibrate the column in terms of $\log_{10} M_r$ versus V_e.

5. Determine the weight-average molecular weights (and other derived averages such as the number and z-averages) and plot a relative mass distribution.

6. Determine the refractive increment, dn/dc (35). Its value is normally

Protocol 5. *Continued*

in the range of 0.18 to 1.19 ml/g for proteins, but can be as low as 0.15 ml/g for heavily glycosylated proteins.

---

[a] Choose SEC columns as appropriate. Molecules like the glycoprotein example of *Figure 13* are at the upper limit of resolution by gel columns. For larger particles, other methods of separation, on-line to the MALLS detector based on field flow fractionation are now available (36,37).

[b] For simultaneous multi-angle detection, the detectors have to be normalized to allow for the differing scattering volumes as a function of angle and the differing responses of the detectors. This is usually performed using a solution of a macromolecule of known M_r (generally $\leq$ 50 000) or for a solution of a larger macromolecule whose R_g is known (e.g. T-500 Dextran).

3. Shape measurement

Although the main thrust of this chapter has been on the estimation from hydrodynamic measurements of the molecular weight of a protein in its native state, the hydrodynamic parameters of a protein are also dependent upon the shape of the molecule. For mol. wt measurement, this can be a complication, although it can be overcome by combining the sedimentation and diffusion coefficients (Equation 9), each of which are affected similarly by the shape. Alternatively, transport methods can be avoided altogether by using either of the thermodynamic equilibrium-based techniques of sedimentation equilibrium and classical light scattering.

On the other hand, hydrodynamic methods provide information about the macromolecular shape. There is the complication that the hydrodynamic shape parameters obtained also depend upon the extent of hydration of the protein (i.e. the amount of aqueous solvent chemically bound or physically entrapped), which is very difficult to measure with any real precision. A further problem is that the more complicated the shape model used, the greater the number of independent parameters needed to specify the model uniquely. For example, to specify uniquely the radius of a spherical model requires only one parameter; for the axial ratio of an ellipsoid of revolution (i.e. an ellipsoid with two equal axes), two parameters are needed; for a general triaxial ellipsoid, with three unequal axes, three parameters are needed (38). All of these approaches are known as 'whole-body' approaches. The most complex way of representing shape is 'hydrodynamic bead modelling', where the protein structure is approximated as an array of spherical beads (39). Problems of the uniqueness of any such model are considerable, however, and this form of modelling is best used for choosing between plausible structures (e.g. subunit arrangements in a multisubunit protein, such as the angle between the two Fab arms of an antibody molecule) or for refining a crystallographic or NMR structure to dilute solution conditions. Segmental flexibility can also, in principle, be modelled using this latter approach.

The choice of hydrodynamic shape parameters is wide:

(a) The 'Perrin' frictional ratio (from the sedimentation or diffusion co-efficients).

(b) The various rotational frictional ratios or relaxation times (from fluorescence depolarization or electro-optic measurements).

(c) The viscosity increment (from measurement of the intrinsic viscosity).

(d) The concentration-dependence of the sedimentation coefficient.

(e) The radius of gyration from solution X-ray scattering (or for proteins of $M_r > 50000$, from classical light scattering).

(f) The molecular co-volume of the protein (from measurements of the non-ideality parameter B in osmotic pressure, sedimentation equilibrium, or classical light scattering measurements).

The viscosity and rotational friction parameters are among the more sensitive but can be correspondingly more difficult to measure. The hydration problem is most effectively dealt with by combining two parameters to give 'hydration-independent' shape parameters.

Whereas the extraction of mol. wt information is relatively straightforward, the extraction of shape information is generally not, and the details are outside the scope of this chapter. The interested reader is referred to a recent article that examines in detail the various approaches and provides the necessary references (8). Suffice here to mention some PC software algorithms for hydrodynamic conformation analysis using either the simpler 'whole-body' or the 'hydrodynamic bead' algorithms.

3.1 Computer programs for conformational analysis

For ellipsoid modelling, we have in-house a suite of algorithms that have been transferred from mainframe FORTRAN to PC (BASIC and FORTRAN). ELLIPS1 (40) evaluates the axial ratio for prolate and oblate ellipsoids for a user-specified value of a hydrodynamic parameter. It is based on polynomial approximations to the full hydrodynamic equations, but the accuracy of this approximation is normally well within the precision of the measurement. ELLIPS2 uses the full hydrodynamic equations for general triaxial ellipsoids to specify the set of hydrodynamic parameters for any given value of the axial ratios. ELLIPS3 and ELLIPS4 carry out the reverse procedure, using a variety of graphical combinations of hydration-independent triaxial shape functions. Elsewhere, the routines HYDRO and SOLPRO developed by J. Garcia de la Torre and colleagues (41,42) are particularly useful for the application of bead models; to facilitate its application, a front-end algorithm (A to B) has been constructed to enable TRV to predict the set of hydrodynamic parameters for a given set of crystal structure co-ordinates (O. Byron, PhD dissertation, 1992, University of Nottingham, UK).

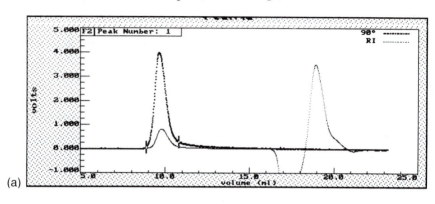

(a)

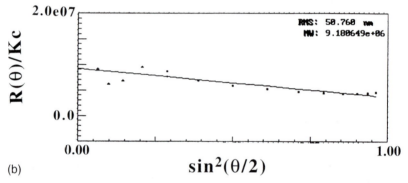

(b)

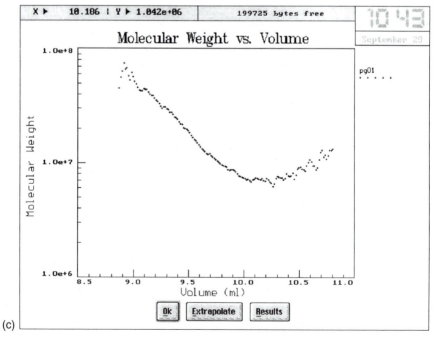

(c)

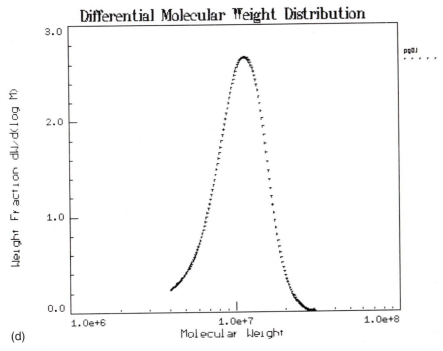

(d)

Figure 13. Extraction of mol. wt distribution of a high mol. wt glycoprotein (a pig gastric mucin preparation '5B1') using SEC-MALLS. (a) Elution profile recorded using the concentration (refractive index) detector (lower profile, lighter dots) and the MALLS detector (only 90° detection shown). The negative and positive peaks at high elution volume correspond to salt elution. (b) 'Debye' plot for a specific value of V_e. (c) Absolute logarithmic calibration plot showing clearly the 'range'of the gel. (d) Mol. wt distribution. The commercial manufacturers software was used for all the analyses: (a–c) ASTRA, (d) EASI. From ref. **34**.

References

1. Ackers, G. (1975). In *The proteins* (3rd edn) (ed. H. Neurath and R. L. Hill), Vol. 1, p. 1. Academic Press, New York.
2. Furth, A. and Moore, R. (1986). *Self assembly of macromolecules*, p. 57. Open University Press, Milton Keynes, UK.
3. Andrews, P. (1965). *Biochem. J.*, **91**, 22.
4. Barth, H. G. (1980). *J. Chromatogr. Sci.*, **18**, 409.
5. Dubin, P. L. and Principi, J. M. (1989). *Div. Polym. Chem. Am. Chem. Soc. Preprints*, **30**, 400.
6. Claes, P., Dunford, M., Kennedy, A., and Vardy, P. (1992). In *Laser light scattering in biochemistry* (ed. S. E. Harding, D. B. Sattelle, and V. A. Bloomfield), p. 66. Royal Society of Chemistry, Cambridge, UK.
7. Wells, C., Molina-Garcia, A. D., Harding, S. E., and Rowe, A. J. (1990). *J. Muscle Res. Cell Motil.*, **11**, 344.

8. Harding, S. E. (1995). *Biophys. Chem.*, **55**, 69.
9. Sanders, A. H. and Cannell, D. S. (1980). In *Light scattering in liquids and macromolecular solutions* (ed. V. Degiorgio, M. Corti, and M. Giglio), p. 173. Plenum, New York.
10. Burchard, W. (1992). In *Laser light scattering in biochemistry* (ed. S. E. Harding, D. B. Sattelle, and V. A. Bloomfield), p. 3. Royal Society of Chemistry, Cambridge, UK.
11. Van Holde, K. E. (1985). *Physical biochemistry* (2nd edn), p. 110. Prentice Hall, Englewood Cliffs, New Jersey.
12. Pusey, P. N. (1974). In *Photon correlation and light beating spectroscopy* (ed. H. Z. Cummins and E. R. Pike), p. 387. Plenum, New York.
13. Johnsen, R. M. and Brown, W. (1992). In *Laser light scattering in biochemistry* (ed. S. E. Harding, D. B. Sattelle, and V. A. Bloomfield), p. 77. Royal Society of Chemistry, Cambridge, UK.
14. Harding, S. E. (1986). *Biotech. Appl. Biochem.*, **8**, 489.
15. Langley, K. H. (1992). In *Laser light scattering in biochemistry* (ed. S. E. Harding, D. B. Sattelle, and V. A. Bloomfield), p. 151. Royal Society of Chemistry, Cambridge, UK.
16. Svedberg, T. and Pedersen, K. O. (1940). *The ultracentrifuge*. Oxford University Press.
17. Giebler, R. (1992). In *Analytical ultracentrifugation in biochemistry and polymer science* (ed. S. E. Harding, A. J. Rowe, and J. C. Horton), p. 16. Royal Society of Chemistry, Cambridge, UK.
18. Harding, S. E., Rowe, A. J., and Horton, J. C. (ed.) (1992). *Analytical ultracentrifugation in biochemistry and polymer science*. Royal Society of Chemistry, Cambridge, UK.
19. Schuster, T. M. and Laue, T. M. (1994). *Modern analytical ultracentrifugation*. Birkhäuser, Boston.
20. Kratky, O., Leopold, H., and Stabinger, H. (1973). In *Methods in enzymology* (ed. C. H. W. Hirs and S. N. Timasheff), Vol. 27D, p. 98. Academic Press, New York.
21. Tombs, M. P. and Peacocke, A. R. (1974). *The osmotic pressure of biological macromolecules*. Oxford University Press, Oxford.
22. Rowe, A. J. (1992). In *Analytical ultracentrifugation in biochemistry and polymer science* (ed. S. E. Harding, A. J. Rowe, and J. C. Horton), p. 394. Royal Society of Chemistry, Cambridge, UK.
23. Gilbert, L. M. and Gilbert, G. A. (1973). In *Methods in enzymology* (ed. C. H. W. Hirs and S. N. Timasheff), Vol. 27, p. 273. Academic Press, New York.
24. Schachman, H. K. (1989). *Nature*, **341**, 259.
25. Creeth, J. M. and Harding, S. E. (1982). *J. Biochem. Biophys. Methods*, **7**, 25.
26. Harding, S. E., Horton, J. C., and Morgan, P. J. (1992). In *Analytical ultracentrifugation in biochemistry and polymer science* (ed. S. E. Harding, A. J. Rowe, and J. C. Horton), p. 275. Royal Society of Chemistry, Cambridge, UK.
27. Correia, J. J. and Yphantis, D. A. (1992). In *Analytical ultracentrifugation in biochemistry and polymer science* (ed. S. E. Harding, A. J. Rowe, and J. C. Horton), p. 231. Royal Society of Chemistry, Cambridge, UK.
28. Teller, D. C. (1973). In *Methods in enzymology* (ed. C. H. W. Hirs and S. N. Timasheff), Vol. 27D, p. 346. Academic Press, New York.
29. Joss, L. A. and Ralston, D. B. (1995). *Anal Biochem.*, in press.

30. Harding, S. E. and Johnson, P. (1985). *Biochem. J.*, **231**, 549.
31. Wyatt, P. J. (1992). In *Laser light scattering in biochemistry* (ed. S. E. Harding, D. B. Sattelle, and V. A. Bloomfield), p. 35. Royal Society of Chemistry, Cambridge, UK.
32. Tanford, C. (1961). *Physical chemistry of macromolecules*. J. Wiley & Sons, New York.
33. Perkins, S. J. (1994). In *Microscopy, optical spectroscopy and macroscopic techniques* (ed. C. Jones, B. Mulloy, and A. H. Thomas), p. 39. Humana Press, New Jersey.
34. Jumel, K., Fiebrig, I., and Harding, S.E. (1995). *Int. J. Biol. Macromol.*, **18**, 133.
35. Huglin, M. B. (ed.) (1972). *Light scattering from polymer solutions*. Academic Press, New York.
36. Arner, E. C. and Kirkland, J. J. (1992). In *Analytical ultracentrifugation in biochemistry and polymer science* (ed. S. E. Harding, A. J. Rowe, and J. C. Horton), p. 209. Royal Society of Chemistry, Cambridge, UK.
37. Adophi, U. and Kulicke, W. M. (1996). *Polymer*, in press.
38. Harding, S. E. (1989). In *Dynamic properties of biomolecular assemblies* (ed. S. E. Harding and A. J. Rowe), p. 32. Royal Society of Chemistry, Cambridge, UK.
39. Garcia de la Torre, J. (1989). In *Dynamic properties of biomolecular assemblies* (ed. S. E. Harding and A. J. Rowe), p. 3. Royal Society of Chemistry, Cambridge, UK.
40. Harding, S. E., Horton, J. C., and Cölfen, H. (1996). *Eur. Biophys. J.*, in press.
41. Garcia de la Torre, J., Navarro, S., Lopez-Martinez, M. C., Diaz, F. G., and Lopez-Cascoles, J. J. (1994). *Biophys. J.*, **67**, 530.
42. Garcia de la Torre, J., Carrasco, B., and Harding, S. E. (1996). *Eur. Biophys. J.*, in press.

10

How to determine the molar absorbance coefficient of a protein

C. NICK PACE and FRANZ X. SCHMID

1. Introduction

It is frequently essential to know the concentration of a protein solution. Protein concentrations are most conveniently and accurately measured by absorbance spectroscopy. The absorbance A is a linear function of the molar concentration c according to the Beer–Lambert relation:

$$A = \varepsilon \times c \times l \qquad [1]$$

where ε is the molar absorbance coefficient, and l is the cell length. Obviously, a value of ε is required to determine c from a measurement of A. This article describes two simple but reliable methods to estimate ε for a globular protein lacking any prosthetic groups.

The absorbance of a protein in the aromatic region between 270 nm and 300 nm is due to only three chromophores: tryptophan (Trp), tyrosine (Tyr), and, to a smaller extent, cystine (i.e. disulfide bonds). The absorbances of Trp and Tyr depend on the microenvironment of their chromophores, and they are slightly red-shifted when transferred from a polar to a non-polar environment (1). As a consequence, in native proteins, the residues that are exposed to solvent and those that are buried will contribute differently to the absorbance coefficient. This is taken into account in different ways in the two methods described here for determining ε.

The first method involves calculations only, and it makes use of the average ε values for the three chromophores in native proteins determined from a statistical analysis of a large set of native proteins with known ε values (2,3). This method works best for proteins that contain at least one Trp residue. The second method requires measuring the absorbance of two solutions with identical protein concentrations: one containing just buffer and one containing buffer plus 6 M GdmCl. This method is sometimes referred to as the Edelhoch method because it uses model compound data from a paper by Edelhoch (1). It is described in more detail by Gill and von Hippel (4) and by

Pace *et al.* (3). This method requires only a small amount of protein and time. Since it is more reliable than the first method, it should be used whenever an accurate ε value is essential.

2. Absorbance coefficients for tryptophan, tyrosine, and the disulfide bond

Only Trp, Tyr, and disulfide bonds contribute to the absorbance of a protein between 270 nm and 300 nm. Absorbance coefficients for these chromophores are compiled in *Table 1*. They are used in the following sections in the procedures to determine the absorbance coefficients of proteins.

3. Calculation of absorbance coefficients from protein reference data

The molar absorbance coefficient of a protein at 280 nm, ε_{280}, is calculated by using Equation 2.

$$\varepsilon_{280} \,[\text{M}^{-1}\text{cm}^{-1}] = 5500 \times n_{\text{Trp}} + 1490 \times n_{\text{Tyr}} + 125 \times n_{\text{SS}} \qquad [2]$$

The molar absorbances from ref. 3 for Trp, Tyr, and the disulfide chromophore (first column of *Table 1*) are used in Equation 2. Alternatively the values from Mach *et al.* (2) can be used. They give ε_{280} values of 5540 $\text{M}^{-1}\text{cm}^{-1}$, 1480 $\text{M}^{-1}\text{cm}^{-1}$, and 134 $\text{M}^{-1}\text{cm}^{-1}$ for Trp, Tyr, and disulfides, respectively. The Trp and Tyr values differ by less than 1% from the values in *Table 1*. The individual steps describing how to calculate ε_{280} for a protein are given in *Protocol 1*. A detailed account of the method and the set of reference data are given in ref. 3.

Table 1. ε Values of Trp, Tyr, and disulfides in the aromatic region used to determine the ε value of a protein in the native and in the denatured state

Chromophore	Average values of ε_{280} ($\text{M}^{-1}\text{cm}^{-1}$) in native proteins[a]	ε Values ($\text{M}^{-1}\text{cm}^{-1}$) for model compounds in 6.0 M GdmCl at wavelength (nm)[b]				
		276	278	279	280	282
Tryptophan	1490	1450	1400	1345	1280	1200
Tyrosine	5500	5400	5600	5660	5690	5600
Cystine	125	145	127	120	120	100

[a] Data taken from ref. 3.
[b] Data for the model compounds *N*-acetyl-L-tryptophanamide, Gly-L-Tyr-Gly, and for cystine in 6.0 M GdmCl, 0.02 M phosphate buffer pH 6.5. The values are taken from *Table 1* in ref. 4. The original data are from ref. 1.

Protocol 1. Calculation of the absorbance coefficient of a protein

1. Confirm that the protein does not contain a tightly bound ligand or prosthetic group that absorbs between 250 nm and 300 nm.[a]

2. Count the number of Trp residues (n_{Trp}) and of Tyr residues (n_{Tyr}) in the protein sequence.

3. Count the number of disulfide bonds. If the disulfide bonding is unknown, make the following assumptions. For cytosolic proteins the number of disulfide bonds (n_{SS}) is zero. For secretory proteins $n_{SS} = n_{Cys}/2$.[b]

4. Use Equation 2 and the numbers from steps 2 and 3 to calculate ε_{280}.

---

[a] In the presence of such a ligand do not use this method. A comparison of the protein spectrum with the absorbance spectra of Tyr and Trp (see Chapter 11) helps to detect non-protein contributions. Tightly-bound nucleotides lead to a decrease in the ratio A_{280}/A_{260}.
[b] Disulfide bonds typically occur only in secreted proteins and often most or all of the thiol groups are engaged in disulfide bonds. Since the molar absorbance of the disulfide bond is small (*Table 1*), errors in the number of disulfide bonds hardly affect the calculated absorbance coefficients.

---

4. Determination of the absorbance coefficient for the unfolded and the folded protein

In this experimental procedure (3,4), the ε value of the denatured protein in 6.0 M GdmCl (ε_{den}) is calculated from the number of tryptophan, tyrosine, and cystine residues, by using the ε values of these chromophores in model compounds (also determined in 6.0 M GdmCl) and Equation 2. Edelhoch (1) showed that the model compound data from 6.0 M GdmCl could accurately reproduce the absorbance spectrum of an unfolded protein in 6.0 M GdmCl. The ε values for the reference compounds are given in *Table 1* for five wavelengths between 276 nm and 282 nm to calculate the ε_{den} value of the unfolded protein either at 280 nm or, preferably, at the absorbance maximum of the native protein. To determine ε value of the native protein (ε_{nat}), solutions of the native and of the unfolded protein with identical concentrations are prepared, and their absorbances, A_{nat} and A_{den}, are measured. Since the concentrations of the native and of the denatured protein are the same, ε_{nat} and ε_{den} are related by Equation 3a (compare with Equation 1), and ε_{nat} is calculated readily from Equation 3b.

$$A_{nat}/A_{den} = \varepsilon_{nat}/\varepsilon_{den} \tag{3a}$$

$$\varepsilon_{nat} = \varepsilon_{den} \times (A_{nat}/A_{den}) \tag{3b}$$

C. Nick Pace and Franz X. Schmid

Protocol 2 outlines the individual steps of the second method for determining the absorbance coefficient of the native protein. It is described in more detail in refs 3 and 4.

Protocol 2. Determination of absorbance coefficients for the unfolded and the native protein

Equipment and reagents

- Double-beam absorbance spectrophotometer
- Two quartz cells, 10 mm long and 4 mm wide[a]
- Buffer A: 0.04 M K phosphate buffer pH 6.5
- Buffer B: 0.02 M K phosphate buffer pH 6.5 (can be made from buffer A by dilution)
- Stock solution with about 1 mg protein in 0.25 ml of buffer B

- 6.6 M GdmCl in buffer A[b] (GdmCl should be 'ultrapure'). Warm gently to dissolve the GdmCl crystals completely. Determine the exact concentration of GdmCl from refractive index measurements (see Chapter 11). Filter all solvents through 0.45 μm filters to remove dust particles.

A. *Preparing the spectrophotometer*

1. Turn the spectrometer and the deuterium lamp on, warm-up for about 30 min.

2. Consult Chapter 11 to check the performance of the spectrometer (if necessary). Thermostatting is not required when room temperature is between 20°C and 30°C.

3. Clean the cells thoroughly and dry them (consult Chapter 11). Do not use organic solvents for cleaning; they may contain UV-absorbing contaminants.

B. *Spectrum of the native protein*

1. Add 910 μl of buffer B to both the sample and the reference cells and scan the absorbance between 360 nm and 240 nm. This baseline should be flat or should deviate by less than 0.005 from the instrument's baseline (with the cells removed). If not, repeat cleaning of the cells.

2. Record or (if possible) store the baseline. In addition, measure individually the absorbance values at 276, 278, 279, 280, and 282 nm.

3. Now add 90 μl of the protein stock solution to the sample cell and 90 μl of buffer B to the reference cell. Mix the solutions before recording the spectrum between 360 nm and 240 nm. The maximum absorbance near 280 nm should be between 0.2 and 1.0, and the absorbance between 320 nm and 360 nm should be zero.[c]

4. Store the spectrum and repeat the absorbance measurement to check the reproducibility.

5. Subtract the baseline.

6. Measure individually the absorbance values at 276, 278, 279, 280, and 282 nm and subtract the respective values of the buffer.

C. *Spectrum of the denatured protein*

1. Add 910 μl of 6.6 M GdmCl to both the sample and the reference cell and scan the absorbance between 360 nm and 240 nm.

2. Store the baseline (or record on the chart) and then measure the individual absorbance values at 276, 278, 279, 280, and 282 nm.[d]

3. Add 90 μl of the same protein stock solution as in part B to the sample cell and 90 μl of buffer B to the reference cell and mix[e] the solutions.

4. Record the spectrum between 360 nm and 240 nm.

5. Store the spectrum, mix the solution again, and repeat the measurement to check for reproducibility and completeness of the unfolding process.

6. Subtract the GdmCl baseline.

7. Measure individually the absorbance values at 276, 278, 279, 280, and 282 nm and subtract the respective values of the GdmCl solution.

8. The procedure should be repeated at least once on a fresh stock solution of protein to check the reliability of the procedure. Careful pipetting is essential.

D. *Data analysis*

1. Calculate the ε_{den} value for the denatured protein. Count the number of tyrosine, tryptophan, and cystine residues in the protein and calculate the values for ε_{den} between 276 nm and 282 nm as outlined in *Protocol 1*, but now using the ε values of these residues at 276–282 nm in 6.0 M GdmCl (*Table 1*).

2. Check for light scattering. In the absence of light scattering, the protein absorbance should be zero above 320 nm, and should not change between 320 nm and 360 nm. A sloping baseline between 320 nm and 360 nm is indicative of light scattering. The contribution of light scattering to the apparent absorbance at 276–282 nm can be obtained by plotting $\log_{10} A$ as a function of $\log_{10}$ wavelength between 320 nm and 360 nm and making a linear extrapolation to the 276–282 nm region. Subtract the resulting values from A_{nat} and A_{den} as measured in this region.

Protocol 2. *Continued*

3. Correct the measured absorbance values according to step 2, if necessary. Then determine the ratios A_{nat}/A_{den} at 276–282 nm and use Equation 3b to calculate ε_{nat} in this wavelength range.

[a] Check whether 900 µl in the cell are sufficient for the light beam. The height of the light beam can be examined visually in the visible range (e.g. near 500 nm) by inserting a strip of white paper into the cell. When 10 × 10 mm cells are used, all volumes have to be multiplied by a factor of 2.5.
[b] A 0.04 M buffer is used, because in a 6.6 M solution GdmCl occupies about half of the volume of the solution. Therefore the final buffer concentration is approx. 0.02 M. If an exact buffer concentration is required, weigh in the desired amounts of GdmCl and buffer and dissolve them in water to the final volume.
[c] If A_{280} is outside this range, make a new protein stock solution, such that A_{280} is about 0.5. Sloping baselines above 320 nm can occur as a result of light scattering, either from dust particles or from the protein itself, if it has a molecular mass greater than about 100 kDa. Corrections for light scattering are described in the data analysis section.
[d] In this procedure, the baseline is measured for 6.6 M GdmCl and not for 6.0 M. This simplifies the procedure; the introduced error is negligible.
[e] Thorough mixing is essential. Otherwise, schlieren patterns can strongly distort the measured absorbance.

5. Concluding remarks

The first method (*Protocol 1*) is simple and fast. Nevertheless, it is reliable, and in a test with 80 proteins it gave an average deviation from measured values of less than 4% (3). It works best for globular proteins that contain at least one Trp residue and show an average distribution of buried and exposed aromatic residues.

In the second method (*Protocol 2*) the absorbances of the native and the denatured protein are measured and thus deviations from the average exposure of the aromatic groups are accounted for experimentally. The second method should be used in addition to the first one when accurate ε values are of crucial importance.

Both methods are less reliable for proteins that contain no Trp residues, only Tyr. Unlike Trp, exposed and buried Tyr residues differ substantially in absorbance at 280 nm (see Chapter 11), the wavelength commonly used for measuring protein concentrations. For such proteins we recommend that the absorbance coefficient be determined not at 280 nm, but at their absorption maxima (275–277 nm), where the absorbances of buried and of exposed Tyr residues are almost identical. The two methods also perform less satisfactorily for very small proteins with an abnormally high content of disulfide bonds (such as insulin). For the rare proteins that contain neither Trp or Tyr residues, the method of Scopes (5) is recommended.

The two methods described in this chapter are much more reliable and much more accurate than the various colorimetric methods, such as the biuret and the Lowry procedures (6), which are commonly used to determine protein concentrations.

References

1. Edelhoch, H. (1967). *Biochemistry*, **6**, 1948.
2. Mach, H., Middaugh, C. R., and Lewis, R. V. (1992). *Anal. Biochem.*, **200**, 4.
3. Pace, C. N., Vajdos, F., Fee, L., Grimsley, G., and Gray, T. (1995). *Protein Sci.*, **4**, 2411.
4. Gill, S. C. and von Hippel, P. H. (1989). *Anal. Biochem.*, **182**, 19.
5. Scopes, R. K. (1974). *Anal. Biochem.*, **59**, 277.
6. Harris, D. A. (1987). In *Spectrophotometry and spectrofluorimetry: a practical approach* (ed. D. A. Harris and C. L. Bashford), p. 49. IRL Press, Oxford.

Optical spectroscopy to characterize protein conformation and conformational changes

FRANZ X. SCHMID

1. Introduction

Proteins absorb light and emit radiation in the UV range of the spectrum. The absorbance is caused by the peptide groups, by the aromatic amino acids, and to a small extent, by disulfide bonds. Fluorescence emission originates from the aromatic amino acids. Some proteins that carry covalently linked cofactors, such as the haem proteins, also show absorbance in the visible range. Proteins also absorb infrared light. The latter two properties will not be dealt with in this chapter. During absorption, light energy is used to promote electrons from the ground state to an excited state. Electrons that participate in delocalized aromatic systems frequently absorb in the near-UV or the visible region. Fluorescence emission is observed when excited electrons revert from the first excited state to the ground state. When a chromophore is part of an asymmetric structure, or when it is immobilized within an asymmetric environment, left-handed and right-handed circularly-polarized light is absorbed to different extents. This phenomenon is called circular dichroism (CD).

The spectral properties of a protein molecule depend upon the molecular environments and upon the mobilities of its chromophores and give information about those aspects. Spectroscopical measurements can be carried out in solution, and only minute amounts of material are required for analysis. The methods are non-destructive, and the samples can be recovered afterwards.

Spectroscopic methods are the methods of choice to investigate changes in the behaviour of a protein under different solvent conditions, and also to compare the properties of related molecules, such as homologous or mutated forms of a protein. In addition, they are widely used to determine protein stability and to follow structural transitions such as unfolding and refolding under a variety of conditions (see Chapter 12). By using rapid mixing techniques, reactions that occur in the millisecond range can be monitored.

Information about the secondary and tertiary structures and about structural changes can be retrieved from the spectral properties. The CD in the amide region is used to estimate the content of secondary structure. The absorbance and fluorescence properties of the aromatic residues depend predominantly on the local environments of these chromophores. Both properties are different in aqueous versus non-aqueous media. Therefore, they are useful to probe the hydrophilic or hydrophobic character of the structure around these chromophores. The CD of the aromatic side chains is sensitive to their precise orientation in a correctly folded protein. It is therefore a good probe for the formation of the native spatial structure.

In this practical guide to spectroscopy of proteins, experimental techniques are described for the measurement of absorbance, fluorescence, and CD. A brief introduction into the spectral properties of amino acids and proteins in different environments is followed by practical aspects and by protocols for performing spectroscopic measurements.

2. Absorbance

Excellent introductions to the fundamental aspects of protein absorbance are available (1–3). Absorbance measurements are most commonly used to determine the concentration of biological macromolecules in solution. Spectrophotometers are standard laboratory equipment, and absorbance can be measured quickly and accurately. The absorbance (A) is related to the intensity of the light before (I_0) and after (I) passage through the protein solution by Equation 1a, and the absorbance depends linearly on concentration, according to the Beer–Lambert relationship (Equation 1b):

$$A = -\log_{10}(I/I_0) \qquad [1a]$$

$$A = \varepsilon \times c \times l \qquad [1b]$$

where c is the molar concentration, l is the pathlength in cm, and ε is the molar absorbance coefficient. The value of ε can be determined by a number of experimental techniques (4) or can be calculated for proteins by adding up the contributions of the constituent aromatic amino acids of a protein (1,5) (see *Chapter 10*).

Apart from small contributions of the peptide bonds, which absorb strongly only below 230 nm, the absorbance of proteins in the 230–300 nm range is caused by the aromatic side chains of tyrosine, tryptophan, and phenylalanine residues. Disulfide bonds display a weak absorbance band around 250 nm. Spectra of the aromatic amino acids are shown in *Figure 1*; their molar absorbance at the respective maxima in the near-UV are compared in *Table 1*. The three aromatic amino acids contribute to the absorbance of a protein to different extents. The molar absorbance of phenylalanine ($\lambda_{max} = 257$ nm) is smaller by an order of magnitude than that of tyrosine and tryptophan, and

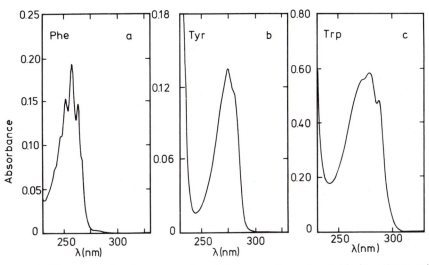

Figure 1. Ultraviolet absorbance spectra of the aromatic amino acids in a 1 cm cell in 0.01 M potassium phosphate buffer pH 7.0 25°C. (a) 1 mM phenylalanine; (b) 0.1 mM tyrosine; (c) 0.1 mM tryptophan.

it is virtually zero at > 270 nm. The spectrum of a protein is therefore dominated by the contributions of Tyr and Trp residues. The Phe residues primarily contribute 'wiggles' in the 250–260 nm region, and they influence the ratio A_{260}/A_{280}, which is often used to identify contaminating nucleic acids in protein preparations. The shape and intensity of a particular protein spectrum depends on the actual number of the three aromatic residues in the molecule. Since the aromatic amino acids do not absorb above 310 nm, solutions containing only protein should not show absorbance at longer wavelengths. Proteins that do not contain Trp residues do not absorb at > 300 nm (see *Figure 2*). A sloping baseline in the 310–400 nm region usually originates from light scattering when large particles, such as aggregates, are present in the solution. The contribution of light scattering to the measured absorbance

Table 1. Absorbance and fluorescence properties of the aromatic amino acids[a]

Amino acid	Absorbance		Fluorescence		Sensitivity
	λ_{max} (nm)	ε_{max} (M^{-1}cm^{-1})	λ_{max} (nm)	φ_F[b]	$\varepsilon_{max} \times \varphi_F$[b] (M^{-1}cm^{-1})
Tryptophan	280	5600	355	0.13	730
Tyrosine	275	1400	304	0.14	200
Phenylalanine	258	200	282	0.02	4

[a] In water at neutral pH; data are from ref. 10.
[b] φ_F, Fluorescence quantum yield.

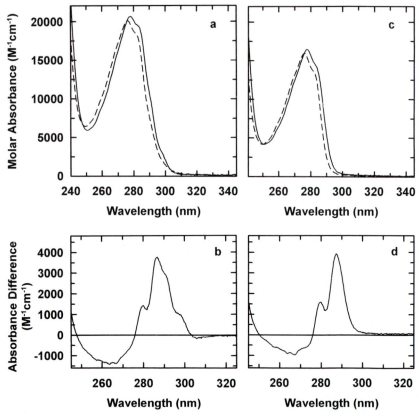

Figure 2. Ultraviolet absorbance specta of (a) the wild-type form and (c) the Trp59Tyr variant of RNase T1. The spectra of the native proteins (in 0.1 M sodium acetate pH 5.0) are shown by the continuous lines. The spectra of the unfolded proteins (in 6.0 M GdmCl in the same buffer) are shown by broken lines. The difference specta between the native and unfolded forms are shown in (b) and (d). Spectra of 15 μM protein samples were measured at 25°C in 1 cm cuvettes in a double-beam instrument with a band width of 1 nm at 25°C. The spectra of the native and unfolded proteins were recorded successively, stored and subtracted.

can be obtained by plotting $\log_{10} A$ as a function of $\log_{10}$ wavelength above 310 nm and extrapolating linearly to shorter wavelengths (1). The absorbance spectra of the wild-type and the Trp59Tyr variant of RNase T1 are shown as examples in *Figure 2a,b*. Wild-type RNase T1 contains one Trp and nine Tyr residues. In the Trp59Tyr variant, the single Trp residue is replaced, so this protein contains ten Tyr residues and no Trp. Accordingly, the absorbance of the Trp59Tyr variant is significantly decreased in the 280–290 nm region (*Figure 2c*), and the absorbance is virtually zero at 300 nm.

The absorbance spectra of the aromatic amino acids depend on their molecular environment, which can produce a broadening of bands, shifts in

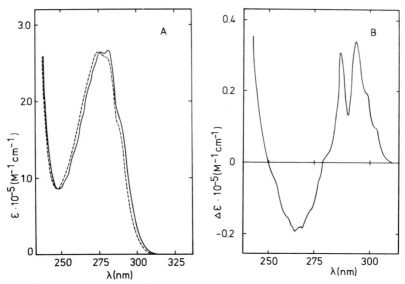

Figure 3. Absorbance spectra of cytoplasmic invertase from yeast. (A) Spectra in the native (—) and in the unfolded states (– – –) at 20°C. Invertase is a large, dimeric protein (M_r 58000 per subunit) and contains 31 Tyr and 16 Trp residues. The native protein (50 μg/ml) was in 0.05 M sodium acetate pH 5.0. The unfolded invertase was in the same buffer plus 6.0 M GdmCl. (B) Difference spectrum between native and unfolded invertase. Native invertase was in the sample cell, and unfolded invertase was in the reference cell. Tandem cells were used for the experiment, as described in the text.

wavelength, and overall changes in intensity. In general, the shift in wavelength predominates: a blue-shift of the spectrum is observed when the polarity of the solvent increases. For example, the maximum of the absorbance of the phenol chromophore is blue-shifted by about 3 nm when the solvent is changed from CCl_4 to water (6). There is a good correlation between this decrease in λ_{max} and the decrease in refractive index of the solvent. The spectral shift leads to peaks in the difference spectra in the descending slope of the original spectrum, that is in the 285–288 nm region for Tyr and around 291–294 nm for Trp (6).

In folded native proteins, some of the aromatic residues are buried in the hydrophobic core of the molecule. Their absorbance spectra are blue-shifted when they become exposed to the aqueous solvent upon unfolding. This is clearly seen for the two forms of RNase T1 in *Figure 2a,c*. The difference spectra in *Figure 2b,d* show that the maximal differences in absorbance occur in the 285–295 nm region. The difference spectrum of the Trp59Tyr variant (*Figure 2d*) shows a prominent maximum at 287 nm, which is typical for proteins that contain Tyr residues only. In the wild-type protein (*Figure 2b*), additional shoulders are observed between 290 nm and 300 nm, which originate from the single Trp residue. The difference spectrum of invertase (*Figure 3*), a large protein with 31 Tyr and 16 Trp residues, displays major peaks near 285 nm (from the Tyr residues) and near 292 nm (from the Trp residues). The differences in the absorbance spectra between the native and the unfolded states of a protein are generally small, but they can be determined with good accuracy (Section 2.4). They are linear with protein concentration and are extremely useful for monitoring conformational changes of a protein.

The folded conformation of a protein can be perturbed in various ways. Widely used are unfolding by heat or by destabilizing agents such as urea or guanidinium chloride (GdmCl). It is important to keep in mind that these variations also influence the spectral properties of the protein solution. With increasing temperature, the refractive index decreases slightly, the solute concentration decreases due to the thermal expansion of the solution, and the ionization of dissociable groups can change. Together, these effects influence protein absorbance only to a minor extent; therefore it is found to be almost independent of temperature in the absence of structural transitions. In contrast, the absorbance of Tyr and Trp at 287 nm and 291 nm, respectively, increase substantially when denaturants are added, even in the absence of structural transitions (*Figures 4* and 5). The observed increases in absorbance at 287 nm and 291 nm reflect a slight red-shift of the spectrum, which originates from the change in refractive index (i.e. the polarity) of the solvent with the concentration of GdmCl or urea. Similar effects are also observed when other denaturants are employed. Consequently, even in the absence of structural changes, protein absorbance is expected to increase with denaturant concentration because of effects on those chromophores that are already exposed to the solvent in the folded protein. The increases shown in *Figures 4*

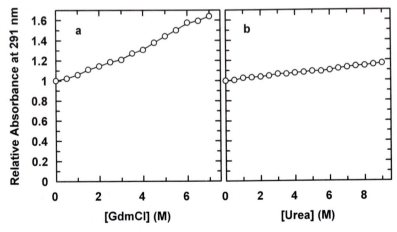

Figure 4. Dependence of the absorbance of tryptophan at 291 nm on (a) GdmCl concentration and (b) urea concentration. Measurements were carried out at a band width of 1 nm employing 0.01 mg/ml solutions the amino acid tryptophan in 0.1 M potassium phosphate buffer pH 7.0 at 25°C. Absorbance values are given relative to the denaturant-free solution.

and 5 occur in the steeply descending slopes of the absorbance spectra (compare with *Figures 1* and *2*) and can depend on the instrumental settings (in particular on the band width).

2.1 Spectrophotometers

For the measurement of difference spectra, it is advantageous to use a double-beam instrument with high sensitivity and an expanded absorbance scale, so that small differences between samples in different environments can be monitored with good accuracy. The development of microprocessor controlled instruments has greatly facilitated the measurement of difference spectra. The measured data can be stored and manipulated later, for example by changing the wavelength and absorbance scales, smoothing the data, addition and subtraction of spectra, and calculation of derivative spectra. Also, difference spectra can be compared with other data already in the computer memory.

Microprocessor controlled single-beam spectrophotometers with data storage facilities are also suitable for the measurement of difference spectra. These instruments must display a high signal stability, however, so that the spectra to be compared can be recorded sequentially and then subtracted.

Diode array spectrophotometers are the instruments of choice when a high spectral resolution is not required. The cell compartment remains freely accessible during the measurements, and changes in absorbance can be measured simultaneously at multiple wavelengths. This is of advantage for measuring difference spectra (e.g. as a function of time). In addition, measuring the absorbance at two wavelengths (e.g. between the absorbances at the

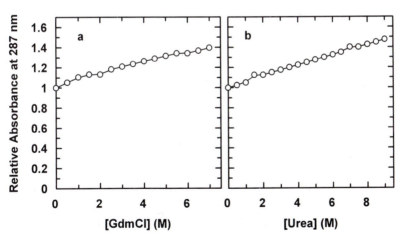

Figure 5. Dependence of the absorbance of tyrosine at 287 nm on (a) GdmCl concentration and (b) urea concentration (measured at a band width of 1 nm). The concentration of the free amino acid tyrosine was 0.04 mg/ml in 0.1 M potassium phosphate buffer pH 7.0 at 25°C. Absorbance values are given relative to the denaturant-free solution.

Table 2. Calibration of the absorbance scale of spectrophotometers with potassium chromate[a]

Wavelength (nm)	Absorbance in a 1 cm cell	Molar absorbance, ε ($M^{-1}cm^{-1}$)
220	0.446	2167
240	0.295	1433
260	0.633	3072
280	0.712	3459
300	0.149	729
320	0.064	308
370	0.987	4795
420	0.124	604
500	0	0

[a] Standard absorbance values for a 0.04 g/litre solution of potassium chromate in 0.05 M KOH at 25 °C are given, together with the respective molar absorbance. The potassium chromate solution is prepared either by dissolving 0.04 g of K_2CrO_4/litre in 0.05 M KOH, or 0.0303 g of $K_2Cr_2O_7$/litre in 0.05 M KOH. The latter may be preferable, because $K_2Cr_2O_7$ can be obtained in a purer state. The absorbance is measured against distilled water as reference (32).

wavelength of maximal difference and at an isosbestic point) exactly at the same time and subtracting them from one another removes all correlated noise, so the signal-to-noise ratio is strongly improved. For a further discussion of spectrophotometers consult ref. 7.

The long-term stability of spectrophotometers must be examined routinely. The wavelength accuracy and the linearity of the absorbance scale should also be checked periodically. Potassium chromate solutions can be used for this purpose. Standard absorbance values at a number of wavelengths in the UV region useful for the calibration of the absorbance scale are given in *Table 2*.

2.2 Samples and cuvettes

2.2.1 Buffers

Good buffers for spectroscopic measurements should ideally not absorb light in the wavelength range of the experiment. For work in the near-UV, buffer absorbance should be small above 230 nm. This criterion is met by most of the commonly used aqueous solvents (see *Table 3*). Buffer absorbance does become a major problem, however, in the far-UV region below 220 nm. Most buffers that contain carboxyl and/or amino groups show considerable absorbance in this wavelength range. Buffers that are fairly transparent in the far-UV include phosphate, cacodylate, and borate (*Table 3*).

Some buffers, such as acetate or glycine, are excellent substrates for micro-organisms. Therefore, these buffers should be kept in the refrigerator, but not for an extended period of time.

Table 3. Absorbance of various salt and buffer substances in the far-UV region[a]

Compound	No absorbance above:		Absorbance of a 0.01 M solution in a 0.1 cm cell at:			
			210nm	200nm	190nm	180nm
NaClO$_4$		170 nm	0	0	0	0
NaF, KF		170 nm	0	0	0	0
Boric acid		180 nm	0	0	0	0
NaCl		205 nm	0	0.02	> 0.5	0.5
Na$_2$HPO$_4$		210 nm	0	0.05	0.3	> 0.5
NaH$_2$PO$_4$		195 nm	0	0	0.01	0.15
Na acetate		220 nm	0.03	0.17	> 0.5	> 0.5
Glycine		220 nm	0.03	0.1	> 0.5	> 0.5
Diethylamine		240 nm	0.4	> 0.5	> 0.5	> 0.5
NaOH	pH 12	230 nm	≥ 0.5	> 2	> 2	> 2
Boric acid, NaOH	pH 9.1	200 nm	0	0	0.09	0.3
Tricine	pH 8.5	230 nm	0.22	0.44	> 0.5	> 0.5
Tris	pH 8.0	220 nm	0.02	0.13	0.24	> 0.5
Hepes	pH 7.5	230 nm	0.37	0.5	> 0.5	> 0.5
Pipes	pH 7.0	230 nm	0.20	0.49	0.29	> 0.5
Mops	pH 7.0	230 nm	0.10	0.34	0.28	> 0.5
Mes	pH 6.0	230 nm	0.07	0.29	0.29	> 0.5
Cacodylate	pH 6.0	210 nm	0.01	0.20	0.22	> 0.5

[a] Buffers were titrated with 1 M NaOH or 0.5 M H$_2$SO$_4$ to the indicated pH values.

2.2.2 Other solvents

Most frequently, GdmCl or urea are employed to unfold proteins, and very high concentrations of these denaturants are required to destabilize protein structures. The purity of these unfolding agents is of utmost importance.

(a) Impurities can lead to irreversible modifications of the unfolded protein chains. The denaturants are present typically in the 1–10 M range, whereas the protein concentration usually is in the range of 10^{-4} to 10^{-7} M. This implies that even impurities present in the denaturant in the 1–10 p.p.m. range are present at a concentration comparable to that of the protein.

(b) Impurities may lead to UV absorbance in the range of protein absorbance and thereby affect the measured difference spectra. Urea and GdmCl are both available in high purity forms ('ultrapure', e.g. from ICN Chemicals). References for purification are given in Chapter 12. Solutions of GdmCl are fairly stable and can be used for several days. Urea slowly decomposes into cyanate and ammonium ions (8). As cyanate can modify amino groups of proteins, it is generally advisable to prepare concentrated solutions of urea fresh daily. Preparation of these solutions and the determination of their concentrations are described in Chapter 12.

To remove dust particles, either filter all solutions through 0.45 μm filters or centrifuge them. Centrifugation is usually preferable because filters or syringes may release small amounts of absorbing or fluorescent material. To pellet dust particles, it is necessary to centrifuge the solution for about 30 minutes in a bench-top centrifuge. The cleared solution should be removed from the tube cautiously with a Pasteur pipette. The dust particles are concentrated at the bottom; therefore the last 10–20% of the solution should be left in the tube.

2.2.3 Cuvettes

A set of matched quartz cells or of quartz tandem cells (if available) is necessary for the measurement of difference spectra in a double-beam spectrophotometer. Tandem cells each contain two compartments of equal pathlength. Alternatively, four individual quartz cuvettes can be used. They should be matched, either all four or pairwise. The cells should be cleaned thoroughly. Washing with a mild non-ionic detergent employing a soft ear cleaner or a bent pipe cleaner has proven fairly effective. Follow by thorough rinsing with distilled water. Cuvette centrifuges (such as the Roto-Vette from Hellma, Germany) are very useful for rapid drying of the cuvettes. This avoids the use of organic solvents, which may contain absorbing impurities. Test the equivalence and the cleanliness of the cuvettes by placing aliquots of the same solution (e.g. a solution of the protein of interest) into the sample and reference cuvettes. The difference in absorbance between both cells should ideally be zero (< 0.5% of the measured absorbance) in the wavelength range of interest.

2.3 Instrumental settings

The instrument should be allowed to warm-up prior to the measurements, because amplifier drifts can cause distortions of measured difference spectra. An adequate test is to record the absorbance of the empty cell compartment, e.g. at 280 nm, until its value is constant. Different instruments may require different times to warm-up.

The major variable that controls the signal-to-noise ratio of the spectrum is the slit width, which determines the spectral band width. To a first approximation, there is a linear correlation between the increase in slit width and in signal-to-noise ratio. However, when the spectral band width becomes larger than 10% of the width of the monitored absorbance bands, the absorbance bands can become distorted and the measured peaks decrease in intensity. An optimal slit width can be found experimentally by recording a particular spectrum with successively larger slit settings. The optimal signal-to-noise ration is obtained with the largest slit setting before a decrease in the measured absorbance is observed.

The signal-to-noise ratio can also be enhanced by repetitive scanning or by increasing the time constant or response time of the instrument. In the latter case, the scanning speed has to be reduced to avoid artificial distortion of the spectrum.

2.4 Experimental protocol for the determination of the difference spectrum between the native and unfolded states of a protein

Protocol 1 describes the measurement of difference spectra with tandem cells in a double-beam spectrophotometer. In the example, 6.0 M GdmCl is used as unfolding agent, which is normally sufficient for complete denaturation of proteins. Of course it is necessary to measure complete unfolding transitions to determine the conditions under which the native and unfolded states are stable. A temperature of 25 °C and pH 7.0 are chosen arbitrarily. For particular proteins, different conditions or buffers can be used. Phosphate buffers should not be used for proteins that have binding sites for phosphate itself or for phosphate-containing molecules (see Chapter 12).

Protocol 1.   Difference spectrum between the native and unfolded state of a protein

Equipment and reagents

- Double-beam absorbance spectrophotometer, with thermostatted cell compartment, and warmed-up for at least 30 min
- Suitable buffer for the native protein (e.g. 0.1 M phosphate buffer pH 7.0)

- A set of matched quartz cells or quartz tandem cells: clean the cells thoroughly and dry them (do not use organic solvents for cleaning; they may contain UV-absorbing contaminants)

A. *Method with tandem cuvettes*

1. Prepare a stock solution of the protein in the buffer with an absorbance of about ten at 280 nm in a 1 cm cell. This stock solution will then be diluted tenfold to prepare the samples of native and of unfolded protein.

2. Prepare a 6.67 M stock solution of GdmCl in the buffer. The buffer used to dissolve GdmCl should be twice as concentrated as the desired final buffer, because about half of the volume of the final solution originates from the dissolved GdmCl. If an exact buffer concentration is important, weigh both the required amounts of buffer and of GdmCl and dissolve them with water to make up the correct final volume of the solution.[a] It is advisable to dissolve the components in slightly less than the nominal final volume in order to be able to control and adjust the pH value of the solution. Note that the pH values of buffers (such as phosphate) can be shifted in the presence of high concentrations of denaturant.

3. Switch on the spectrophotometer and set the bath that controls the

---

Protocol 1. *Continued*

temperature of the cell holders so that the temperature in the cuvettes (filled with water) is 25°C (as measured conveniently with a flexible thermistor). Wait until the absorbance at 280 nm and the temperature are constant. Record an air/air or water/water baseline from 350 nm to 240 nm.

4. Prepare solutions of native and unfolded protein of equal protein concentration by a tenfold dilution of the protein stock solution (step 1) with buffer and with 6.67 M GdmCl, respectively. A 6.0 M GdmCl solution should also be prepared by an appropriate dilution of the 6.67 M stock solution of GdmCl. These solutions together with an aliquot of buffer should be centrifuged or filtered to remove dust particles.

5. Fill the left compartment of the sample cell (*Figure 6*) with the native protein and the right side of the reference cell with buffer. Record the spectrum from 350 nm to 240 nm at a band width of about 0.5 nm. The absorbance maximum should be about 0.4–0.5 in a 0.5 cm cell. In the absence of light scattering, the absorbance should be zero between 310 nm and 350 nm. Now open the slit gradually, until the spectrum begins to change in shape. Select the largest slit width at which the spectrum did not show any change. If the absorbance coefficient is known, this spectrum is used to calculate the actual protein concentration in the tandem cell.

6. Fill the left compartment of the reference cell (*Figure 6*) with native protein and the right compartment of the sample cell with buffer, and record the absorbance again. If the cells are matched properly, the signal should coincide with the water/water baseline. This is the reference baseline for the following measurement of the difference spectrum.

7. Now replace the native protein in the reference cell by unfolded protein in 6.0 M GdmCl and the buffer in the right compartment of the sample cell by 6.0 M GdmCl solution. Scan the difference spectrum between 350 nm and 240 nm. To get a good representation of the difference spectrum, it is advisable to shift the baseline to the centre of the chart and to use an appropriately expanded absorbance scale.

8. The measured absorbance difference, ΔA, can be converted into a molar change, $\Delta \varepsilon$, by use of the known protein concentration, c, (step 5) and the optical pathlength of the tandem cell, l, by Equation 2:

$$\Delta \varepsilon = \Delta A / (c \times l). \qquad [2]$$

B. *Comments*

1. Solutions should be allowed to stand in the cuvettes in the instrument for about 10 min to warrant adequate thermal equilibration.

2. Cells should be stoppered to avoid evaporation of water.

3. The protein concentration can be varied, depending on the sensitivity of the instrument and the magnitude of the difference spectrum.

4. Instead of tandem cells, a matched set of four individual cuvettes can be used.

C. *Alternative method without tandem cuvettes*

1. Use a matched set of 1 cm quartz cuvettes. Put the native protein into the sample cuvette and the buffer into the reference cuvette. Record and store the spectrum of the native protein (spectrum 1).

2. Empty the cuvettes (save the solution of the native protein for step 4). Clean the cuvettes, and then put the denatured protein (in 6.0 M GdmCl) into the sample cuvette and the 6.0 M GdmCl solution into the reference cuvette. Record the spectrum of the denatured protein (spectrum 2).

3. Subtract spectrum 2 from spectrum 1 to get the difference spectrum.

4. After the experiment, the spectrum of the native protein (step 1) should be rescanned to ascertain that the instrument's baseline and sensitivity remained stable throughout the measurements.

[a] Warm gently to dissolve the GdmCl crystals completely. Determine the exact concentration of GdmCl by the refractive index (see Chapter 12). Filter the solution through 0.45 μm filter to remove dust particles.

2.5 Data interpretation

The size and shape of the difference spectrum depends on the kind and number of aromatic amino acids, as well as on the degree of burial of their side chains in the interior of the native protein. The contribution of Phe residues to the difference spectrum is very small. It is sometimes apparent as a ripple structure in the 250–260 nm region (*Figure 2b,d*). Proteins that lack Trp display a Tyr difference spectrum with a prominent positive peak at 287 nm and a minor peak at around 278 nm (*Figure 2d*). Proteins that contain both Tyr and Trp show an additional prominent peak around 292 nm that originates from the buried Trp residues in the folded protein (*Figures 2b and 3b*).

A quantitative analysis of the magnitude of the observed peaks, e.g. in terms of numbers of aromatic side chains that become exposed during unfolding, is not warranted for the following reasons:

(a) The absorbance of both the native and the unfolded proteins increase with the concentration of denaturant (*Figure 7*). Therefore the difference in absorbance between 0 M and 6 M GdmCl is an underestimate of the real absorbance change that occurs in the course of unfolding. To measure this change, it is necessary to monitor the entire unfolding

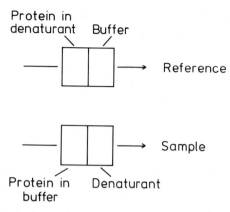

Figure 6. Tandem cell arrangement to measure difference specta.

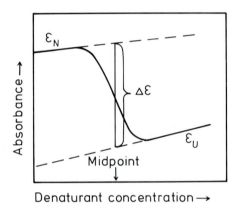

Denaturant concentration →

Figure 7. Sketch of a protein unfolding transition, as meausred by UV difference spectro-scopy. ε_N and ε_U are the absorbances of the native and the denatured form, respectively, which both depend on the concentration of denaturant. $\Delta\varepsilon$ is the molar change in absorbance, observed at the midpoint of the transition; it is greater than ε (0 M denatu-rant) $-\varepsilon$ (6 M denaturant).

transition, draw the respective baselines for the absorbance of the native and of the unfolded state, and determine the absorbance difference at the midpoint of the transition as shown in *Figure 7*. Methods to measure and analyse such unfolding transitions are described in Chapter 12.

(b) The magnitude of the absorbance change during unfolding depends strongly on the environment of the respective chromophores in the native protein. The absorbance change upon transfer of a particular aromatic residue to an aqueous environment depends on the hydrophobicity of its environment in the folded protein, as well as on its polarizability, and on the proximity of other residues. Rough estimates for the change in absorb-ance produced by the transfer of a residue from the protein interior to water were given by Donovan (9). The respective numbers are -700 $M^{-1}cm^{-1}$ at 287 nm for Tyr and -1600 M^{-1} cm^{-1} at 292 nm for Trp. These values were derived assuming that the transfer from the interior of a protein to an aqueous environment is equivalent to a hypothetical transfer from 120% ethylene glycol to water. This assumption may be questionable, but the above values are reasonable first approximations.

Despite the limitations encountered in the mechanistic interpretation of difference spectra, the measurement of spectral changes that accompany unfolding transitions provides a very powerful technique to determine the stability of proteins (Chapter 12) and to follow the kinetics of conformational changes. Absorbance spectrophotomers are standard laboratory equipment, absorbance changes linearly with concentration and can be measured with very high accuracy and reproducibility, and absorbance changes can be followed directly as a function of time in the cuvette following either manual or stopped-flow mixing.

3. Fluorescence

Fluorescence emission is observed when, after excitation, an electron returns from the first excited state back to the ground state. In the excited state, some energy is always lost by non-radiative processes (such as vibrational transitions). Therefore the energy of the emitted light is always less than that of the absorbed light, and the fluorescence of a chromophore occurs at greater wavelengths than its absorbance. A good introduction to the basic principles of fluorescence of biological samples is found in refs 3 and 10–13. Fluorescence emission is much more sensitive to changes in the environment of the chromophore than is light absorbance. As the lifetime of the excited state is long, a broad range of interactions or perturbations can influence this state and thereby the emission spectrum. Fluorescence is thus an excellent probe to investigate conformational changes of proteins.

3.1 Fluorescence of the aromatic amino acids

The fluorescence of proteins originates from Phe, Tyr, and Trp residues. Emission spectra for the three aromatic amino acids are shown in *Figure 8*, and their absorbance and emission properties are summarized in *Table 1*. The excitation spectra correspond to the respective absorbance spectra (*Figure 1*).

3.2 Fluorescence of proteins

In proteins that contain all three aromatic amino acids, fluorescence is usually dominated by the contribution of the Trp residues because both their absorbance at the wavelength of excitation and their quantum yield of emission are considerably greater than the respective values for Tyr and Phe. This

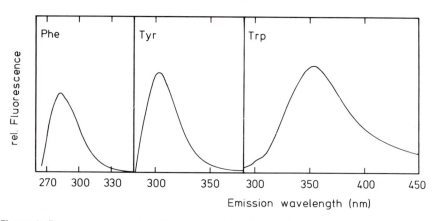

Figure 8. Fluorescence spectra of the aromatic amino acids in 0.01 M potassium phosphate buffer pH 7.0 at 25 °C. For the measurements, 100 μM phenylalanine, 6 μM tyrosine, and 1 μM tryptophan were used, with excitation at 257 nm, 274 nm, and 278 nm, respectively.

is expressed by the 'sensitivity' parameter (*Table 1*), which is 730 for tryptophan and 200 for tyrosine. Phe fluorescence is not observed in native proteins because its sensitivity of four is very low (*Table 1*). The other factor in fluorescence is transfer of energy between residues. For example, Phe fluorescence is also not observed because its emission is efficiently quenched by energy transfer to the other two aromatic amino acids. Tyr and Trp residues absorb strongly around 280 nm, where Phe emits fluorescence. In proteins that contain both Tyr and Trp residues, fluorescence from Tyr is barely detectable because Trp emission is strong (see above), because in folded proteins emission from Trp is frequently shifted to smaller wavelengths towards that of Tyr, and because non-radiative energy transfer can occur from Tyr to Trp residues in the compact native state (10–13).

Changes in protein conformation, such as unfolding, very often lead to large changes in the fluorescence emission. In proteins that contain Trp residues, both shifts in wavelength and changes in intensity are generally observed upon unfolding. The tryptophan emission of a native protein can be greater or smaller than the emission of free tryptophan in aqueous solution. Consequently both increases and decreases in fluorescence can occur upon protein unfolding. The emission maximum is usually shifted from smaller wavelengths to about 350 nm in the unfolded state, which corresponds to the fluorescence maximum of tryptophan in aqueous solution. The exact location of this maximum depends to some extent on the nature and concentration of the buffer. In a hydrophobic environment, such as in the interior of a folded protein, Trp emission occurs at smaller wavelengths (indole shows an emission maximum of 320 nm in hexane (15)). As an example, the emission spectra of native and of unfolded RNase T1 are shown

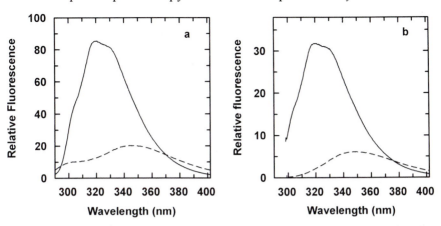

Figure 9. Fluorescence emission spectra of native (—) and unfolded (– – –) RNase T1. Native RNase T1 (1.4 μM) was in 0.1 M sodium acetate pH 5.0; the sample of unfolded protein contained 6.0 M GdmCl in addition. Fluorescence was excited at (a) 278 nm and (b) 295 nm. The band widths were 3 nm for excitation and 5 nm for emission. Spectra were recorded at 25 °C in 1 × 1 cm cells in a Hitachi F-4010 fluorimeter.

in *Figure 9*. RNase T1 contains nine Tyr residues and only one Trp (Trp59), which is inaccessible to solvent in the native protein (16).

The fluorescence of the Trp residues can be investigated selectively by excitation at wavelengths greater than 295 nm. Because of the red-shift and the increased intensity of the absorbance spectrum of Trp when compared with Tyr (see *Figures 1* and *2*), protein absorbance above 295 nm originates almost exclusively from Trp residues. A comparison of the emission observed after excitation at 280 nm and at 295 nm gives information about the contribution of the Trp and the Tyr residues to the observed fluorescence spectra. The data for RNase T1 in *Figure 9* show that the shapes of the fluorescence spectra observed after excitation at 278 nm and 295 nm are virtually identical. The measured emission originates almost completely from the single Trp59, which is inaccessible to solvent and hence displays a strongly blue-shifted emission maximum near 320 nm. Tyrosine emission is barely detectable in the spectrum of the native protein, because of energy transfer to Trp. Unfolding by GdmCl results in a strong decrease in Trp fluorescence and a concomitant red-shift of the maximum to about 350 nm. The distances between the Tyr residues and Trp59 increase, and energy transfer becomes less efficient. As a consequence, the Tyr fluorescence near 303 nm becomes visible in the spectrum of the unfolded protein when excited at 278 nm (*Figure 9a*), but not when excited at 295 nm (*Figure 9b*). The examples in *Figure 9* indicate that, unlike absorbance, the changes in fluorescence upon folding can be very large, and the contribution of the Trp residues can be studied selectively by changing the excitation wavelength. These features, together with the very high sensitivity of fluorescence measurements, make

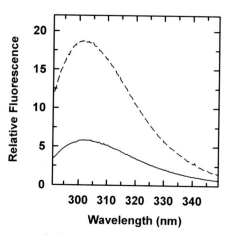

Figure 10. Fluorescence emission spectra of 1.5 μM native (—) and unfolded (– – –) Trp59Tyr RNase T1. This protein contains ten Tyr residues. The native protein was in 0.1 M sodium acetate pH 5.0; the unfolded sample contained 6.0 M GdmCl in addition. Fluorescence was excited at 278 nm; the spectra were recorded as in *Figure 9*.

them extremely useful as sensitive probes of conformational changes in proteins. Multiple emission bands do not necessarily originate from different Trp residues of a folded protein. *Figure 9* shows that even single Trp residues, such as Trp59 of RNase T1, can give rise to several emission bands.

The fluorescence maximum of a Tyr residue remains around 303 nm, irrespective of its molecular environment. Therefore the unfolding of proteins that contain Tyr, but no Trp, is often accompanied by changes in the intensity, but not in the wavelength, of emission. As an example, the fluorescence emission spectra of the Trp59Tyr variant of RNase T1 in the native and unfolded states are shown in *Figure 10*. Decreased Tyr fluorescence in the native state (as in *Figure 10*) is frequently observed. It is thought to originate from hydrogen bonding of the tyrosyl hydroxyl group and/or the proximity of quenchers, such as disulfide bonds, in the folded state (14).

3.3 Environmental effects on Tyr and Trp emission

The fluorescence of the exposed aromatic amino acids of a protein depends on the solvent conditions even in the absence of conformational changes. This effect can be used to probe the solvent accessibility of the aromatic residues of a native protein by fluorescence quenching techniques (10,17). It is also important for the evaluation of fluorescence difference spectra between native and unfolded proteins, such as those shown in *Figures 9* and *10*. The most commonly used unfolding agents are temperature, GdmCl, and urea. The dependence of the emission of the two aromatic amino acids on these variables is described in this section.

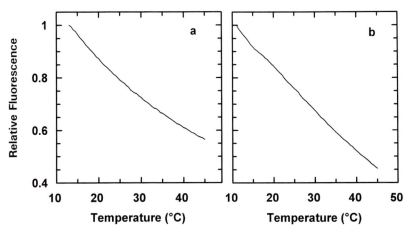

Figure 11. Dependence on temperature of the relative fluorescence of (a) tyrosine (6 μM) at 303 nm (excitation at 274 nm) and of (b) tryptophan (1 μM) at 355 nm (excitation at 278 nm). The free amino acids were in 0.01 M potassium phosphate pH 7.0. Spectra were recorded as described in *Figure 9*.

3.3.1 Temperature

The fluorescence intensity generally decreases with increasing temperature (*Figure 11*). This decrease is substantial; to a first approximation, Tyr emission decreases $\geq 1\%$ per degree increase in temperature. The dependence on temperature of the fluorescence of Trp is even more pronounced (*Figure 11*).

3.3.2 GdmCl and urea

Both denaturants exert a significant influence on the fluorescence of Tyr and Trp residues. The dependences on the concentration of GdmCl and urea of tryptophan and tyrosine emission are displayed in *Figures 12* and *13*. These curves depend to some extent on the instrument's geometry and on features such as light beam focus and slit settings. For proteins that contain both Tyr and Trp residues, the emission of the unfolded protein can be represented by an appropriate mixture of the two aromatic amino acids. The dependence of the emission of this mixture on denaturant concentration is often very useful to determine the baseline equivalent to the unfolded protein for the quantitative analysis of unfolding transition curves.

3.4 Measurement of protein fluorescence

3.4.1 Instruments

Most spectrofluorimeters operate in a split-beam mode, where a small portion of the incident light is directed to a reference photomultiplier to correct the fluorescence signal for inherent instabilities of the light source. To avoid baseline drifts, some instruments interrupt the light beam periodically or employ pulsed lamps. The dark periods are used to adjust the baseline during

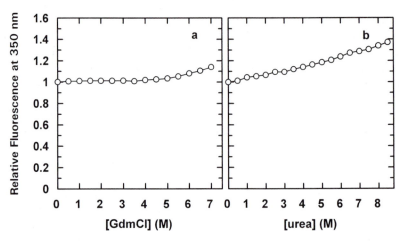

Figure 12. Dependence of the relative fluorescence of tryptophan at 355 nm (excitation at 278 nm) on (a) GdmCl concentration and (b) urea concentration. The amino acid concentration was 2 μM in 0.1 M potassium phosphate buffer pH 7.0 and the indicated concentration of denaturant. Spectra were recorded as described in *Figure 9*.

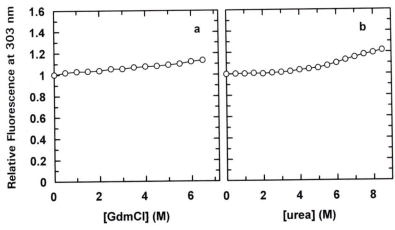

Figure 13. Dependence of the relative fluorescence of tyrosine at 303 nm (excitation at 274 nm) on (a) GdmCl concentration and (b) urea concentration. The amino acid concentration was 10 μM. Other conditions were as in *Figure 12*.

the measurement. This increases the long-term constancy of the measured fluorescence. For further description of instruments, consult refs 7 and 11 and the respective instruments' manuals.

3.4.2 Cuvettes

Emission is usually observed at a right angle to the excitation beam; therefore fluorescence cuvettes are manufactured with polished surfaces on all four sides. Quartz cells are required for work in the UV region. For routine work rectangular cuvettes (0.4 × 1 cm) are convenient. Since the exciting beam illuminates only the central part of the cell, the observed fluorescence intensity does not necessarily decrease when cuvettes with reduced cross-sections are used. The advantage is that smaller sample volumes are required. To measure the minimal volume required, determine the actual size and location of the exciting light beam by introducing a narrow strip of white paper into the dry cuvette and observing with an excitation wavelength in the visible range (e.g. around 500 nm).

Fluorescence is an extremely sensitive technique, so it is mandatory to avoid contamination of cuvettes and glassware with fluorescing substances. Use deionized and quartz-distilled water, and avoid plastic containers, which may leach out fluorescent additives. Also, laboratory detergents usually contain strongly fluorescing substances. Fluorimeter cells should be cleaned with 50% nitric acid. If detergents are used for cleansing, extensive rinsing with distilled water is required. Before measurements, the cleanliness of the cell (and the distilled water) should be checked routinely by filling the cuvette with water and recording a blank in the wavelength range of interest. There should be no emission except the Raman peak of water (Section 3.4.4).

3.4.3 Solvents and buffers

Any solvent can be used for fluorescence measurements, provided that it does not absorb in the spectral regions of the excitation and emission of the fluorophore to be examined. For protein work, this implies that the buffers or solvent additives should not absorb (or fluoresce) at wavelengths greater than about 250 nm. This criterion is met by most of the commonly used buffers (*Table 3*) or denaturants. Prepare buffers with quartz-distilled water that does not contain fluorescent impurities. Also check the containers used to store buffers for fluorescent materials. A convenient way to detect such contaminants is to rinse the container with a small volume of water and to measure the emission properties of this rinse water.

3.4.4 Raman peak of water

In water, a Raman scattering peak is observed that is separated from the incident radiation by a fixed energy difference. This energy is equivalent to an O-H vibrational mode of the H_2O molecule. The Raman peak of water is weak and occurs at a constant frequency difference, not at a constant wavelength difference, from the exciting light. The nomogram in *Figure 14* gives the position of the Raman peak as a function of the excitation wavelength. The Raman peak frequently overlaps with the emission spectrum of proteins. When instruments with data storage facility are used, the spectrum of the solvent with the Raman peak can be stored and subtracted from the sample spectrum. Otherwise the subtraction has to be done manually. In some

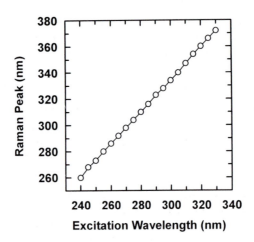

Figure 14. Position of the Raman peak of water as a function of the excitation wavelength.

instances, a shift of the excitation wavelength can alleviate the interference of the Raman peak with the protein emission.

The Raman peak of water is very useful to check the performance of the fluorescence spectrometer. It should be recorded under fixed instrumental settings at regular time intervals. A decrease in the signal-to-noise ratio usually indicates that the lamp needs readjustment or replacement. Detailed procedures and performance criteria should be described in the instrument's manual.

3.4.5 Instrument setting

i. General

Since fluorescence is strongly dependent on temperature, good temperature control of the cell is mandatory. Thermostatted cuvette holders are usually supplied by the manufacturers. The instrument should be allowed to warm-up until the signal is constant. This can easily be examined by recording the intensity of the water Raman peak with time.

ii. Choice of wavelengths: excitation

The shape of the emission spectrum of a fluorophore is independent of the excitation wavelength, provided that it remains within the respective absorbance bands. Since fluorescence is a very sensitive technique, excitation at wavelengths where absorbance is fairly weak still results in emission signals of good quality. This flexibility in the choice of excitation wavelength offers a number of advantages for practical work.

(a) A shift in excitation wavelength can be employed to minimize interference of the Raman peak of water (see *Figure 14*) with the emission maximum of the protein under investigation.

(b) If the solution contains fluorescent impurities, it is sometimes possible to minimize their contribution to the emission by choosing an excitation wavelength where the absorbance of the impurity is minimal.

(c) If the absorbance of the protein sample at its absorbance maximum is too high (Section 3.4.6) for fluorescence experiments, a different excitation wavelength can be used, where the absorbance is weaker.

(d) To investigate selectively the Trp emission of proteins that also contain Tyr residues, excitation should be performed at wavelengths greater than 295 nm, where the absorbance of Tyr is almost zero (see *Figure 1*).

(e) Upon protein unfolding, absorbance changes in a complex manner (Section 2). This implies that in fluorescence experiments the emission of native and unfolded protein molecules are excited at most wavelengths to a different extent, thus leading to differences in the fluorescence signals. To avoid this effect, use an excitation wavelength where the native and the unfolded molecules do not show a difference in absorbance. The precise location of such isosbestic points depends on the individual proteins under investigation. They are frequently found near 270 nm.

The slit width used for excitation can be varied. Wide slit openings should be used for very dilute solutions to improve the signal-to-noise ratio of emission. A narrow slit setting should generally be used when the quality of the signal is sufficient. With a narrow excitation slit, light intensity is reduced and the risk of photochemical decomposition of the sample is decreased. The excitation slit should always be closed between measurements to minimize the exposure of the sample to light.

iii. Choice of wavelengths: emission

Protein emission spectra are broad, and without detectable fine structure; changes upon structural transitions frequently involve the entire fluorescence spectrum. Therefore a variety of wavelengths and wide emission slits can be used to follow conformational changes of proteins.

3.4.6 Samples for fluorescence measurements

i. Preparation of samples

Because of the high sensitivity of fluorescence measurements, avoid impurities and dust as much as possible. Consult Section 3.4.3 on buffers and solvents. Generally, samples should be filtered or, preferably, centrifuged to remove dust particles. A single dust particle moving slowly through the light beam can cause severe distortions of the signal due to light scattering. It is helpful to stir the sample solution continuously during the fluorescence measurement, using tiny magnetic stirring bars placed at the bottom of the cuvette. Magnetic stirrers that fit underneath the cuvette in the sample compartment of the instrument and cuvettes with specially designed bottoms for magnetic bars are commercially available. To avoid distortion of the signal, the stirring bar must not interfere with the exciting light beam. If this cannot be achieved, it helps to mask the lowest part of the cuvette (where the bar is rotating) with black tape. Stirring has three beneficial effects.

(a) It keeps residual dust particles in rapid motion and thereby minimizes their distorting effect on the signal.

(b) It continually transports new protein molecules into the small volume of the cell that is illuminated by the excitation light beam, and thereby averages out photochemical decomposition reactions over the entire sample.

(c) Thermal equilibration within the cuvette is improved.

ii. Concentration of samples: inner filter effect

There is not a linear relationship between fluorophore concentration and emission intensity. The actual form of the concentration-dependence varies with the optical geometry of the fluorimeter and the cell.

The non-linearity is caused by inner filter optical effects. Between the entrance window of the cell and the volume element of the sample from which fluorescence emission is collected, the exciting light is attenuated by absorbance by the protein and the solvent. This attenuation depends on the absorbance of the sample at the wavelength of excitation. Its extent can therefore be varied by a shift in the excitation wavelength. In strongly absorbing samples (i.e. at high concentration), the emission intensity is actually decreased, as almost the entire incident light is absorbed before it can reach the centre of the cuvette (7). Further attenuation originates from reabsorbance of fluorescent light by protein and solvent molecules. This effect is usually small, because absorbance of proteins and solvents is very weak beyond 300 nm. An approximately linear relationship between fluorescence emission and concentration is obtained when the absorbance of the sample at the excitation wavelength is low. It should generally be smaller than 0.1 absorbance units. When fluorescence methods are used to measure concentrations, it is generally necessary to determine the concentration-dependence of the emission in a calibration curve, in the same solvent and the same cuvettes as used for the measurements.

Protocol 2. Performing a fluorescence measurement

Equipment
- Fluorescence spectrometer
- Fluorescence cuvette

Method
1. Prepare dust-free samples of the protein and of the buffer used as outlined in Section 3.4.3 and 3.4.6. The absorbance of the sample around 280 nm should be well below 0.1.

---

2. Switch on the instrument and allow for sufficient warm-up until the signal (e.g. the Raman peak of the buffer) remains constant. Adjust the background and the recorder to zero as described in the instrument's manual.

3. The magnitude of the fluorescence signal depends on the photomultiplier voltage, the amplifier gain, and the excitation and emission slit widths. To select optimal instrument settings, insert the protein sample and dial the excitation and emission wavelengths to their respective maxima. Set the photomultiplier voltage to an optimal value (consult the manual). The excitation slit should preferably be kept fairly narrow (usually < 5 nm) to minimise photochemical reactions; the emission slit can be opened fairly wide (e.g. 10 nm). After setting the slits, adjust the maximal signal to about 70% deflection on the recorder scale or on the screen.

4. Record the excitation spectrum, keeping the emission wavelength constant, usually at the fluorescence maximum. Repeat the experiment with the buffer cuvette and subtract the blank spectrum (which should basically be flat, apart from the Raman peak) from the protein spectrum. The excitation spectrum in the 240–300 nm region should resemble the absorbance spectrum of the respective fluorophores (only Trp, or the sum of Tyr and Trp, depending on the wavelength of emission).

5. Record the emission spectrum. The following criteria are relevant for the selection of an appropriate excitation wavelength. If maximal intensity and an excitation of both Tyr and Trp are desired, excitation should be near 280 nm. In the case of high absorbance, wavelengths away from the maximum can be used, where the absorbance is lower (Section 3.4.5). For the selective measurement of Trp emission, the excitation wavelength should be greater than 295 nm (Section 3.4.5). Again, blank spectra of the buffer should be recorded under the same instrumental conditions and subtracted from the protein emission spectra.

Comments:

(a) Proper thermostatting is mandatory in fluorescence spectroscopy. Allow at least 10–15 min for thermal equilibration of the cell prior to measurements.

(b) The intense excitation light may damage the sample. Therefore select a narrow excitation slit and keep the excitation shutter closed between measurements. In the case of extended (e.g. kinetic) measurements, the shutter should be closed intermittently.

(c) The long-term stability of some fluorimeters poses a problem in quantitative measurements. In such cases it is advisable to prepare a reference solution (e.g. of the protein that is investigated) in a sealed cuvette, measure its emission intensity at regular time intervals during the experiment, and use these data to correct the obtained results for intensity fluctuations.

(d) For long-term experiments, seal the cuvette to avoid evaporation of solvent during the experiment.

3.5 Fluorimetric determination of the Trp content of proteins

The number of Trp residues in a protein can be determined by a comparison of its emission spectrum with the emission of tryptophan solutions of known concentration (18). For this comparison the Trp residues have to be normalized by prior unfolding of the protein chain. The individual steps of the analysis are described in *Protocol 3*.

Protocol 3. Determination of the Trp residue content of a protein

Equipment
- Fluorescence spectrometer
- Fluorescence cuvette

Method

1. Denature the protein (0.01–0.05 mg/ml) by incubation in 6.0 M Gdmcl pH 7.0 for 60 min. See *Protocol 1* for how to make the solutions of GdmCl and of the denatured protein. For the reduction of disulfide-containing proteins, 30 mM 2-mercaptoethanol (or 10 mM dithiothreitol) should be added.

2. Record the emission spectrum of this solution and of the solvent between 300 nm and 400 nm. Excitation should be at 295 nm, to excite Trp fluorescence selectively.

3. Under identical experimental and solvent conditions, record spectra of solutions of the free amino acid tryptophan of varying and known concentrations and compare them to the spectrum of the protein.

4. The Trp residue concentration in the protein solution is equivalent to the concentration of free tryptophan in that reference solution that gives the best fit of the respective emission spectra. The number of Trp residues, n_{trp}, is equal to $c_{trp}/(1.1 \times c_P)$, where c_P is the molar protein concentration and c_{trp} is the molar free tryptophan concentration that gives the closest fit of the spectra. The empirical factor 1.1 was found by a calibration of this method with proteins of known Trp content (18).

This method can also be used to validate mutations involving Trp residues. Note that in native proteins the contributions of the individual Trp residues to the measured fluorescence can differ enormously.

4. Circular dichroism

Circular dichroism (CD) and optical rotary dispersion (ORD) are two inter-related phenomena that are used to measure the optical activity of asymmetric molecules in solution. ORD is the ability of a molecule to rotate the plane of linearly polarized light as a function of the wavelength. CD gives information about the unequal absorption of left- and right-handed circularly-polarized light by optically-active molecules. A good introduction to the basic principles of the two methods is provided by refs 3,12,19,20. Because of the availability of high precision instruments, CD has become the standard technique to measure the optical activity of proteins. This chapter will therefore deal only with practical aspects of CD measurements.

CD signals are observed in the same spectral region as the absorbance bands of a particular chromophore, provided that the chromophore or its molecular environment are asymmetric. CD bands of proteins occur in two spectral regions. The far-UV or amide region (170–250 nm) is dominated by contributions of the peptide bonds, whereas CD bands in the near-UV region (250–300 nm) originate from the aromatic amino acids. In addition, disulfide bonds give rise to several CD bands. The two spectral regions give different kinds of information about protein structure.

The CD in the amide region reports on the backbone (i.e. the secondary) structure of a protein and is used to characterize the secondary structure and changes therein. In particular the α helix displays a strong and characteristic CD spectrum in the far-UV region. The spectral contributions of the other elements of secondary structure are less well defined. Several methods exist for determination of the secondary structural elements of a protein from its CD spectrum in the amide regions; they are summarized in Section 4.5.

CD bands in the near-UV region are observed when, in a folded protein, aromatic side chains are immobilized in an asymmetric environment. The CD of the aromatic residues is very small in the absence of ordered structure (e.g. in short peptides). The signs, magnitudes, and wavelengths of the aromatic CD bands cannot be calculated; they depend on the immediate structural and electronic environment of the immobilized chromophores. Therefore the individual peaks in the very complex near-UV CD spectrum of a protein usually cannot be assigned to transitions in the vicinity of specific amino acid side chains. However, the near-UV CD spectrum represents a highly sensitive criterion for the native state of a protein. It can thus be used as a fingerprint of the correctly folded conformation. It is only surpassed by NMR in its sensitivity.

4.1 CD spectra of native and unfolded proteins

CD is very useful for assessing the structure of a protein in dilute aqueous solution and for monitoring structural transitions. The CD differences between native and unfolded proteins are usually very large in both spectral regions. The aromatic CD is virtually zero in denatured proteins. As an example, the CD spectra of RNase A in the amide region and in the aromatic region are shown in *Figure 15* for both the native and the unfolded proteins. Although RNase A contains only 15% α helix, the far-UV spectrum of this protein is dominated by the contribution of this type of secondary structure, with prominent bands at 222 nm and at 208 nm. The aromatic CD contains contributions of the six Tyr residues and the four disulfide bonds (21).

4.2 Measurement of circular dichroism

Note that the difference in absorbance of right- and left-handed circularly-polarized light of a protein sample is extremely small. In the far-UV region it is in the range of 10^{-4} to 10^{-6} absorbance units in samples with a total absorbance of about 1.0. This requires that less than 0.1% of the absorbance signal be measured accurately and reproducibly. Therefore highly sensitive instruments are necessary, and careful sample preparation is important.

4.2.1 Instruments

Modern CD spectropolarimeters use a high frequency photoelectric modulator to generate alternatively the two circularly-polarized light components. This leads to an alternating current contribution to the photomultiplier signal that is proportional to the CD of the sample. A detailed description of the principles of CD measurements is found in refs 19 and 20 and usually in the instrument's manual. Depending on the instrument, the signal is recorded either directly as difference in absorbance $\Delta A = A_L - A_R$, or as ellipticity expressed in millidegrees (mdeg). The interrelationship between the two is given in Section 4.4.

i. Instrument calibration

CD instruments are operated in a similar way as single-beam absorbance spectrophotometers. It is important to check the expansion of the CD scale periodically. A suitable compound for the calibration of CD instruments is d-10-camphorsulfonic acid, which shows a molar ellipticity of 7800 deg × $dmol^{-1}$ × cm^2 at 290.5 nm in aqueous solution (19,22). Since commercial camphorsulfonic acid may contain water, recrystallization from acetic acid (22) and drying under vacuum prior to use is advisable. The actual concentration of camphorsulfonic acid solutions should be determined spectrophotometrically by using $\varepsilon_{285} = 34.5$ $M^{-1}cm^{-1}$. Camphorsulfonic acid also has a negative CD band at 192.5 nm with a molar ellipticity of 15 840 deg × $dmol^{-1}$ × cm^2, which can be employed to check instrument performance in the far-

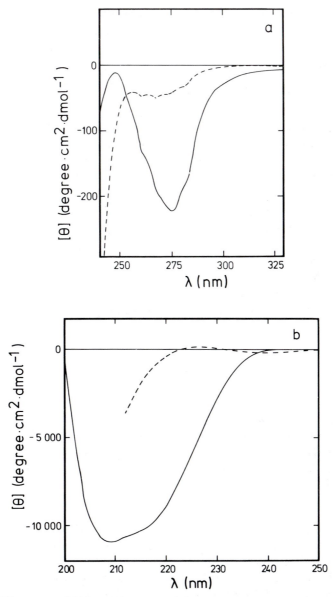

Figure 15. CD spectra of RNase A from red deer at 20°C in the native (—) and in the unfolded states (– – –). The native protein was in 0.1 M sodium cacodylate/HCl pH 6.0; the unfolded protein was in the same buffer plus 6.0 M GdmCl. (a) CD in the aromatic region; the protein concentration was 78 μM in a 1 cm cell. (b) CD in the peptide region, using 28 μM protein in a 0.1 cm cell.

Table 4. Standards to calibrate the ellipticity scale of CD instruments

Compound	Concentration	CD in a 1 cm cuvette
epi-Androsteron [3-β-Hydroxy-5α-androstan-17-on]	0.5 mg/ml in dioxan	192.4 mdeg[a] at 304 nm
D(−)Pantoyl-lactone [D(−)2-hydroxy-3,3-dimethyl-γ-butyrolactone]	0.15 mg/ml in water	190 mdeg at 219 nm
D(+)-Camphorsulfonic acid monohydrate	0.6 mg/ml in water	188 mdeg at 290 nm

[a] mdeg = millidegrees.

UV (19). Compounds that are useful for the calibration of the ellipticity scale in the UV range are given in *Table 4*. The maxima of these compounds can also be employed to examine the wavelength accuracy of the instrument.

ii. Stability and baseline

Since the measured CD bands are generally very small, an extremely high stability of the CD signal is required. In general the instrument should be allowed to warm-up for 30–60 min. Determine the stability by recording the signal, e.g. of a blank solution, in the wavelength region of interest for an extended time. The signal drift should be less than about 0.5–1 mdeg/h. The spectral baseline of most CD instruments can be flattened either manually by a multipot system (in old instruments) or by computer. The baseline with and without the cuvette should be checked regularly.

iii. Slit width, time constant, and scanning speed

Slit width and time constant are the two major instrumental variables that determine the signal-to-noise ratio of the CD signal. An increase in slit width improves it. The upper limit in slit width is set by the requirement that the fine structure of the spectrum not be averaged out by an excessively large slit, and by the appearance of artefacts due to fluorescence or stray light effects at wide slit openings and high absorbance of the sample. The best way to determine the optimal slit width is to record spectra with successively wider slits and then to select the maximum value that does not give distortion of the measured CD spectrum.

An optimal combination of time constant and scanning speed can be found in a similar way. At a given scanning speed, record the spectrum first with a low time constant (e.g. 0.5 sec) and then increase the time constant progressively to a value above which the CD spectrum becomes distorted. The product of the scanning rate and the time constant should be smaller than 0.33 nm. The signal-to-noise ratio increases with the square root of the selected time constant. If the quality of the spectrum is still not satisfactory, decrease the scanning speed and increase the time constant. Alternatively the spectrum may be scanned repetitively and averaged. Here again the signal-to-noise ratio increases with the square root of the number of scans. Most modern CD instruments allow repetitive scanning, provide post-run smoothing routines, or both. If variations in time constant, scanning speed, and/or repetitive scanning do not lead to adequate quality of the spectrum, modifications in solvent, protein concentration, and/or path length should be considered (Sections 4.2.2 and 4.2.4).

4.2.2 Cells for CD measurements

Preferably use only fused quartz cuvettes, specifically manufactured for CD measurements. They display high transparency below 200 nm and are not birefringent. Examine the quality of the CD cells by recording the CD of the thoroughly cleaned empty cuvette. It should coincide with the instrument's baseline. CD cells are usually circular and are available in a wide variety of pathlengths. The actual length of the very thin cuvettes can be determined by placing a coloured solution of known concentration (see *Table 2*) into the cuvette and measuring its absorbance in an absorbance spectrophotometer. For good temperature control, water-jacketed cells should be used. Thermostatted holders for cylindrical cells usually do not provide efficient thermal equilibration. For cleaning procedures see Section 2.2.3.

4.2.3 Solvents

The contributions of buffers and salts to the total absorbance of the sample should be as small as possible. This poses severe restrictions on the choice of solvents for CD measurements in the far-UV region (see Section 2.2.1). Absorbance values in the far-UV region for several salts and buffers are given in *Table 3*. A table with the transparency of non-aqueous solvents is given in ref. 19. Common denaturants, such as urea and GdmCl absorb strongly in the far-UV. Therefore their use is restricted to the wavelength region above about 210 nm. Use only highly purified samples of these denaturants, to avoid interference from absorbing contaminants. Oxygen absorbs light below 200 nm, so degas solvents for measurements in this region prior to the CD measurement. The optics of the instrument should always be purged with nitrogen to avoid the production of ozone, which damages the optics and the experimentalist. The cell compartment has to be purged with nitrogen only when measurements below 190 nm are performed.

4.2.4 Properties of samples

Proper choices of protein concentration, pathlength, and solvent are essential for obtaining good CD spectra and for avoiding artefacts. Since the CD instrument measures very small differences in the transmitted light, the total absorbance of the sample in the desired spectral region is of utmost importance. A good signal-to-noise ratio is achieved when the absorbance is around

1.0; it should generally stay below 2.0. The magnitude of the CD depends on the protein concentration; therefore its contribution to the total optical density of the sample should be as high as possible, and the absorbance of the solvent should be as small as possible. Most buffers are transparent in the near-UV, so that pathlengths of 1 cm or more can be selected in this spectral region. For measurements in the far-UV region, short pathlengths (0.2 to 0.01 cm) and increased protein concentrations should be employed to minimize the contributions of the solvent to the total absorbance (Section 4.2.3 and *Table 3*). As a rule, 0.1 cm cells can be used for the spectral region down to about 200 nm. For measurements extending below 200 nm, 0.01 cm cells are advisable. Protein absorbance is about tenfold higher in the far-UV than in the aromatic region; therefore the optical density of the protein solution in the amide region can be estimated from the absorbance at 280 nm. To a first approximation, protein concentrations of 0.1% and 0.01% (w/v) are suitable for measurements with 0.01 and 0.1 cm pathlength, respectively. To get high quality spectral data over an extended wavelength range, it is sometimes helpful to use different pathlengths for the various wavelength regions. In practice, measurements at different pathlengths and protein concentrations should be used as a test for artefacts that can arise easily, e.g. from excessive absorbance of the sample (absorption flattening) or from protein aggregation.

Filter or centrifuge solutions to be employed for CD measurements to remove dust or other solid suspended particles (Section 2.2.2). The presence of light scattering in CD measurements leads to anomalous, red-shifted spectra (3). When the protein samples are exposed to UV light around 220 nm for an extended time, as is frequently the case when measuring far-UV CD, photochemical damage can occur. It is advisable to examine some functional and/or structural properties of the protein under investigation after long-term CD experiments.

4.3 Measurement of CD spectra of proteins

Protocol 4 provides an experimental guide for measuring the CD spectrum of a protein.

Protocol 4. Measurement of protein CD spectra

Equipment
- CD spectrometer
- CD cuvettes: 1 cm, 0.1 cm, or 0.01 cm cells, depending on the wavelength range

A. Preparation of the instrument and the cells
1. Allow 30–60 min for the instrument to warm-up.
2. Check the long-term stability, the instrument's baseline, and the calibration of the CD scale at regular intervals (Section 4.2.1).

---

3. Insert the empty, dry cell and record its CD at high sensitivity ($\leqslant 1$ mdeg/cm) in the relevant wavelength range. It should coincide with the instrument's baseline. Deviations occur if the cell is birefringent, or, more likely, if traces of impurities, such as residual proteins, cover the glass surfaces. In this case, clean the cell thoroughly (Section 2.2.3).

4. Use the same cuvette in the same orientation for measuring the sample spectrum and the baseline.

B. *Preparation of samples*

1. The absorbance of the samples should be around unity for a good signal-to-noise ration of the CD signal; it should not exceed 2.0 (Section 4.2.4).

2. Record absorbance spectra of the samples prior to the CD experiment to determine the protein concentration necessary for the calculation of the molar CD (Section 4.4), and to select the optimal conditions for the CD measurement.

3. Adjust the desired total optical density of the sample by varying the protein concentration and/or optical pathlength. Protein CD in the aromatic region is generally small and buffers usually do not absorb in this spectral region. Use cells with pathlengths of 1 cm or more in the near-UV region.

4. For measurements in the far-UV amide region, select a buffer to keep the absorbance of the solvent as low as possible and use cells with pathlengths of 0.1 cm or less (Section 4.2.4).

5. All samples and buffers should be passed through 0.45 μm filters or centrifuged.

C. *Scanning the CD spectrum*

1. Select an adequate combination of spectral band width, scan speed, and time constant.

2. If possible, use repetitive scanning as outlined in Section 4.2.1 to obtain an optimal signal-to-noise ratio.

3. After placing the sample into the cell, allow 10–15 min for thermal equilibration at the desired temperature.

4. Record the CD spectrum of the protein sample and the baseline of the buffer successively under identical instrumental settings in the same CD cell.

5. Subtract the buffer spectrum manually or (if the spectra were stored in the instrument's memory) by computer from the protein spectrum.

6. When spectra in the far-UV region are monitored, special attention

Protocol 4. *Continued*

should be paid to the increase of the photomultiplier voltage with decreasing wavelength. An excessive multiplier gain indicates that the absorbance of the sample is too high. This high absorbance causes a progressive increase in the noise level, and, more seriously, can lead to an artificial decrease of the ellipticity at low wavelengths because of absorbance flattening. A return of the measured CD to the baseline with decreasing wavelength is indicative of such an artefact.

7. When repetitive scanning is performed for an extended time, single spectra should be recorded at the beginning and at the end to examine the stability of the sample.

4.4 Evaluation of data

Depending on the CD instrument, the data are recorded either as the difference in absorbance of right- and left-handed circularly-polarized light, $\Delta A = A_L - A_R$, or as ellipticity, Θ_{obs}, in degrees or millidegrees. Data in both formats can be converted to the molar values, i.e. to the differential molar circular dichroic extinction coefficient, $\Delta\varepsilon = \varepsilon_L - \varepsilon_R$ and to the molar ellipticity, $[\Theta]$, which are interrelated by Equation 3.

$$[\Theta] = 3300 \times \Delta\varepsilon \qquad [3]$$

It should be noted that the concentration standards are different for $[\Theta]$ and $\Delta\varepsilon$. $\Delta\varepsilon$ is the differential absorbance of a 1 mol/l solution in a 1 cm cell, whereas $[\Theta]$ is the rotation in degrees of a 1 dmol/cm^3 solution and a path-length of 1 cm.

CD of proteins in the amide region is nearly always given as mean residue ellipticity, $[\Theta]_{MRW}$, which is based on the concentration of the sum of the amino acids in the protein solution under investigation. If the molar protein concentration and the number of amino acids are known, the concentration of residues is simply the product of both numbers. If the protein concentration is known in mg/ml, the concentration of amino acids can be calculated by assuming a mean residue weight (MRW) of 110 per amino acid residue.

For data in the aromatic region, different units are used in the literature. Frequently, aromatic CD is given as $\Delta\varepsilon$, but $[\Theta]$, based on the protein concentration, and $[\Theta]_{MRW}$, based on the residue concentration, are also found.

4.4.1 Differential molar CD extinction coefficient

When $A_L - A_R$ is measured by the instrument, $\Delta\varepsilon$ is calculated by the use of Equation 4, which is basically the Beer–Lambert relation of Equation 1:

$$\Delta\varepsilon = \varepsilon_L - \varepsilon_R = \frac{A_L - A_R}{c \times l} \qquad [4]$$

where $A_L - A_R$ is the measured differential CD absorbance, c is the concentration in mol per litre, and l is the pathlength in cm. $\Delta\varepsilon$ has the unit of litre $\times$ mol^{-1} $\times$ cm^{-1} or litre $\times$ (mol residues)$^{-1}$ $\times$ cm^{-1}.

4.4.2 Molar or residue ellipticity

The molar ellipticity, $[\Theta]$, or the residue ellipticity, $[\Theta]_{MRW}$, are calculated from the measured Θ (in degrees) by using Equations 5 or 6:

$$[\Theta] = \frac{\Theta \times 100 \times M_r}{c \times l} \qquad [5]$$

$$[\Theta]_{MRW} = \frac{\Theta \times 100 \times MRW}{c \times l} \qquad [6a]$$

$$[\Theta]_{MRW} = \frac{\Theta \times 100 \times M_r}{c \times l \times N_A} \qquad [6b]$$

where Θ is the measured ellipticity in degrees, c is the protein concentration in mg/ml, l is the pathlength in cm, and M_r and MRW are the protein molecular weight and the mean residue weight, respectively. N_A is the number of amino acids per protein. $[\Theta]$ and $[\Theta]_{MRW}$ have the units degrees $\times$ cm^2 $\times$ dmol^{-1}. The factor 100 in Equations 5 and 6 originates from the conversion of the molar concentration to the dmol/cm^3 concentration unit.

The relation between $[\Theta]$ and $\Delta\varepsilon$ given in Equation 3 allows an immediate transformation of raw Θ data into ΔA values and vice versa by the relationship: $\Theta = 33(A_L - A_R)$. This implies that a measured ellipticity of 10 mdeg is equivalent to a ΔA of only 0.0003 absorbance units.

4.5 Determination of protein secondary structure by CD

CD data in the far-UV region can be used to determine the relative amounts of the different secondary structural elements of a protein. A number of different procedures are available for such analyses (19,22). They are based on the approximation that a protein CD spectrum in the amide region can be represented as a linear combination of the contributions of the different elements of secondary structure, according to Equation 7.

$$[\Theta(\lambda)] = f_\alpha \times [\Theta_\alpha(\lambda)] + f_\beta \times [\Theta_\beta(\lambda)] + f_t \times [\Theta_t(\lambda)] + f_n \times [\Theta_n(\lambda)] \qquad [7]$$

where $[\Theta_\alpha(\lambda)]$, $[\Theta_\beta(\lambda)]$, $[\Theta_t(\lambda)]$, and $[\Theta_n(\lambda)]$ are the basis spectra for α helical, β sheet, turn, and non-regular structures, and the f_i are the respective fractions of these structural elements. The coefficients f_i are obtained by solving Equation 7 simultaneously at a set of selected wavelengths.

The various procedures for the calculation of the fractions of secondary structure differ in the number and the choice of basis spectra $[(\Theta_i(\lambda)]$ in the wavelength range where spectral data are required and in computational aspects, such as the requirement to restrict Σf_i to unity. The major uncertainties arise from the choice of basis spectra. Early procedures, based on the CD of poly-amino acids that form extended secondary structures were found not to be satisfactory. More recently developed procedures generally use different sets of native proteins, whose CD spectra and three-dimensional structures are known. From the deconvolution of their CD spectra according to Equation 7 and the known f_i values, individual basis spectra $[(\Theta_i(\lambda)]$ can be derived that are a better representation of the secondary structural elements in globular proteins. Component spectra for the different types of secondary structure and for the average aromatic contribution to the far-UV CD are found in *Figure 6* of ref. 23. They are very helpful for a first, visual inspection of a measured protein CD spectrum.

The method of Provencher and Glöckner (24) avoids the problem of assigning basis spectra to the different types of secondary structure. It uses a linear combination of the CD spectra of reference proteins with known secondary structure for a direct analysis of the CD spectrum of a protein.

Secondary structure prediction from CD was significantly improved by using methods of variable selection of reference spectra (25–27). In these procedures, proteins with similar CD spectra as the unknown protein are employed as reference data in the analysis. Sreerama and Woody (27) provide a good description of the different methods for selecting adequate reference proteins, and they compare their performance when they are combined with the established procedures of secondary structure estimation.

Methods to determine the secondary structure of proteins and references to computer programs are given in refs 23–31. Recent critical evaluations of the methods are found in refs 27 and 31. The most pertinent points are summarized below.

(a) α Helices show a strong and characteristic CD spectrum, so f_α can be determined very well.

(b) The CD of β structures is weak; it depends on the length and twist of the respective β strands. This makes the estimates of f_β less reliable. The same holds for the CD of β turns.

(c) The aromatic side chains and disulfide bonds display CD bands in the region of the peptide CD as well. The magnitude and the sign of these contributions cannot be computed. They depend on the microenvironment of these chromophores. Aromatic contributions can invalidate secondary structure estimates for proteins with a low helix content and a large proportion of aromatic residues.

References

1. Wetlaufer, D. B. (1962). *Adv. Protein Chem.*, **17**, 303.
2. Gratzer, W. B. (1967). In *Poly-α-amino acids* (ed. G. D. Fasman), p. 177. Marcel Dekker Inc., New York.

3. Cantor, C. and Timasheff, S. N. (1982). In *The Proteins* (3rd edn) (ed. H. Neurath), Vol. V, p. 145. Academic Press, New York.
4. Peterson, G. L. (1983). In *Methods in enzymology* (ed. C. H. W. Hirs and S. N. Timasheff), Vol. 91, p. 95. Academic Press, New York.
5. Pace, C. N. and Schmid, F. X. This volume, Chapter 10.
6. Yanari, S. and Bovey, F. A. (1960). *J. Biol. Chem.*, **235**, 2818.
7. Bashford, C. L. (1987). In *Spectrophotometry and spectrofluorimetry: a practical approach* (ed. D. A. Harris and C. L. Bashford), p. 1. IRL Press, Oxford.
8. Stark, G. R. (1965). *Biochemistry*, **4**, 1030.
9. Donovan, J. W. (1973). In *Methods in enzymology* (ed. C. H. W. Hirs and S. N. Timasheff), Vol. 27, p. 497. Academic Press, New York.
10. Eftink, M. R. (1991). In *Methods of biochemical analysis* (ed. C. H. Suelter), Vol. 35, p. 127. J. Wiley, New York.
11. Penzer, G. R. (1980). In *An introduction to spectroscopy for biochemists* (ed. S. B. Brown), p. 70. Academic Press, London.
12. Cantor, C. and Schimmel, P. R. (1980). *Biophysical chemistry*, Vol. II. W. H. Freeman, San Francisco.
13. Brand, L. and Witholt, B. (1967). In *Methods in enzymology* (ed. C. H. W. Hirs and S. N. Timasheff), Vol. 11, p. 748. Academic Press, New York.
14. Cowgill, R. W. (1967). *Biochim. Biophys. Acta*, **140**, 37.
15. Teale, F. W. J. (1960). *Biochem. J.*, **76**, 381.
16. Heinemann, U. and Saenger, W. (1982). *Nature*, **299**, 27.
17. Eftink, M. R. and Ghiron, C. A. (1981). *Anal. Biochem.*, **114**, 199.
18. Payot, P. (1976). *Eur. J. Biochem.*, **63**, 263.
19. Johnson, W. C. (1985). In *Methods of biochemical analysis* (ed. D. Glick), Vol. 31, p. 61. J. Wiley, New York.
20. Bayley, P. M. (1980). In *An introduction to spectroscopy for biochemists* (ed. S. B. Brown), p. 148. Academic Press, London.
21. Strickland, E. H. (1972). *Biochemistry*, **11**, 3465.
22. Yang, J.-T., Wu, C.-S. C., and Martinez, H. M. (1986). In *Methods in enzymology* (ed. C. H. W. Hirs and S. N. Timasheff), Vol. 130, p. 208. Academic Press, New York.
23. Perczel, A., Park, K., and Fasman, G. D. (1992). *Proteins: Struct. Funct. Genet.*, **13**, 57.
24. Provencher, S. W. and Glöckner, J. (1981). *Biochemistry*, **20**, 33.
25. Manavalan, P. and Johnson, W. C. Jr. (1987). *Anal. Biochem.*, **167**, 76.
26. van Stokkum, I. H. M., Spoelder, H. J. W., Bloemendal, M., van Grondelle, R. and Groen, F. C. A. (1990). *Anal. Biochem.*, **191**, 110.
27. Sreerama, N. and Woody, R. W. (1994). *J. Mol. Biol.*, **242**, 497.
28. Chang, C. T., Wu, C.-S. C., and Yang, J.-T. (1978). *Anal. Biochem.*, **91**, 12.
29. Hennessey, W. C. Jr. and Johnson, W. C. Jr. (1981). *Biochemistry*, **20**, 1085.
30. Compton, L. A. and Johnson, W. C. Jr. (1986). *Anal. Biochem.*, **155**, 155.
31. Venyaminov, S. Y., Baikolov, I. A., Wu, C.-S. C., and Yang, J.-T. (1991). *Anal. Biochem.*, **198**, 250.
32. Haupt, G. W. (1952). *J. Res. Natl. Bureau Standards*, **48**, 414.

12

Measuring the conformational
stability of a protein

C. NICK PACE and J. MARTIN SCHOLTZ

1. Introduction

The stability of proteins, especially enzymes, has long been a practical concern (1), because this is usually the factor that most limits their usefulness. There are two very different aspects of protein stability. One is the chemical stability of the covalent structure, which involves covalent changes and is usually irreversible. The other is the conformational stability of the folded state, in the absence of covalent changes (2–8). The latter is the subject of this chapter, which will describe the simplest methods available for measuring how much more stable is the folded conformation of a protein than its unfolded conformations.

Measuring the conformational stability requires determining the equilibrium constant and the free energy change, ΔG, for the reaction:

$$\text{Folded (F)} \longleftrightarrow \text{Unfolded (U)} \qquad [1]$$

We will refer to the value of ΔG at 25 °C in the absence of a denaturant, $\Delta G(H_2O)$, as the conformational stability of a protein.[1] Measurements of the conformational stabilities of proteins are needed for a variety of purposes and have become especially important now that we can construct proteins to order. Studies of proteins differing slightly in structure are helping us gain a better understanding of the forces that determine the conformations of proteins and to optimize their stabilities (9,10).

We will describe how to determine and analyse thermal, urea, and guanidinium chloride (GdmCl) denaturation curves. These are relatively simple experiments that can be done in almost any laboratory. We will show how to use this information to estimate $\Delta G(H_2O)$, to determine the stability curve for a protein (11), and to measure differences in stability among proteins. These experiments sometimes reveal additional features of a protein such as the existence of domains or the presence of stable folding intermediates.

[1]Energies are given here in units of calories (cal); to convert to joules (J), multiply by 4.18 J/cal.

2. Selecting a technique to follow unfolding

You must first decide which technique to use to follow unfolding. The techniques used most often are UV absorbance spectroscopy, fluorescence, and circular dichroism (CD), which are described in Chapter 11. Other techniques used are biological activity measurements, nuclear magnetic resonance (NMR), viscosity, and other hydrodynamic methods (Chapters 8 and 9). Only spectroscopic techniques will be discussed here; biological activity measurements present special problems (12). To decide on a technique, the spectra of the folded and unfolded conformations of the protein of interest should be determined. As examples, the fluorescence and UV absorbance spectra of folded and unfolded ribonuclease T1 (RNase T1) are shown in *Figures 1a* and *1b*. Three features of the spectra are important in deciding on a technique to follow unfolding.

First, the magnitude of the response may be of crucial importance. The two absorbance spectra in *Figure 1b* required considerably more RNase T1 than the fluorescence spectra in *Figure 1a*. This is generally true; when fluorescence or far-UV CD can be used, less protein will be required than with other techniques. UV absorbance spectroscopy and near-UV CD generally require greater amounts of material, although the amounts depend on the wavelength chosen to follow unfolding. NMR generally requires the greatest amount of protein. In return, however, NMR will generally yield considerably more detailed structural information than the other techniques. Thus, the technique you choose may be limited by the amount of protein that you have available for the experiments.

Secondly, it is necessary to pick a technique and a wavelength for which the spectra of the folded and unfolded conformations differ significantly. With fluorescence, we chose 320 nm where the intrinsic fluorescence of folded RNase T1 is approximately sixfold greater than that of the unfolded protein (*Figure 2a*). With UV absorbance, we chose 286 nm and 292 nm which are maxima in the difference spectrum (*Figure 2b*). In general, pick the wavelength where the properties of the folded and unfolded conformations differ most.

The third factor that must be considered is the signal-to-noise ratio; the greater its value, the more accurate the measurements will be.

The spectral changes observed upon unfolding often depend upon different features of protein structure. For example, fluorescence and UV absorbance spectroscopy respond to changes in the environment of the tryptophan (Trp) and tyrosine (Tyr) residues, and hence to changes in tertiary structure, while CD measurements below 250 nm depend mainly on changes in the secondary structure of a protein. This may also be a consideration in determining the technique you should use to follow unfolding. These topics are discussed in more detail in Chapter 11.

Finally, two practical matters should be mentioned. Fluorescence measure-

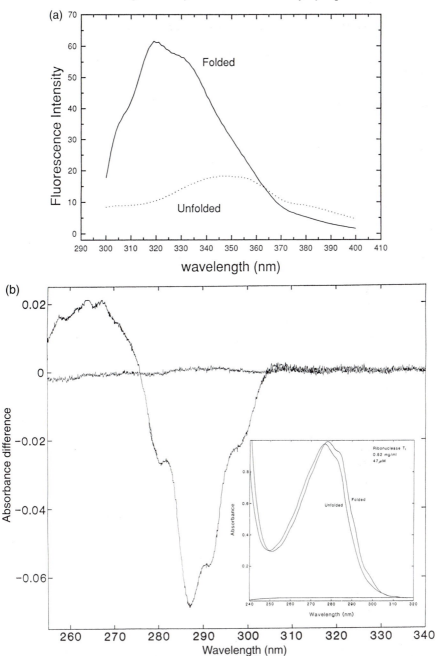

Figure 1. Spectra of folded and unfolded (8 M urea) RNase T1. (a) Intrinsic fluorescence emission spectra with excitation at 278 nm. (b) Difference in absorbance between the two absorbance spectra shown in the inset. The RNase T1 concentration was 0.01 mg/ml for (a) and 0.5 mg/ml for (b).

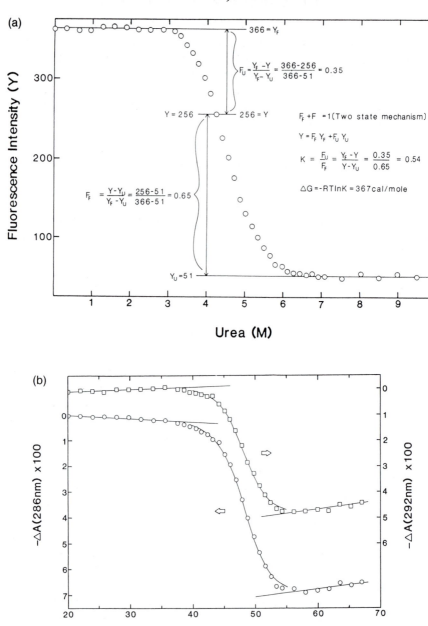

Figure 2. Urea- and heat-induced unfolding transitions of RNase T1. (a) Fluorescence intensity as a function of urea concentration at 25°C in 30 mM Mops buffer pH 7. The equations used to analyse an unfolding curve for a two-state mechanism are given. (b) Absorbance difference as a function of temperature in 30 mM Mops buffer pH 7. From ref. 48.

ments are less useful for following thermal unfolding, because the pre- and post-transition baselines are steep and temperature-sensitive. UV absorbance measurements are less convenient for following urea and GdmCl unfolding, and the pre- and post-transition baselines are steeper than with most of the other techniques. In general, unfolding curves with steeper pre- and post-transition baselines lead to larger errors in the parameters determined in the analysis (see Section 4).

3. Determining unfolding curves

Figure 2a and *b* show typical urea and thermal unfolding curves. In these cases, fluorescence was used to follow urea unfolding and UV absorbance was used to follow thermal unfolding, but we will refer to the physical parameter used to follow unfolding as y for the discussion that follows. The curves can be conveniently divided into three regions:

(a) The pre-transition region, which shows how y for the folded protein, y_F, depends upon the denaturant.

(b) The transition region, which shows how y varies as unfolding occurs.

(c) The post-transition region, which shows how y for the unfolded protein, y_U, varies with denaturant.

All of these regions are important for analysing unfolding curves. As a minimum, we recommend determining four points in the pre- and post-transition regions, and five points in the transition region. Of course, the more points determined the better defined the curve. In general, points at the corners, between regions, are less useful in the analysis of the results.

3.1 Equilibrium and reversibility

Since we are dealing with thermodynamic measurements, it is essential that the unfolding reaction has reached equilibrium before measurements are made and that the unfolding reaction is reversible. The time required to reach equilibrium can vary from seconds to days, depending upon the protein and the conditions. For example, unfolding of RNase T1 reaches equilibrium in minutes at 30°C, but requires hours at 20°C. For unfolding, the time to reach equilibrium is longest at the midpoint of the transition and decreases in both the pre- and post-transition regions. To ensure that equilibrium is reached, y is measured as a function of time to establish the time required to reach equilibrium.

To test the reversibility of unfolding, allow a solution to reach equilibrium in the post-transition region and then, by cooling or dilution, return the solution to the pre-transition region and measure y. The value of y measured after complete unfolding should be identical to that determined directly. In general, urea and GdmCl unfolding are more likely to be completely reversible than thermal unfolding. In fact, we have left RNase T1 in 6 M GdmCl solutions for three months and found that the protein will refold completely on dilution. The thermal unfolding of RNase T1 is not completely reversible, and the degree of irreversibility increases the longer the protein is exposed to unfolding conditions and the higher the temperature. Similar observations have been made with many other proteins. For this reason, thermal denaturation curves should be determined as quickly as possible.

Proteins that contain free sulfydryl groups present special problems. If the protein contains only free –SH groups and no disulfide bonds, then a reducing agent such as 10 mM dithiothreitol (DTT) can be added to ensure that no disulfide bonds form during the experiments. For proteins containing both free –SH groups and disulfide bonds, disulfide interchange can occur and this may lead to irreversibility. Disulfide interchange can be minimized by working at low pH (Chapter 7).

3.2 Urea and GdmCl unfolding

Each point shown in *Figure 2a* was determined on a separate solution. These solutions were prepared volumetrically, using the best available pipettes, by mixing a fixed volume of protein stock solution with the appropriate volumes of a buffer solution and a urea or GdmCl stock solution. The protein and buffer solutions are prepared by standard procedures. The urea or GdmCl stock solution must be prepared with considerable care; some suggestions are given below (*Table 1* and *Protocol 1*).

Protocol 1. Example of preparation of a urea stock solution[a]

1. Add approx. 60 g of urea to a tared beaker and weigh (59.91 g).[b] Now add 0.69 g of Mops buffer (sodium salt), 1.8 ml of 1 M HCl, 52 ml of distilled water, and weigh the solution again (114.65 g).[b]

2. Allow the urea to dissolve and check the pH. If necessary, add a weighed amount of 1 M HCl to adjust to pH 7.0.

3. Prepare a 30 mM Mops buffer pH 7.0.

4. Determine the refractive index of the urea stock solution (1.4173) and of the buffer (1.3343). Therefore, $\Delta N = 1.4173 - 1.3343 = 0.0830$.[b]

5. Calculate the urea molarity from ΔN using the equation given in *Table 1*: $M = 10.08$.[b]

6. Calculate the urea molarity based on the recorded weights. The density is calculated with the equation given in *Table 1*: weight fraction urea $(W) = 59.91/114.65 = 0.5226$; therefore $d/d_0 = 1.148$. Therefore volume $= 114.65/1.148 = 99.88$ ml. Therefore urea molarity $= 59.91/60.056/.09988 = 9.99$ M.[b]

---

7. The molarities calculated in steps 5 and 6 differed by less than 1%, so this solution was used to determine a urea unfolding curve.

[a] This describes the preparation of $\sim$ 100 ml of $\sim$ 10 M urea stock solution containing 30 mM Mops buffer pH 7.0. We use a top loading balance with an accuracy of about $\pm$ 0.02 g.
[b] Experimental result obtained in this example.

The method for determining an unfolding curve is given in *Protocol 2*. After the solutions for measurement have been prepared, they are incubated until equilibrium is reached at the temperature chosen for determining the unfolding curve. After the measurements have been completed, it is good practice to measure the pH of the solutions in the transition region. If the amount of protein is limited, instruments have been developed for following the unfolding of a protein simultaneously with several spectral techniques using a single protein solution (13). In addition, it is possible to use titration methods to determine urea or GdmCl denaturation curves (14).

Protocol 2.   Determining a urea or GdmCl unfolding curve

1. Prepare three solutions: a denaturant stock solution as described in *Protocol 1*, a protein stock solution, and a buffer solution.

2. Prepare the solutions on which measurements will be made volumetrically (e.g. with Rainin EDP2 pipettes) in clean, dry test tubes. Typical solutions used in determining a urea unfolding curve with fluorescence measurements are shown below. Only two solutions are shown for the pre- and post-transition regions. In the actual experiment, a total of 32 solutions were prepared and their fluorescence was measured.

3. Allow these solutions to equilibrate at the temperature chosen for the experiment until they reach equilibrium. (This is best determined in a separate experiment as described in the text.)

4. Measure the experimental parameter being used to follow unfolding on the solutions in order of ascending denaturant concentration. Leave the cuvette in the spectrophotometer, and do not rinse between samples. Simply remove the old solution carefully with a Pasteur pipette with plastic tubing attached to the tip, and then add the next sample. Leaving the cuvette in position improves the quality of the measurements, and the error introduced by the small amount of old solution is negligible.

5. Plot these results to determine if any additional points are needed. If so, prepare the appropriate solutions and make the measurements just as for the original solutions.

6. Measure the pH of the solutions in the transition region.

Protocol 2. *Continued*

7. Examples of the experimental results obtained are given in *Table 2*.

8. Analyse the results as described in *Protocol 5*.

Both urea and GdmCl can be purchased commercially in highly purified forms (e.g. from United States Biochemical). However, some lots of GdmCl are found to contain fluorescent impurities, and some lots of urea contain significant amounts of metallic impurities. Methods are available for checking the purity of GdmCl and for recrystallization when it is necessary (15). A procedure for purifying urea has also been described (16). GdmCl solutions are stable for months, but urea solutions slowly decompose to form cyanate and ammonium ions (17) in a process accelerated at high pH. The cyanate ions can react with amino groups on proteins (18). Consequently, a fresh urea stock solution should be prepared for each unfolding curve and used within one day.

Table 1 summarizes useful information for preparing urea and GdmCl stock solutions. We prepare urea stock solutions by weight, and then check the concentration by refractive index measurements using the equation given in *Table 1*. If the concentrations agree within 1%, we use the solution for determining an unfolding curve. The preparation of a typical urea stock solution is outlined in *Protocol 1*. Since GdmCl is quite hygroscopic, it is more difficult to prepare stock solutions by weight. Consequently, the molarity of GdmCl stock solutions is generally based on refractive index measurements and the equation given in *Table 1*.

Table 1. Information for preparing urea and GdmCl stock solutions

Property	Urea	GdmCl
Molecular weight	60.056	95.533
Solubility (25 °C)	10.49 M	8.54 M
$d/d_0{}^a$	$1 + 0.2658W + 0.0330W^2$	$1 + 0.2710W + 0.0330W^2$
Molarity^b	$117.66(\Delta N) + 29.753(\Delta N)^2$ $+ 185.56(\Delta N)^3$	$57.147(\Delta N) + 38.68(\Delta N)^2$ $-91.60(\Delta N)^3$

Grams of denaturant per gram of water to prepare:

6 M	0.495	1.009
8 M	0.755	1.816
10 M	1.103	–

^{a} W is the weight fraction denaturant in the solution, d is the density of the solution, and d_0 is density of water (19).
^{b} ΔN is the difference in refractive index between the denaturant solution and water (or buffer) at the sodium D line. The equation for urea solutions is based on data from ref. 20, and the equation for GdmCl solutions is from ref. 15.

3.3 Thermal unfolding

The time spent initially to ensure that the temperature of your cell can be measured accurately and maintained will be repaid many times over. Modern instruments may employ thermoelectric devices, such as the Peltier design, that allow for accurate and reliable control over the sample temperature. If the thermoelectric cells are not available, many instruments provide devices that allow water from a constant temperature bath to be circulated to maintain the temperature of the cell. Alternatively, a variety of cells can be purchased that allow water from a constant temperature bath to be circulated around the solution on which measurements are to be made. It is essential that great care be taken in securing all the tubing connections. Water at 90 °C spewing from a loose tube can be a major disaster. It is generally a good idea to insulate the tubing leading from the water-bath to the instrument. How accurately the temperature must be maintained depends mainly on the value of the enthalpy change, ΔH, for the unfolding of your protein and on whether you wish to determine the heat capacity change, ΔC_P, for unfolding (see Section 4.2). The greater the magnitude of ΔH, the more the equilibrium constant will fluctuate with temperature. In general, maintaining the temperature within $\pm$ 0.05 °C is adequate.

One way to monitor the temperature is to insert a thermistor directly into the sample cell, and then seal the cell to minimize evaporation. We use a probe and a telethermometer manufactured by Yellow Springs Instrument Co. for this purpose. Since the scale on our instrument is small, we attach a voltmeter (Micronta Digital Multimeter) and use this to read the output of the telethermometer. The system is calibrated using a National Bureau of Standards certified thermometer. With this approach, it is essential that the temperature is homogeneous throughout the cell, because the thermistor is generally in the top of the cell and measurements are made on solution near the centre of the cell. Consequently, the solution should be stirred. An alternative procedure is to calibrate the cell temperature against the bath temperature with your thermometer or thermistor in a separate experiment. Then it is necessary only to record the bath temperature while measurements are in progress and use the calibration curve to determine the cell temperature.

Only a single solution is needed to determine a thermal unfolding curve (*Protocol 3*). Consequently, less protein is required than for urea and GdmCl unfolding experiments. It is a good idea to filter (e.g. with a 0.65 μm filter) the protein solution before the experiment. This will remove any dust and often improves the signal-to-noise ratio. Since it may require several hours to determine a thermal unfolding curve, it is imperative that a very stable measuring instrument be used. Also, it is essential that blanks be run before and after the experiment to ensure that the instrument has not drifted significantly during the course of the experiment. After the solution has been cooled and the reversibility of unfolding checked, it is good practice to remeasure the pH of the solution.

Protocol 3. Determining a thermal unfolding curve

1. Select a buffer with a small ΔH of ionization so that the variation of the pK of the buffer and, hence, the pH of the solution, with temperature will be minimized. Near pH 7, Mops is excellent, and acetate and formate buffers are good at lower pH values. In general, ΔH is small for carboxyl groups and large for amino groups.

2. Make certain the instrument to be used to follow unfolding is thoroughly warmed-up and that the system for maintaining and monitoring the temperature of the cell is operating properly.

3. Prepare the protein solution and measure the pH and protein concentration. If the solution is not completely clear, filter it with a 0.65 μm filter.

4. Carefully measure the experimental parameter, y, for the blank solution (all components except the protein) at the lowest temperature that will be used in the unfolding curve.

5. Replace the blank solution with your protein solution and measure y at sufficient temperatures in the pre-transition region to determine y_F.

6. Measure y at temperatures through the transition region. These measurements should be made as quickly as possible, but it is essential that the system reach thermal and chemical equilibrium at each new temperature before the measurement is made. Generally the equilibration time is best determined in a separate experiment designed for this purpose. If only T_m and ΔH_m are needed, it is only necessary to make measurements at about six temperatures: three above and three below the T_m. If you are attempting to determine ΔC_P, measurements should be made with great care and at more temperatures in the transition region.

7. Measure y at sufficient temperatures in the post-transition region to determine y_U.

8. At the highest temperature used, replace the protein solution with a blank solution and remeasure y. If y for the blank is identical to that determined at the lowest temperature, your experiment is probably successful. If your instrument is not very stable, it is necessary to monitor the blank more carefully during the course of the experiment. These blank measurements are very important, especially with single-beam instruments.

9. Cool the sample back down to the starting temperature and measure

---

y again. If *y* has changed appreciably, this is evidence for a lack of reversibility of the thermal folding transition.

10. Remove the sample and check the pH of the protein solution after the experiment.

11. Analyse the data as described in *Protocol 6*. *Figure 2b* shows a typical thermal unfolding curve and *Protocol 6* shows an analysis of thermal unfolding data.

4. Analysing unfolding curves

The unfolding of many small globular proteins has been found to approach closely a two-state folding mechanism, such as that shown in Equation 1. *Protocol 4* gives a summary of methods that can be used to gain more information about the folding mechanism.

Protocol 4. Investigating the unfolding mechanism

1. If an unfolding curve shows more than one transition, unfolding is more complex than a two-state mechanism and the analysis of the data is more complicated (21). This behaviour is frequently observed with multidomain proteins where the domains unfold independently (13,22,23).

2. If a single-stage unfolding curve is observed, it does not prove that unfolding is a two-state mechanism (24,25).

3. Further insight can be gained by using different techniques or probes to following unfolding. Non-coincidence of plots of f_U as a function of temperature or denaturant concentration determined by different techniques indicates that significant amounts of intermediates are present at equilibrium and hence a simple two-state mechanism can not be used in analysing the data (12,13,24). Unfortunately, coincidence of the unfolding curves is only consistent with, but does not prove, that unfolding follows a two-state mechanism (26).

4. The best evidence that thermal unfolding follows a two-state mechanism is to show that ΔH determined by the van't Hoff relationship (Equation 7) is identical to that determined calorimetrically, where a specific folding mechanism is not required to determine ΔH_{cal}. When $\Delta H_{vH} < \Delta H_{cal}$, it is clear evidence that significant concentrations of intermediates are present at equilibrium (6,26–28).

5. The mechanism of folding of globular proteins is an area of great interest to biochemists and has been reviewed recently (29). For our purposes here, we need not be concerned with kinetic folding intermediates, only those intermediates present at equilibrium.

C. Nick Pace and J. Martin Scholtz

A two-state folding mechanism will be assumed for the discussion here. Consequently, for any of the points shown in *Figure 2a* or *b*, only the folded and unfolded conformations are present at significant concentrations, and $f_F + f_U = 1$, where f_F and f_U represent the fraction protein present in the folded and unfolded conformations, respectively. Thus, the observed value of y at any point will be $y = y_F f_F + y_U f_U$, where y_F and y_U represent the values of y characteristic of the folded and unfolded states, respectively, under the conditions where y is being measured. Combining these equations gives:

$$f_U = (y_F - y)/(y_F - y_U). \qquad [2]$$

The equilibrium constant, K, and the free energy change, ΔG, can be calculated using:

$$K = f_U/f_F = f_U/(1 - f_U) = (y_F - y)/(y - y_U) \qquad [3]$$

and

$$\Delta G = -RT \ln K = -RT \ln [(y_F - y)/(y - y_U)] \qquad [4]$$

where R is the gas constant (1.987 cal mol^{-1} K^{-1}) and T is the absolute temperature. Values of y_F and y_U in the transition region are obtained by extrapolating from the pre- and post-transition regions, as illustrated in *Figure 2a* and *b*. Generally, y_F and y_U are found to be linear functions of temperature or denaturant concentration and a least-squares analysis can be used to determine the linear expressions for y_F and y_U.

The calculation of f_U, K, and G from data such as those shown in *Figure 2a* and *b* is illustrated in *Protocols 5* and *6*. Values of K can be measured most accurately near the midpoints of solvent or thermal denaturation curves, and the error becomes substantial for values outside the range 0.1–10. Consequently, we generally only use ΔG values within the range ± 1.5 kcal/mol.

4.1 Urea and GdmCl unfolding

Protocol 5. Analysis of a urea unfolding curve

1. Data points from the transition region of an unfolding curve are listed in *Table 2*. Values of f_U, K, and ΔG were calculated using Equations 2, 3, and 4, respectively, and the y_F and y_U values given by the equations below for the pre- and post-transition regions.[a]

2. A least-squares analysis of this data yields the following for Equation 5: $\Delta G = 5680–(1218) \times$ [urea]. Thus, $\Delta G(H_2O) = 5.68$ kcal mol^{-1}, $m = 1218$ cal mol^{-1} M^{-1}, and [urea]$_{1/2} = 4.66$ M.

3. All of these parameters should be reported to characterize the results from a urea denaturation curve.

[a] $y_F = 112 + 19 \times$ [urea], and $y_U = 765$.

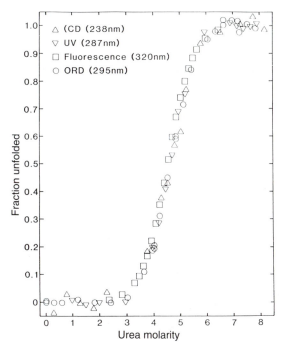

Figure 3. Fraction of RNase T1 unfolded, f_U, as a function of urea concentration. The values of f_U were calculated from data such as those in *Figure 2a* using Equation 2. From ref. 48.

Data such as those shown in *Figure 2a* have been analysed as described in *Protocol 5* and typical results are shown in *Figures 3* and *4*, and *Table 3*. In *Figure 3*, f_U is shown as a function of urea concentration. When comparing results from studies of related proteins that have similar pre- and post-transition baselines it is often useful to show just f_U as a function of denaturant in the transition region rather than a complete unfolding curve.

It can be seen in *Figure 4* that ΔG varies linearly with denaturant concentration in the limited region where ΔG can be measured. Similar results

Table 2. Data of a urea unfolding curve

[Urea] (M)	y	f_U	K	ΔG(cal/mol)
3.64	244	0.109	0.123	1242
3.94	293	0.186	0.229	873
4.03	302	0.199	0.248	825
4.24	370	0.312	0.453	469
4.54	452	0.449	0.815	121
4.85	536	0.594	1.462	−225
5.15	607	0.716	2.525	−549
5.45	677	0.840	5.261	−983

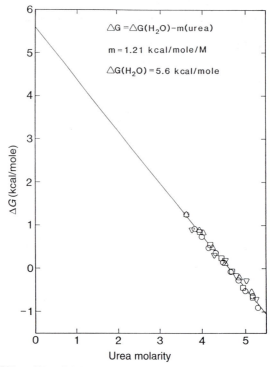

$$\triangle G = \triangle G(H_2O) - m(urea)$$

$$m = 1.21 \; kcal/mole/M$$

$$\triangle G(H_2O) = 5.6 \; kcal/mole$$

Figure 4. ΔG for RNase T1 unfolding as a function of urea concentration. The experimental values of ΔG were calculated from data such as those in *Figure 2a* using Equation 4. The solid line represents Equation 5 with $\Delta G(H_2O) = 5.6$ kcal mol^{-1} and $m = 1220$ cal mol^{-1} M^{-1}. From ref. 48.

have been obtained with many other proteins (30,31). The simplest method of estimating the conformational stability in the absence of urea, $\Delta G(H_2O)$, is to assume that this linear dependence continues to zero concentration and to use a least-squares analysis to fit the data to the following equation:

$$\Delta G = \Delta G(H_2O) - m \, [D] \qquad [5]$$

where m is a measure of the dependence of ΔG on denaturant concentration, [D]. Note also that the denaturant concentration at the midpoint of the unfolding curve, $[D]_{1/2} = \Delta G(H_2O)/m$. We strongly recommend that values of $\Delta G(H_2O)$, m, and $[D]_{1/2}$ be given in any study of the unfolding of a protein by urea or GdmCl.

If the linear extrapolation method is assumed for the analysis, it allows a convenient method to be used for analysing a urea or GdmCl denaturation curve (32). The transition region can then be characterized by two parameters, m, which measures the steepness of the transition, and $[D]_{1/2}$, which measures the midpoint of the transition, and the pre- and post-transition

312

Table 3. Example of experimental results obtained with a urea unfolding experiment

Urea (ml)[a]	Buffer (ml)[b]	Protein (ml)[c]	Urea (M)[d]	I[e]
0.45	2.55	0.2	1.41	135
0.90	2.10	0.2	2.83	136
1.20	1.80	0.2	3.77	115
1.25	1.75	0.2	3.93	113
1.30	1.70	0.2	4.09	105
1.35	1.65	0.2	4.25	99
1.40	1.60	0.2	4.40	87
1.45	1.55	0.2	4.56	81
1.50	1.50	0.2	4.72	69
1.55	1.45	0.2	4.87	62
1.60	1.40	0.2	5.03	53
1.65	1.35	0.2	5.19	47
1.70	1.30	0.2	5.34	40
2.05	0.95	0.2	6.45	21
2.50	0.50	0.2	7.87	22

[a] 10.07 M urea stock solution (30 mM Mops pH 7.0).
[b] 30 mM Mops buffer pH 7.0.
[c] 0.10 mg/ml RNase T1 stock solution (30 mM Mops pH 7.0).
[d] Urea molarity = 10.07 {(ml urea)/(ml urea + ml protein + ml buffer)} = 10.07 (ml urea)/3.2 = 3.1469 (ml urea).
[e] Fluorescence intensity (278 nm excitation and 320 nm emission) measured with a Perkin–Elmer MPF 44B spectrofluorometer.

regions can also be characterized by two parameters, y_F, and y_U for the intercepts, and m_F and m_U for the slopes. Consequently, the following equation can represent the entire denaturation curve:

$$y = \{(y_F + m_F[D]) + (y_U + m_U [D]) \times \exp[m \times ([D]-[D]_{1/2})/RT]\}/ (1 + \exp[m \times ([D]-[D]_{1/2})/RT]). \qquad [6]$$

The best fit of the six parameters and their confidence intervals can be determined by using the non-linear least-squares program described by Johnson and Frasier (33). This method can also be used to analyse thermal denaturation curves.

Other extrapolation methods can be used to estimate $\Delta G(H_2O)$. At present, there is no good reason for using these more complicated procedures, and they will not be discussed here (see ref. 21 for more information). In general, the estimates resulting from the other methods do not differ substantially when urea is the denaturant, but they may with GdmCl.

4.2 Thermal unfolding

We will now consider how to analyse a thermal unfolding curve, such as the one shown in *Figure 2b*, in order to determine the conformation stability, $\Delta G(H_2O)$ (see ref. 11 for an excellent discussion of this subject). This requires extrapolating ΔG measurements from the narrow temperature range where unfolding occurs to a reference temperature, T, such as 25 °C. *Protocol 6* illustrates the calculation of K and ΔG from results such as those in *Figure 2b*.

Protocol 6. Analysis of a thermal unfolding curve

1. The data are from a thermal unfolding curve for ribonuclease T1 determined under the same conditions as the urea unfolding curve analysed in *Protocol 5*. Values of f_U, K, and ΔG were calculated using Equations 2, 3, and 4 respectively (see *Table 4*). A least-squares analysis of the pre- and post-transition regions gave $y_F = 87.2 + 0.66 \times T$ and $y_U = 646$.

2. The slope of the plot of ΔG versus T is -300.3 cal mol^{-1} K^{-1}, and $T = T_m = 48.3$°C at $\Delta G = 0$. Since $\Delta G = 0$ at T_m, $H_m = T_m \times \Delta S_m$. Therefore, $\Delta H_m = (-1) \times (48.3 + 273.2) \times (-300.3) = 96.5$ kcal mol^{-1}. The slope of the van't Hoff plot (Equation 7) is $-48\,631$. Therefore, $\Delta H_m = -(1.987) \times (-48\,631) = 96.6$ kcal mol^{-1}. The van't Hoff plot shows no curvature over the temperature range of the transition, so clearly these data can not be used to determine ΔC_P with Equation 8.

3. To estimate ΔG at 25 °C use Equation 9. If it is assumed that $\Delta C_P = 0$, the second term in the equation is 0, and $\Delta G(25°C) = 7.0$ kcal mol^{-1}. If the average of eight experimental values is used, $\Delta C_P = 1460$ cal mol^{-1} K^{-1} and $\Delta G(25°C) = 5.7$ kcal mol^{-1}. If ΔC_P is calculated using Equation 10, $\Delta C_P = 1680$ cal mol^{-1} K^{-1} and $\Delta G(25°C) = 5.5$ kcal mol^{-1}. Note the excellent agreement with the value of $\Delta G(25°C) = 5.7$ kcal mol^{-1} from the analysis of a urea unfolding curve in *Protocol 5*.

The familiar method used to obtain the enthalpy change, ΔH, from these measurements is with the van't Hoff equation:

$$d(\ln K)/d(1/T) = -\Delta H/R \qquad [7]$$

The van't Hoff plots ($\ln K$ versus $1/T$) of protein unfolding transitions are found to be non-linear provided the transition covers a wide temperature range. This indicates that ΔH varies with temperature, which is expected when the heat capacities of the products and reactants differ:

$$d(\Delta H)/d(T) = C_P(U) - C_P(F) = \Delta C_P \qquad [8]$$

where $C_P(U)$ and $C_P(F)$ are the heat capacities of the unfolded and folded conformations, respectively, and ΔC_P is the change in heat capacity that accompanies protein unfolding. Therefore, both ΔC_P and ΔH are required to calculate ΔG as a function of temperature. Since ΔH is needed at only a single temperature, the best temperature to use is T_m, the midpoint of the thermal unfolding curve where $\Delta G(T_m) = 0 = \Delta H_m - T_m \Delta S_m$. Now with these

Table 4. Thermal unfolding data for ribonuclease T1

T (°C)	y	f_U	K	ΔG (cal/mol)
16.2	98.1			
21.0	100.9			
25.6	103.7			
30.2	107.4			
45.4	221.3	0.197	0.245	890
46.3	263.9	0.277	0.383	610
47.2	313.9	0.371	0.589	337
48.1	367.6	0.472	0.894	72
49.0	422.2	0.575	1.353	−193
49.9	474.1	0.673	2.061	−464
50.8	518.5	0.757	3.123	−733
51.7	555.5	0.828	4.805	−1013
61.2	645.4			
65.5	646.3			
69.8	646.3			
73.8	645.4			

parameters, the equation used to calculate ΔG at any temperature T, $\Delta G(T)$, is:

$$\Delta G(T) = \Delta H_m(1 - T/T_m) - \Delta C_P[(T_m - T) + T \ln (T/T_m)]. \qquad [9]$$

Thus, we need T_m, ΔH_m, and ΔC_P in order to calculate $\Delta G(T)$. The simplest method to determine T_m and ΔH_m is to plot ΔG as a function of temperature and then to use $T_m = T$ at $\Delta G = 0$, and $\Delta H_m = -1 \times (T_m$ in K$) \times$ (slope at T_m). Calculations based on this approach and on the van't Hoff equation are illustrated in *Protocol 6*.

The determination of ΔC_P is more difficult. One approach is to use a calorimeter to determine ΔC_P. In our best cases, ΔC_P can be measured directly to about ± 10% with a differential scanning microcalorimeter. A non-calorimetric approach that has been used successfully with T4 lysozyme (11), the chymotrypsinogen family (34) and other proteins (35), is to measure T_m and ΔH_m as a function of pH. T_m usually varies with pH with a concomitant variation in ΔH_m. The slope of the plot of ΔH_m versus T_m will provide ΔC_P, as shown by Equation 8. This assumes that ΔH and ΔC_P do not vary significantly with pH, which appears to be the case normally (26,33,34). This approach can also be used to determine ΔC_P with a calorimeter (27). In favourable situations, ΔC_P can be determined directly using data such as those shown in *Figure 2b*. Equations 7 and 8 show that this requires taking the second derivative of the experimental data; this is only possible if the experimental data are exceptionally good. We had success with this approach in a favourable case (36), but we have not been able to determine a reasonable and reproducible value of ΔC_P for RNase T1 using this method. Brandts used this approach to determine ΔC_P for chymotrypsinogen in his pioneering studies in this area (37). All the methods described here give comparable values of ΔC_P for chymotrypsinogen (38).

Another useful technique for determining ΔC_P without a calorimeter has been described by Pace and Laurents (39). In this method, the $\Delta G(H_2O)$ values from urea denaturation experiments at low temperatures are combined with ΔG values from the transition region of a thermal unfolding curve. A least-squares fit of the $\Delta G(T)$ data to Equation 9 provides a measure of ΔH_m, T_m, and ΔC_P. This method is illustrated in *Figure 5* for a small protein, HPr (40). In favourable cases, such as this, when the temperature of maximum stability is above 0°C, the ΔC_P value can be determined with an accuracy of 5–10%. If the temperature of maximum stability is below 0°C, there is more un-certainty in the ΔC_P value determined from this method; however, since this method does not employ a calorimeter, almost any laboratory can use this method to determine ΔC_P using the simple methods discussed in this chapter.

For several proteins, estimates of ΔC_P in reasonable agreement with ex-perimental values can be calculated using model compound data and the amino acid composition of the protein (41). A recent report (42) provides a simple way to estimate the ΔC_P for the protein unfolding reaction using only information from the primary sequence of the protein:

$$\Delta C_P \approx 172 + 17.6 \times N - 164 \times SS \text{ (in cal mol}^{-1}\text{ K}^{-1}) \qquad [10]$$

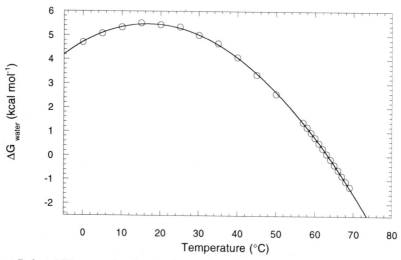

Figure 5. A stability curve for the small bacterial protein HPr using the method of Pace and Laurents (39). The data at $T \le 50°C$ are $\Delta G(H_2O)$ values from urea denaturation curves, while the data at higher temperatures are from a thermal unfolding curve in the absence of denaturant. The line represents the best fit to Equation 8 with T_m = 63.5 ± 0.2°C, ΔH_m = 70.4 ± 1.4 kcal mol^{-1}, and C_P = 1.45 ± 0.08 kcal mol^{-1} K^{-1}. The data are from Nicholson and Scholtz (40).

where N is the number of amino acid residues and SS is the number of disulfide cross-links in the protein. If the structure of your protein is known, there are several other equations that can be used to provide even better estimates for ΔC_P (see ref. 42 for a comparison of the various empirical methods). We have used this approach in *Protocol 6* to obtain a value of ΔC_P for RNase T1 folding to illustrate the use of Equation 9.

5. Determining differences in stability

It is frequently of interest to determine differences in conformational stability among proteins that vary slightly in structure. The structural change might be a single change in amino acid sequence achieved through site-directed mutagenesis, or a change in the structure of a side chain resulting from chemical modification. *Table 5* presents results from urea and thermal unfolding studies of wild-type RNase T1 and a mutant that differs in amino acid sequence by one residue. Three different methods of calculating the differences in stability, $\Delta(\Delta G)$, are illustrated.

The midpoints of urea, $[\text{urea}]_{1/2}$, and thermal, T_m, unfolding curves can be determined quite accurately and do not depend to a great extent on the unfolding mechanism (21). In contrast, measures of the steepness of urea, m, and thermal, $-\Delta S_m$, unfolding curves cannot be determined as accurately, and deviations from a two-state folding mechanism will generally change these values. Consequently, differences in stability determined by comparing the $\Delta G(H_2O)$ values can have large errors. However, when comparing completely different proteins or forms of a protein that differ markedly in stability, no other choice is available. This approach is illustrated by the first column of $\Delta(\Delta G)$ values in *Table 5*. *Table 6* illustrates the dangers of trying to draw conclusions about the conformational stabilities of unrelated proteins based solely on the midpoints of their unfolding curves. Lysozyme and myoglobin have similar $\Delta G(H_2O)$ values at 25 °C, but a much higher concentration of GdmCl is needed to denature lysozyme because the m value is much smaller. Likewise, lysozyme and cytochrome c unfold at about the same temperature, even though lysozyme has a much larger value of $\Delta G(H_2O)$ at 25 °C.

A second approach is to compare the proteins in the presence of urea by taking the difference between the $[D]_{1/2}$ values and multiplying this by the average of the m values. This is illustrated by the $\Delta(\Delta G)$ values in the last column of *Table 5*. The rationale here is that the error in measuring the m values should generally be greater than any differences resulting from the effect of small changes in structure on the m value. However, substantial differences in m values between proteins differing by only one amino acid in sequence have been observed with some proteins (43).

Becktel and Schellman have provided a simple method for the analysis of

Table 5. Analysis of urea unfolding curves (at 25°C) and thermal unfolding curves for wild-type RNase T1 and a mutant which differs by one amino acid

Protein	$\Delta G(H_2O)^a$	$\Delta(\Delta G)^b$	m^c	$[Urea]_{1/2}^d$	$\Delta[Urea]_{1/2}^e$	$\Delta(\Delta G)^f$
Wild-type	6.41		1210	5.30		
Tyr11Phe	4.52	1.9	1270	3.56	1.74	2.2

Protein	ΔS_m^g	ΔH_m^h	T_m^i	$\Delta(T_m)^j$	$\Delta(\Delta G)^k$
Wild-type	339	110	50.9		
Tyr11Phe	317	101	44.9	6.0	2.0

[a] From Equation 5, in kcal mol^{-1}.
[b] Difference between the $\Delta G(H_2O)$ values in kcal mol^{-1}.
[c] From Equation 5, in cal mol^{-1} M^{-1}.
[d] Midpoint of the urea unfolding curve in M.
[e] Difference between the $[urea]_{1/2}$ values in M.
[f] From $[urea]_{1/2} \times 1240$ (the average of the two m values) in kcal mol^{-1}.
[g] Negative of the slope of ΔG versus T at T_m in cal mol^{-1} K^{-1}.
[h] $\Delta H_m = T_m(K) \times (\Delta S_m)$ in kcal mol^{-1}.
[i] Midpoint of the thermal unfolding curve in °C.
[j] Difference between the T_m values.
[k] $\Delta(\Delta G) = (\Delta T_m) \times \Delta S_m(wt) = \Delta T_m \times [\Delta H_m(wt)/T_m(wt)]$, where ΔT_m is the difference in the midpoints of the thermal transitions and $\Delta S_m(wt)$ and $\Delta H_m(wt)$ are the values for the better characterized, usually wild-type, protein (see ref. 11).

thermal denaturation curves in order to determine differences in stability (11). We have used their suggestions in calculating the $\Delta(\Delta G)$ values from thermal unfolding in *Table 5*. For these proteins, the estimates of $\Delta(\Delta G)$ are in good agreement, which is reassuring. This good agreement, however, is not always observed. We think it is prudent to measure $\Delta(\Delta G)$ by both urea and thermal unfolding studies whenever possible.

An ingenious method for determining a urea denaturation curve in a single experiment using electrophoresis (44) is described in Chapter 8. *Figure 6a* shows a urea gradient gel for RNase T1. $[D]_{1/2}$ values can be estimated from these gels reasonably well using a ruler. In *Figure 6b*, we show that $[D]_{1/2}$ values determined by urea gradient gel electrophoresis are in surprisingly good agreement with values determined by conventional urea unfolding

Table 6. Unfolding curve midpoints of different proteins

Protein	$[GdmCl]_{1/2}$ (M)	$\Delta G(H_2O)$ (kcal mol^{-1}) at 25°C
Myoglobin	1.7	12
Lysozyme	3.1	12

Protein	T_m (°C)	$\Delta G(H_2O)$ (kcal mol^{-1}) at 25°C
Cytochrome c	80	8
Lysozyme	80	12

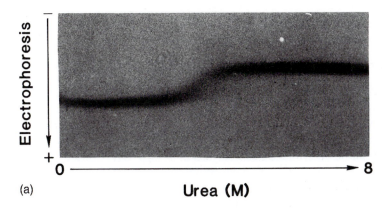

(a) **Urea (M)**

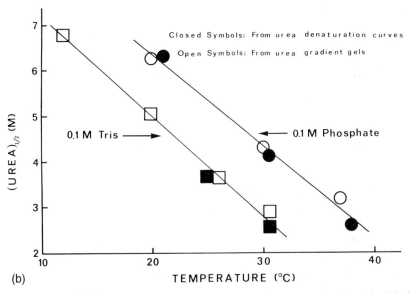

(b)

Figure 6. Unfolding of RNase T1 measured by urea gradient electrophoresis. (a) An example of a urea gradient gel (see Chapter 8). (b) Comparison of the unfolding measured by urea gradient gels (open symbols) and by conventional urea unfolding curves (closed symbols), using two different buffers. The midpoints of the urea denaturation curves are given at various temperatures.

curves. These results show that RNase T1 is much more stable in a phosphate buffer than in a Tris buffer; they also show the marked dependence of the stability of RNase T1 on temperature. Urea gradient gels are useful for a variety of purposes and provide a quick means of assessing possible changes in stability. In addition, they can be used to obtain a rough estimate of $\Delta G(H_2O)$ (45).

6. Concluding remarks

In the past, we preferred results from urea and GdmCl unfolding over thermal unfolding studies for several reasons: the product of unfolding seemed to be better characterized, unfolding was more likely to closely approach a two-state mechanism, and unfolding was more likely to be completely reversible. However, in the cases that have been carefully investigated, the two techniques seem to give estimates of $\Delta G(H_2O)$ that are in good agreement (48). After getting the equipment set-up and learning the procedures, either type of unfolding curve can be determined and analysed in a single day. Consequently, the safest course is to determine both types of unfolding curves whenever possible. Also, we generally prefer urea over GdmCl because salt effects can be investigated, and sometimes reveal interesting information (46). However, GdmCl is a more potent denaturant, and this is sometimes very useful.

Studies of mutant proteins over the past few years have improved our understanding of the forces that contribute to the conformational stability of globular proteins (10). In addition, our understanding of how to make proteins more stable and more useful is improving, as is our ability to design new proteins (47). However, much remains to be learned and the techniques described in this chapter are essential tools for doing so.

Finally, we should emphasize that this is a 'how to do it chapter' and we ignored or only briefly mentioned some interesting and unresolved questions. See refs 2–11 for more information on the topics discussed here.

Acknowledgements

We thank Janis Chmiel for her excellent assistance in preparing this chapter. We also thank the NIH (GM 37039 and GM 52483), the Robert A. Welch Foundation (A-1060 and A-1281), the American Cancer Society Junior Faculty Research Award program (JFRA-577), and the Tom and Jean McMullin Professorship for financial support of the work in our laboratories.

References

1. O'Sullivan, C. and Tompson, F. W. (1890). *J. Chem. Soc.*, **57**, 834.
2. Kauzmann, W. (1959). *Adv. Protein Chem.*, **14**, 1.
3. Tanford, C. (1968). *Adv. Protein Chem.*, **23**, 121.
4. Tanford, C. (1970). *Adv. Protein Chem.*, **24**, 1.
5. Pace, C. N. (1975). *Crit. Rev. Biochem.*, **3**, 1.
6. Privalov, P. L. (1979). *Adv. Protein Chem.*, **33**, 167.
7. Pace, C. N. (1990). *Trends Biochem. Sci.*, **15**, 14.
8. Creighton, T. E. (ed.) (1992). *Protein folding*. Freeman, New York.
9. Shortle, D. (1992). *Q. Rev. Biophys.*, **25**, 2.
10. Matthews, B. W. (1993). *Adv. Protein Chem.*, **46**, 249.

11. Becktel, W. J. and Schellman, J. A. (1987). *Biopolymers*, **26**, 1859.
12. Wong, K. P. and Tanford, C. (1973). *J. Biol. Chem.*, **248**, 8518.
13. Saito, Y. and Wada, A. (1983). *Biopolymers*, **22**, 2105.
14. Scholtz, J. M. (1995). *Protein Sci.*, **4**, 35.
15. Nozaki, Y. (1972). In *Methods in enzymology* (ed. C. H. W. Hirs and S. N. Timasheff), Vol. 26, p. 43. Academic Press, New York.
16. Prakash, V., Loucheux, C., Scheufele, S., Gorbunoff, M. J., and Timasheff, S. N. (1981). *Arch. Biochem. Biophys.*, **210**, 455.
17. Hagel, P., Gerdling, J. J. T., Fieggen, W., and Bloemendal, H. (1971). *Biochim. Biophys. Acta*, **243**, 366.
18. Stark, G. R. (1965). *Biochemistry*, **4**, 1030.
19. Kawahara, K. and Tanford, C. (1966). *J. Biol. Chem.*, **241**, 3228.
20. Warren, J. R. and Gordon, J. A. (1966). *J. Phys. Chem.*, **67**, 1524.
21. Pace, C. N. (1986). In *Methods in enzymology* (ed. C. H. W. Hirs and S. N. Timasheff), Vol. 131, p. 266. Academic Press, New York.
22. Waheed, A., Qasim, M. A., and Salahuddin, A. (1977). *Eur. J. Biochem.*, **76**, 383.
23. Azuma, T., Hamaguchi, K., and Migita, S. (1972). *J. Biochem.*, **72**, 1457.
24. Kuwajima, K. (1989). *Proteins: Struct., Funct., Genet.*, **6**, 87.
25. Price, N. C. (1994). In *Mechanisms of protein folding* (ed. R. H. Pain), p. 160. IRL Press, Oxford.
26. Lumry, R., Biltonen, R., and Brandts, J. F. (1966). *Biopolymers*, **4**, 917.
27. Privalov, P. L. and Khechinashvili, N. N. (1974). *J. Mol. Biol.*, **86**, 665.
28. Freire, E. and Biltonen, R. L. (1978). *Biopolymers*, **17**, 463.
29. Kim, P. S. and Baldwin, R. L. (1990). *Annu. Rev. Biochem.*, **59**, 631.
30. Greene, R. F. and Pace, C. N. (1974). *J. Biol. Chem.*, **249**, 5388.
31. Ahmand, F. and Bigelow, C. C. (1982). *J. Biol. Chem.*, **257**, 12935.
32. Santoro, M. M. and Bolen, D. W. (1989). *Biochemistry*, **27**, 8063.
33. Johnson, M. L. and Frasier, S. G. (1985). In *Methods in enzymology* (ed. C. H. W. Hirs), Vol. 117, p. 301. Academic Press, New York.
34. Shiao, D. F., Lumry, R., and Fahey, J. (1971). *J. Am. Chem. Soc.*, **93**, 2024.
35. Swint, L. and Robertson, A. D. (1993). *Protein Sci.*, **2**, 2037.
36. Pace, C. N. and Tanford, C. (1986). *Biochemistry*, **7**, 198.
37. Brandts, J. F. (1964). *J. Am. Chem. Soc.*, **86**, 4291.
38. Jackson, W. M. and Brandts, J. F. (1970). *Biochemistry*, **9**, 2294.
39. Pace, C. N. and Laurents, D. V. (1989). *Biochemistry*, **28**, 2520.
40. Nicholson, E. M. and Scholtz, J. M. (1996). *Biochemistry*, **35**, 11369.
41. Edelhoch, H. and Osborne, J. C. Jr. (1976). *Adv. Protein Chem.*, **30**, 183.
42. Myers, J. K., Pace, C. N., and Scholtz, J. M. (1995). *Protein Sci.*, **4**, 2138.
43. Shortle, D. (1995). *Adv. Protein Chem.*, **46**, 217.
44. Creighton, T. E. (1986). In *Methods in enzymology* (ed. C. H. W. Hirs and S. N. Timasheff), Vol. 131, p. 156. Academic Press, New York
45. Hollecker, M. and Creighton, T. E. (1982). *Biochim. Biophys. Acta*, **701**, 395.
46. Pace, C. N. and Grimsley, G. G. (1988). *Biochemistry*, **27**, 3242.
47. Betz, S. F., Bryson, J. W., and DeGrado, W. F. (1995). *Curr. Opin. Struct. Biol.*, **5**, 457.
48. Thomson, J. A., Shirley, B. A., Grimsley, G. A., and Pace, C. N. (1989). *J. Biol. Chem.*, **264**, 11614.

13

Immunochemical analysis of protein conformation

BERTRAND FRIGUET, LISA DJAVADI-OHANIANCE, and
MICHEL E. GOLDBERG

1. Introduction

Antibodies can be used to study protein conformation in two ways. One, already outlined in Chapter 3, is to take advantage of their specificity and of the sensitivity of immunochemical detection methods to trace the protein after an appropriate analytical procedure. This approach is particularly useful when the biological activity of the protein is not easy to test. This is the case for instance with engineered proteins that have lost some of their original properties. Using immunochemical detection, it is possible to determine, even with crude extracts, such physicochemical parameters as molecular weight (by electrophoresis; see Chapters 1 and 8), diffusion coefficient (by gel filtration; see Chapter 9), or sedimentation coefficient (by sucrose gradient), ligand fixation (by immunoprecipitation of the ligand-bound protein), and stability (by irreversible heat precipitation). In this approach, polyclonal as well as monoclonal antibodies can be used, as long as the antibodies are able to detect the protein with sufficient efficiency and specificity.

The second approach is to use antibodies as real conformational probes aimed at analysing conformational changes in a protein. Such an analysis requires a quantitative estimate of the 'strength' of the association between the antibody and the various forms of the antigen. This is why polyclonal antibodies as obtained, without fractionation, from an immune serum can usually not be used as conformational probes. Their heterogeneity in terms of the antigenic sites they recognize on their target protein precludes the observation of a precise region of the molecule, thus failing to detect local changes. Even if rendered more specific for a given region of the antigen by a rigorous purification on an appropriate immunoadsorbent, they will usually remain heterogeneous in terms of binding constants and thus cannot be used as a quantitative probe. This is why the heterogeneous antibodies of a polyclonal immune serum should be used only in preliminary studies aimed at deciding qualitatively whether or not immunochemical changes do occur in a protein. For instance, the remarkable finding of Crumpton (1) that an immune serum raised against apomyoglobin expels the haem from haem-saturated metmyoglobin is a perfect illustration of the capacity of a polyclonal serum to reveal a conformational change, but of its limitations in providing even a qualitative understanding of the nature, the amplitude, and the rate of this change. Hence the need for a quantitative immunochemical probe adapted to the study envisaged. Properly selected monoclonal antibodies can be used as such quantitative probes, because they are homogeneous in terms of both their antigenic site and their affinity.

In this chapter, our purpose is to develop some aspects of the immuno-chemical analysis of protein conformation with monoclonal antibodies (mAb).

The strategies and procedures for the preparation and cloning of hybrido-mas are described in different reviews (2–4). The one we often use success-fully is the mass fusion protocol described by P. Legrain *et al.* (5). We wish however to emphasize that, depending on the problem to be solved, it is important before the immunization to define the antibodies needed as con-formational probe. For example, has the immunization to be done with the whole protein, with short fragments of this protein, such as synthetic polypeptides, or with large fragments, such as functional or folding domains? Moreover, many questions arise concerning the behaviour of the protein during the injection and the fate of this protein in the immunized animal. These problems are, however, outside the scope of this article, but some are considered in Chapter 3.

This chapter first describes experiments that must be designed to select and characterize mAb appropriate for conformational probes. It then describes how to determine a quantitative parameter, the affinity in solution, permit-ting use of mAb as conformational probes.

2. Antibody selection: how to obtain the appropriate antibodies

Before the hybridomas are obtained, it is necessary to devise a reliable tech-nique to detect the production of specific antibodies. Often hundreds of hybridoma supernatants are tested, so the technique used must be both powerful and simple. Commonly used are immunoprecipitation, using a radiolabelled antigen, or a solid phase immunoassay. Monoclonal antibodies are monospecific, so direct immunoprecipitation of the antigen by the mAb cannot be expected, and a double antibody precipitating system would be required. Therefore, the solid phase immunoassay, using commercially avail-able enzymatic immunoconjugates, or radiolabelled antibody, or protein A, appears the most convenient technique. *Protocol 1* gives a comprehensive description of a typical enzyme-linked immunosorbent assay (ELISA), point-ing out the pitfalls of the technique.

Protocol 1.   Indirect ELISA procedure

Equipment and reagents

- ELISA plate spectrophotometer or fluorimeter (Titertek, Dynatech, SLT Labinstruments)
- Plate washer (e.g. Titertek Handiwash, Nunc Immunowash)
- 96-well flat bottom microtitration plates and plate sealers
- Repeater pipettes and the tips (e.g. Eppendorf Multipette and Combitips)
- 5 ml glass tubes
- Substrate solution 1 for alkaline phosphatase:[b] dissolve 20 mg of disodium *p*-nitrophenylphosphate (PNPP) in 10 ml DEA solution (1 M diethanolamine adjusted with HCl to pH 9.8 and supplemented with 1 mM magnesium sulfate)—prepare this solution fresh just before use
- Substrate solution 2 for β-galactosidase:[b] prepare 0.4% (w/v) *o*-nitrophenyl-β-D-galactopyranoside (ONPG) stock solution in PM₂ buffer (70 mM disodium phosphate, 30 mM monosodium phosphate, 1 mM magnesium sulfate, 0.2 mM manganese sulfate, 2 mM EDTA magnesium salt pH 7.0 with HCl). To prepare a working solution just before use, add 5 ml of 0.4% (w/v) ONPG stock solution to 20 ml PM₂ buffer and then add 18 µl 2-mercaptoethanol. The ONPG stock solution can be kept in the dark at 4°C until it becomes yellowish.
- Substrate solution 3 (fluorogenic substrate solution for β-galactosidase):[b] prepare a 20 mM MUG (methyl-umbelliferyl-β-galactopyranoside) stock solution by dissolving 34 mg of MUG in 5 ml dimethylformamide in a glass test tube. If necessary, warm gently by passing the tube quickly on the pilot-light of a Bunsen burner until the MUG is dissolved. Keep the stock solution in the dark at –20°C. To prepare a working solution, mix 0.25 ml of the 20 mM MUG stock solution to 25 ml PM₂ and add 18 µl 2-mercaptoethanol.
- Immunoconjugate: antibody directed against mouse immunoglobulins linked to alkaline phosphatase or β-galactosidase[a] (e.g. from Biosys, Southern Biotechnology Associates Inc., Promega)
- Coating buffer: 50 mM carbonate buffer pH 9.6—dissolve 1.59 g Na_2CO_3 and 2.93 g $NaHCO_3$ in 1 litre of distilled water final volume
- Washing buffer: PBS pH 7.4 supplemented with 0.05% (w/v) Tween 20—dissolve 8 g NaCl, 0.2 g KH_2PO_4, 2.8 g $Na_2HPO_4.12 H_2O$, and 0.2 g KCl in 1 litre of distilled water final volume
- Specific antibodies (at the desired concentrations in the buffer which will be used to determine the K_D in *Protocol 7*)
- Antigen 1 µg/ml in coating buffer

Method

Assays must be done in triplicate, plus a control with a non-coated well (see legend of *Figure 1*).

1. Add 0.1 ml of 1 µg/ml antigen solution (in coating buffer) to wells + of an ELISA microtitration plate following the scheme in *Figure 1*. Do not coat the first vertical line of wells, which will become blank wells containing only the substrate solution.

2. Incubate the plate, covered with a plate sealer, for at least 3 h at room temperature or overnight at 4°C. Usually a blocking step by overcoating with BSA (0.5%, w/v) is not needed.

3. Empty the plate by turning it upside down and shaking it over a sink, then hitting it strongly several times on a pile of paper towels.

4. Wash out all wells with washing buffer three times, allowing a 3 min incubation between washes at the chosen temperature.

5. After the third wash, add to three coated wells and one non-coated

Protocol 1. *Continued*

well 0.1 ml of each antibody solution. Incubate the covered plate for 30 min at the chosen temperature.

6. Wash the wells three times, as in steps 3 and 4, at room temperature.

7. Add to each well 0.1 ml of immunoconjugate previously diluted in washing buffer. Depending on the immunoconjugate supplier, use a 500- to 1000-fold dilution. Incubate for 30 min at room temperature.

8. Wash the wells three times, as in steps 3 and 4, at room temperature.

9. Add to each well 0.1 ml of the appropriate substrate solution.

10. Cover the plate with an adhesive plate sealer to avoid substrate evaporation, which can lead to erroneous results. Incubate at room temperature. If the reaction is too slow at room temperature, incubate the plate at 37 °C.

11. Follow the appearance of the product by measuring (without the plate sealer) the absorbance or fluorescence at the relevant wavelength for each substrate. Thus, for substrate solution 1 or 2, measure the change in absorbance at 405 nm. When using substrate solution 3, measure the fluorescence at 480 nm (excitation wavelength 355 nm).

12. If needed, stop the reaction by adding 0.05 ml of the appropriate stopping solution; 1 M sodium phosphate pH 7 for substrate solution 1; 1.43 M Na_2CO_3 for substrate solution 2 or 3.

[a] These two enzymes are preferred to peroxidase because of their linearity in the assay response.
[b] Three substrate solutions are listed, one for use with an immunoconjugate linked to alkaline phosphatase, and two for β-galactosidase-linked immunoconjugates. Prepare only the appropriate substrate solution.

2.1 Important remarks: what should be remembered?

For antibody selection by the indirect ELISA procedure just described, the antigen must be highly pure. Fortunately, the ELISA procedure requires only very small amounts of pure antigen, and only the antibody selection step must be highly specific. Therefore, in the case of proteins difficult to purify, a partially purified protein can be used for immunization. Antibodies directed against specific regions or subunits of the protein can be selected directly during the screening of the hybridoma if highly purified subunits or fragments are available and used in the coating step of the ELISA.

As already pointed out, solid phase immunoassays such as ELISA deal with a protein adsorbed on to the plastic wells. Depending upon the protein, two primary phenomena can occur:

(a) When coated on to the plastic, the protein may undergo complete or partial denaturation, which will expose some areas normally hidden in

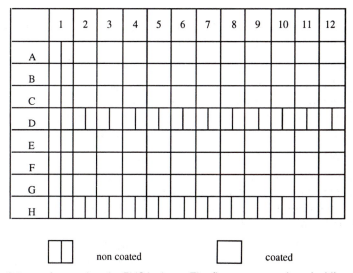

Figure 1. Scheme for coating the ELISA plates. The first non-coated vertical line A1 to H1 will become the blank wells containing only the substrate solution when measuring absorbances in the spectrophotometer. The wells used for assays (triplicate plus control in a non-coated well) are A2 to D2 for the first sample, A3 to D3 for the second one, etc.

the native protein. This may lead to the selection of antibodies directed against a non-native form of the protein (6).

(b) Some proteins, such as those binding to membranes or to other proteins, may adopt specific orientations when interacting with the plastic wells. Consequently, important antigenic determinants will be hidden on the coated protein.

To avoid problems encountered by the direct adsorption of the protein on the plastic, modified ELISA methods have been proposed (7,8) in which the protein is immobilized via a capture protein such as protein–avidin–biotin system or polyclonal antibodies.

One should, however, know that polyclonal antiserum contains antibodies directed against the denatured form of the antigen; consequently, denatured forms of the protein that could be present in the antigen preparation can be captured. In this case, the screening will lead also to the selection of some antibodies against non-native forms of the protein.

An ELISA method that seems to offer a reliable way to select, during the screening, mAbs recognizing the native antigen has been described by Smith and Wilson (9). In this method, the medium containing the mAb is first incubated with the native antigen in solution. After this first incubation, the antibody–antigen mixture is transferred to the wells of a microtitration plate coated with polyclonal antibodies against the same antigen, and the complex is detected by enzyme conjugated anti-mouse immunoglobulin. The presence of the complex reflects the ability of the antibody to bind the native antigen during the incubation in solution. One should even incubate for a shorter time (20 min instead of 1 h) the antibody–antigen mixture in solution to avoid any binding by the antibody of transient unfolded forms of the protein.

If the screening of antibodies has been performed by indirect ELISA, then a competition ELISA described in *Protocol 4* must be performed to discriminate between the anti-native and anti-denatured antibodies.

3. Antibody characterization

After hybridoma cloning, antibody can be produced either *in vitro* from cultures of the clones or *in vivo* by inducing ascitic fluids that contain a high yield of monoclonal immunoglobulins. Before using these antibodies as conformational probes of the protein structure, it is necessary to purify and characterize them.

3.1 Antibody purification

Pure antibodies are obtained by performing successively ammonium sulfate precipitation of the immunoglobulins (*Protocol 2*) and ion exchange chromatography with diethylaminoethyl (DEAE) anion exchanger (*Protocol 3*).

Protocol 2. Ammonium sulfate precipitation of mAb

Equipment and reagents
- Refrigerated centrifuge
- Centrifuge tube
- Magnetic stirrer and stirring bar
- Spectrophotometer
- Saturated ammonium sulphate solution, pH 7
- 40 mM sodium phosphate, pH 8

Method

1. Prepare saturated ammonium sulfate solution (76 g $(NH_4)_2SO_4$ into 100 ml distilled water). Adjust to pH 7.
2. Centrifuge the ascitic fluid at 5000 g for 30 min.
3. Transfer the supernatant to an appropriate beaker and determine the volume. Add stirring bar and place on a magnetic stirrer.
4. While the antibody solution is stirring gently at room temperature, add slowly an equal volume of the saturated ammonium sulfate solution to bring the final concentration to 50% saturation.
5. 20 min after the last ammonium sulfate addition, incubate the suspension at 4°C overnight, or for at least 6 h.
6. Centrifuge the suspension (precipitate plus solution) at 8000 g for 20 min at 4°C.
7. Discard the supernatant and suspend the pellet in a 50% saturated

---

ammonium sulfate solution in the centrifuge tube. Centrifuge again at 8000 g for 20 min at 4°C.

8. Discard the supernatant and dissolve the pellet in 40 mM sodium phosphate buffer pH 8 (the buffer to be used in the purification step on DEAE).[a] The buffer volume used should be about one-third of the volume of the ascitic fluid treated.

9. Transfer the antibody solution to dialysis tube and dialyse at 4°C at least overnight against three changes of 40 mM sodium phosphate buffer pH 8.[a]

10. Remove the antibody solution from the dialysis tubing.

11. Centrifuge at 8000 g for 20 min at 4°C to remove aggregates.

12. Measure the absorbance at 280 nm of the antibody solution to estimate the protein concentration. The extinction coefficient of most immunoglobulin G is close to 1.5 mg^{-1}.ml.cm^{-1}.

[a] If you do not need further purification of these antibodies on the DEAE column, suspend the pellet in PBS and dialyse against PBS. Store at –20°C in small aliquots.

Protocol 3. Anion exchange chromatography

Equipment and reagents
- 5 to 10 mL disposable columns (Biorad)
- DEAE resin (e.g. DEAE-Trisacryl from IBF)
- Spectrophotometer
- Dialysis tube
- 40 mM sodium phosphate, pH 8

Method

1. Equilibrate a DEAE gel such as DEAE–cellulose or DEAE–Trisacryl (IBF) with 40 mM sodium phosphate buffer pH 8. Change the buffer several times until both the pH and the conductivity of the buffer do not change.

2. Pack a column with 2 ml of equilibrated anion exchanger gel for each 1 ml of antibody solution. For best results, do not exceed a total volume of 10 ml.

3. Apply the antibody solution to the gel and start elution immediately with the equilibration buffer used in step 1; collect 0.5 ml fractions.

4. Measure the absorbance at 280 nm of the eluted fractions, and pool the most concentrated fractions.

5. Extensively dialyse the pooled fractions that contain antibody at 4°C against a buffer such as PBS pH 7.4.

6. Check the purity of the antibody solution by gel electrophoresis. Store the antibody solutions at –20°C in small aliquots.

Generally, this easy two-step purification procedure leads to an almost pure mAb preparation, as observed with SDS–PAGE under reducing conditions. Better results can be obtained when antibodies are eluted from the DEAE–matrix column by increasing the ionic strength of the column buffer.

In this case, after the ammonium sulfate precipitation, the antibody solution is dialysed against 10 mM Tris pH 8.5, and the DEAE column equilibrated in 10 mM Tris pH 8.5. The antibody is eluted by increasing NaCl concentrations in 10 mM Tris pH 8.5.

3.2 Isotype determination

The affinity of an immunoglobulin G (IgG) for protein A and its susceptibility to papain cleavage depend on its isotype, i.e. its constant heavy chain composition. It is then important to determine the isotype of the specific mAb.

Immunoconjugates specific to IgG_1, IgG_{2a}, IgG_{2b}, IgG_3 are commercially available and permit isotype determination. The same ELISA protocol as described in *Protocol 1* can be followed for each mAb, but the four anti-class immunoconjugates, instead of the classical anti-immunoglobulin immunoconjugate, must be used for each antibody. This experiment usually gives a clear-cut determination of the subclass of the antibody studied. The response might, however, be positive for two different classes, as happened for one of the antibodies isolated in our laboratory. It was subsequently found to be a hybrid antibody containing two different heavy chains, γ_1 and γ_{2b} (10). Consequently, this antibody is heterogeneous although monoclonal, and is present under three different arrangements: $IgG_{2b}/_{2b}$ specific for the antigen, hybrid $IgG_1/_{2b}$, which is also specific because of its γ_{2b} moiety, and $IgG_1/_1$, which does not bind the antigen.

This example shows how crucial it is to check the isotype homogeneity of mAb before starting any quantitative immunochemical analysis.

3.3 Discrimination between anti-native and anti-denatured antibodies

As already mentioned, the screening of hybridoma by a solid phase immunoassay may also lead to the selection of antibodies directed against a non-native form of the protein, since the protein may undergo some denaturation upon coating.

To determine the capacity of antibodies to recognize the native form of the protein, we devised an ELISA competition test (6) (*Protocol 4*) based on the following rationale. The mAbs and an excess of the native antigen are first incubated in solution for a short period of time, to allow 'rapid' binding of only antibodies recognizing an antigenic determinant (epitope) of the native protein. The free antibody remaining in solution is then monitored by a classical indirect ELISA test as described in *Protocol 1*.

Protocol 4. Competition ELISA

Equipment and reagents
- As in protocol 1

Method

1. Prepare 2 ml of a 2 μg/ml antibody solution (or hybridoma supernatant) in the incubation buffer.

2. Prepare 2 ml of a solution of the native antigen in incubation buffer at a tenfold higher molar concentration than the antibody solution.

3. Make two aliquots of 1 ml of the antibody solution. Add 1 ml of incubation buffer to one and 1 ml of the antigen solution to the other. Also prepare an incubation buffer control.

4. Incubate these mixtures for 15 min at room temperature.

5. Add 0.1 ml of each mixture per well in a plate coated and washed as described in *Protocol 1*. Assays must be done in triplicate. Cover the plate and incubate for 30 min at room temperature.

6. Follow the indirect ELISA procedure described in *Protocol 1*, starting from step 6.

If the antibody tested is anti-native, the absorbance obtained in wells that contain the antigen–antibody mixture will be much lower (even equal to zero) than that obtained in wells that contain antibody without antigen.

If the antibody tested did not bind the native protein in solution, the absorbance obtained with the antigen–antibody mixture will be similar to that of the antibody alone (see *Table 1* as an example).

Soluble denatured forms of the protein can be used in this competition test to cross-check the behaviour of the antibodies with these denatured forms of the antigen (6).

3.4 Do the antibodies recognize different epitopes?

The method commonly used to determine whether two mAbs recognize identical, overlapping, or distant epitopes is the competition immunoassay. In this assay, one of the mAbs is purified and labelled and the capacity of the other mAb preparation to block the binding of this labelled antibody to the immobilized antigen is tested (see ref. 11). This method requires a labelling step for each new mAb to be tested. To avoid this, we set-up an ELISA in which the ability of a pair of unlabelled mAbs to bind simultaneously on the immobilized antigen is tested (12). The antibodies are added separately or together, at saturating concentration, to the coated antigen, and the amount

Table 1. Competition assays of eight mAb with the native β2 subunit of *E. coli* tryptophan synthase

mAb	Absorbance[a]		Native β2
	No competitor	Native β2	Recognition
68–1	1.47	0.06	+
93–6	0.94	0.11	+
46–9	0.20	0.03	+
B3B$_5$	0.34	0.48	−
D$_4$B$_6$	0.80	0.64	−
69–4	0.63	0.65	−
169–3	1.28	1.29	−
172–3	0.97	1.07	−

[a] Values obtained after stopping the β-galactosidase reaction in the ELISA. Each experiment was repeated three times, and the results were reproducible within ± 5%.

of bound antibodies is then detected quantitatively by the usual indirect ELISA described in *Protocol 1*.

A sandwich ELISA using unlabelled mAbs has been devised by Kenett (13). In this case one mAb is adsorbed on to a microtitration plate as a capture antibody for the antigen, and binding of the second mAb added on the antigen is measured.

It can be necessary, for some purposes, to determine precisely the epitopes recognized by the different mAbs. Then an epitope mapping must be performed. Epitope mapping involves:

(a) Chemical or enzymatic cleavage of the antigen, followed by purification of the resulting peptides.

(b) Chemical synthesis of a large number of peptides.

(c) Synthetic peptides on pins.

(d) Random fragment expression library.

(e) Site-directed mutagenesis.

(f) *In vitro* expression of cDNA fragments.

4. The affinity, a quantitative parameter to analyse conformational changes

The affinity of an antibody for its antigen depends on the structural complementarity between the two molecules, so changes in the antigen conformation are expected to be reflected by changes in affinity for the appropriate antibodies. Therefore, the affinity of the antibody–antigen complex is an appropriate quantitative parameter for detecting and characterizing conformational changes in proteins. Hence the need for convenient and precise experimental methods for measuring affinity values. Such a method is described in Section 5. Before describing this method, difficulties due to the multivalency of antibodies and/or antigens will be considered. The determination of the affinity constant of an antibody–antigen complex requires that the stoichiometry of the antigen–antibody association reaction be known. Immunoglobulins are invariably multivalent molecules whose association with oligomeric proteins could lead to the formation of circular antigen–antibody complexes, which may result in an affinity enhancement. Consequently, the use of homogeneous monovalent immunoglobulins, i.e. Fab fragments from mAb, is often preferable.

In the case of an oligomeric protein, the antigen carries as many potential antigenic determinants as it has protomers. However, since even Fab fragments are large molecules, steric hindrance may prevent simultaneous saturation of all the epitopes present on the protein. Therefore, it is necessary to determine how many copies of a given epitope of the oligomer can be saturated by the corresponding Fab at saturation.

For all these reasons, we describe experimental procedures for Fab preparation and for determining the stoichiometry of binding.

4.1 Preparation of 'fragments antigen binding' (Fab)

To nick an IgG into two Fab fragments and one Fc fragment, the proteolytic enzyme papain (commercially available) is commonly used. Prior to large scale digestion, determine the optimum papain to immunoglobulin (w/w) ratio using assays with small aliquots, as described in *Protocol 5*.

Protocol 5. Limited proteolysis of antibodies with papain

Equipment and reagents

- Gel filtration column and liquid chromatography system (e.g. Superose 12 and FPLC from Pharmacia)
- SDS-PAGE equipment and reagents
- Thermostated waterbath
- Papain (Boehringer Mannheim)
- 100 mM potassium phosphate buffer, pH 7.2 supplemented with 2 mM 2-mercaptoethanol and 2 mM EDTA
- Iodoacetamide (Sigma)

Method

1. Prepare 1 ml of the mAb solution at about 5 mg/ml in 100 mM potassium phosphate buffer pH 7.2. Add 2-mercaptoethanol to a final concentration of 2 mM and EDTA to 2 mM.

2. Divide into five 0.2 ml aliquots. Add papain to (w/w) ratios of between 1/20 to 1/200. Seal the tubes, mix well, and place them in a 37°C water-bath.

3. Incubate for appropriate time. Usually, adequate proteolysis is

---

Protocol 5. *Continued*

obtained in 1–3 h without significant further proteolytic cleavage of the fragments.

4. Follow the proteolysis from 15 min by analysing the disappearance of intact IgG and the appearance of both Fab and Fc fragments, using either gel filtration (see Chapter 9) or SDS–PAGE (see Chapter 1).

5. With the papain to immunoglobulin ratio found to give an efficient proteolysis in 1–2 h, carry out proteolysis of the total mAb solution prepared as in step 1.

6. Stop the papain at the appropriate time by adding iodoacetamide to a final concentration of 40 mM, in 40 mM NaH_2PO_4 pH 8.0, and chill immediately in ice.

7. Extensively dialyse the nicked immunoglobulins against 40 mM sodium phosphate buffer pH 8 at 4°C.

To obtain pure Fab, it is necessary to remove the papain, the Fc fragments, and the remaining uncleaved immunoglobulins. This can be done by the following successive chromatographic steps.

The first step consists of anion exchange chromatography like that described in *Protocol 3*. Fab and uncleaved IgG are not retained by the column while Fc is. If desired, Fc can be eluted with 2 M NaCl. The second chromatographic step consists of gel filtration to separate the remaining uncleaved IgG (M_r = 150 000) and Fab (M_r = 50 000). After each chromatographic step, concentrate the eluted protein by ammonium sulfate precipitation (see *Protocol 2*) or ultrafiltration. Dialyse the final Fab preparation at 4°C against a buffer such as PBS, and store at 4°C for no longer than a few weeks.

4.2 Stoichiometry determination

The stoichiometry of binding of pure Fabs to a protein is important only with oligomeric proteins. Because steric hindrance can prevent the simultaneous fixation of Fab molecules to all the sites of same specificity, the number of Fab molecules able to bind to an oligomeric antigen must be known before undertaking any quantitative analysis of the antigen–antibody interaction. This can be determined by titrating the antibody binding sites on the antigen with fluorescence measurements (14,15) or by liquid chromatography such as FPLC (16,17).

For FPLC, follow the disappearance of the free Fab or free antigen, and the appearance of antigen–Fab complexes, at different antigen to Fab ratios using gel filtration or ion exchange columns. A constant amount of antigen is incubated with various amounts of Fab (or vice versa), then injected into the FPLC system. A sample of 50 μl injected requires protein concentrations of about 10^{-5} M for satisfactory detection by the absorbance at 280 nm. The areas of the peaks give the titration curve of antigen by Fab (or vice versa). To verify the stoichiometry found by titration, the Fab–antigen complexes can also be separated and analysed by SDS–PAGE (see Chapter 1), followed by quantitative scanning densitometry of the Coomassie blue stained gel.

The molecular weight of the Fab–antigen complex and, consequently, the number of Fab molecules bound to the antigen, can be also determined by sedimentation-diffusion equilibrium centrifugation (18) (see Chapter 9).

4.3 Are monovalent Fabs always required for affinity measurements in solution?

The method we developed (see below) to determine the true affinity constant of antigen–antibody complexes in solution (14) can be used not only with Fabs but also with intact bivalent antibody molecules, either by working under conditions where most of their sites are liganded, or by using an appropriate mathematical treatment of the data (19). This should not be attempted, however, with decavalent IgM molecules.

5. Affinity measurements in solution by competition ELISA

Determining a quantitative parameter like the equilibrium dissociation constant of an antibody–antigen complex requires a convenient and reliable method. Equilibrium dialysis, the method of choice for haptens and dialysable antigens, is not applicable to macromolecular antigens. Several convenient methods such as ELISA or surface plasmon resonance, with the Ag or the mAb immobilized on the titration plate or on the 'sensor chip' (e.g. gold covered with a carboxymethylated dextran hydrogel), are often used to provide values of the affinity. Such measurements yield real values of K_D only rarely. One reason is that immobilization often results in some denaturation of the protein, thus modifying its binding properties (see refs 20 and 21 for review). Another reason is that K_D is defined in solution, with both the mAb and antigen diffusing and rotating freely in solution, while in the solid phase assay one of the partners is immobile. This can result in estimates of K_D by solid phase assay being orders of magnitude different from the real affinity in solution.

These considerations have led us to set-up a convenient method aimed at measuring the affinity constant in solution of a monoclonal antibody for its antigen.

5.1 Theoretical aspects

The antibody site to antigenic site association reaction can be written as follows:

$$\text{antibody} + \text{antigen} \longleftrightarrow \text{complex}$$

with the concentrations of free antibody sites, free antigen sites, and complex at equilibrium given as [Ab], [Ag], and [x] respectively.

The concentration of free antibody sites [Ab] and antigen sites [Ag] at equilibrium are related to the total antibody sites $[Ab_t]$ and the total antigen sites $[Ag_t]$ by:

$$[Ab] = [Ab_t] - [x] \qquad [1]$$

$$[Ag] = [Ag_t] - [x]. \qquad [2]$$

K_D, the dissociation constant is defined by:

$$K_D = [Ag][Ab]/[x].$$

If $[Ag_t]$ is varied while $[Ab_t]$ is kept constant:

$$K_D = [Ag] \, ([Ab_t] - [x])/[x])$$

consequently,

$$[x]/[Ab_t] = [Ag]/([Ag] + K_D) \qquad [3]$$

Several linear plots of the Equation 3 have been proposed, the most commonly used of which are the following.
Scatchard equation:

$$[x]/[Ag] = ([Ab_t] - [x])/K_D. \qquad [4]$$

Klotz equation:

$$[Ab_t]/[x] = K_D/[Ag] + 1. \qquad [5]$$

If Ab_t and Ag_t are known, the experimental determination of K_D (or $K_A = 1/K_D$) requires precise measurement of only one of the three concentrations [Ab], [Ag], or [x].

5.2 Rationale

The method we have developed requires two steps:

(a) In the first step, antibody at a constant concentration and antigen at various concentrations are incubated in solution until equilibrium is reached.

(b) In the second step, the concentration of antibody that remains free at equilibrium is measured by a classical indirect ELISA in which the antigen is coated on the microtitration plate.

The state (native or partially denatured upon coating) of the coated antigen and whether or not it is recognized by the mAb differently from the soluble antigen is not important so long as the coated antigen can specifically and quantitatively trap the free antibody.

5.3 Requirements for the determination of K_D

For correct determination of free antibody concentration at equilibrium, several requirements must be fulfilled.

(a) The absorbance obtained in the last step of the indirect ELISA, which reflects free antibody concentration, must be proportional to the antibody concentration tested.

(b) Only a small percentage of free antibody molecules (i.e. less than 10%) must bind to the coated antigen, to prevent any significant disruption of equilibrium in the liquid phase.

(c) Since the dissociation constant K_D is generally dependent on the temperature, the temperature must be kept constant throughout.

To satisfy requirements (a) and (b), it is necessary to determine (using the indirect ELISA procedure described in *Protocols 1* and *2*) the total (i.e. initial) antibody concentration range that must be used, the concentration of coated antigen, and the optimal incubation time of the antibody solutions in the coated wells.

Protocol 6.   Quantification of the amount of antibody trapped on to the coated antigen

Equipment and reagents
- As in *Protocol 1*

Method

1. Prepare two coated plates as described in *Protocol 1*, step 1, using an antigen concentration of 1 μg/ml.

2. Prepare antibody solutions at different concentrations (e.g. from 10^{-7} to 10^{-11} M) in the buffer used for the antigen, supplemented with 0.02% (w/v) bovine serum albumin.

3. Wash the first coated plate as described in *Protocol 1*.

4. Add 0.1 ml (per well) of each antibody solution, prepared in step 2, to three coated wells and one non-coated well of the first plate (the assays are carried out in triplicate to have enough solution to make duplicates in step 7). Incubate 30 min at the temperature chosen for the affinity measurement.

5. Carefully pipette out of each well the antibody solution and pool the triplicate samples corresponding to the same antibody concentration. Follow the ELISA procedure described in *Protocol 1* from step 6 for this first plate.

6. Wash the second coated plate as in step 3.

Protocol 6. *Continued*

7. Add to the wells 0.1 ml of each of the pooled antibody solutions that were recovered from the first plate. Carry out the assays in duplicate. Incubate for the same time and temperature as in step 4.

8. Follow the ELISA described in *Protocol 1* from step 6 for the second plate. In the two plates, the enzymatic reaction must be stopped after the same time of incubation. Plot the absorbance read in each plate versus the initial antibody concentrations used for the first plate.

The plot is linear over a limited concentration range of antibody and gradually reaches a plateau. The constant concentration of antibody to be used for the determination of K_D must be chosen in the linear portion of the plot. In this linear range (see *Figure 2*), the fraction of antibody retained on the coated antigen in the first plate is deduced from the ratio $(S_1-S_2)/S_1$, where S_1 is the slope obtained with the first plate and S_2 with the second plate. In the case described in *Figure 2*, the amount of antibody retained on the coated antigen represents 6% of the total amount of antibody incubated in the plate.

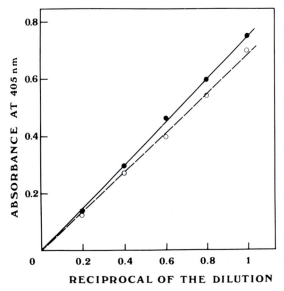

Figure 2. Determination of the amount of antibody retained on the coated antigen. The mAb 46–9 directed against the β2 subunit of *E. coli* tryptophan synthase was incubated for 1 h at 20°C in plates previously coated with β2 subunit (1 μg/ml). The range of antibody concentration was 1×10^{-10} M to 7×10^{-10} M. After incubation, the liquid in each well was transferred to a second coated plate and incubated under identical conditions (see Section 5.3). The β-galactosidase enzymatic reaction was stopped after 2.5 h at 37°C. Closed and open symbols correspond, respectively, to the absorbance values obtained with the first and the second plate.

If the amount of antibody retained is greater than 10% of the total amount of antibody, it is absolutely necessary to repeat the experiment with a lower concentration of antigen in the coating step 1 and/or to reduce the time of incubation of the antibody solutions on the plate in the step 4, until conditions are found under which the two slopes differ by no more than 10%.

5.4 Determination of K_D

Binding equilibrium studies require that the total concentration of antibody (Ab_t) should be close to, or lower than the value of the dissociation constant (K_D). Since the dissociation constant is a priori not known, the total antibody concentration should be chosen as small as possible. Choose the lowest antibody concentration that gives an absorbance of one for the enzymatic reaction after a reasonable time. Do not forget that this antibody concentration should be in the linear part of the plot established in *Protocol 2*.

The results are not satisfactory if the total antibody concentration (Ab_t) is higher than the K_D. The sensitivity of the immunoenzymatic assay with chromogenic substrate limits the minimum total antibody concentration to about 10^{-10} M. Thus for an antibody with very high affinity, the experiment should be done with the β-galactosidase or alkaline phosphatase immuno-conjugate and their fluorogenic substrates, which increases the sensitivity of the assay and permits working with lower antibody concentration.

Taking into account these considerations determine, the K_D as described in *Protocol 7*.

Protocol 7. Determination of the dissociation constant (K_D)

Equipment and reagents
• As in *Protocol 1*

Method
1. Prepare a coated plate as described in *Protocol 1*, step 1, with antigen at a concentration of 1 μg/ml, or the one determined in *Protocol 6*.
2. Prepare 3 ml of the antibody solution at twice the concentration (since it will be diluted twice with the antigen) determined in *Protocol 6*. The antibody and antigen solutions are prepared in the buffer usually required for the antigen and supplemented with 0.02% (w/v) bovine serum albumin.
3. Prepare the antigen solutions (0.3 ml for each concentration) at ten different concentrations (e.g. by serial dilutions from 10^{-8} to 10^{-11} M). The value of [Ag_t] corresponds to the total concentration of antigenic sites (for oligomeric proteins, it is the concentration of protomers if each protomer binds one antibody molecule).

339

Bertrand Friguet et al.

Protocol 7. *Continued*

4. Prepare 11 tubes, ten containing 0.25 ml of the antigen at different concentrations (prepared in step 3) and one containing 0.25 ml of incubation buffer (0 antigen).

5. Add 0.25 ml of the antibody solution prepared in step 2 to each tube and incubate at desired constant temperature until equilibrium is reached (e.g. overnight).

6. When equilibrium is reached, wash as described in *Protocol 1*, steps 3 and 4, the coated plate prepared in step 1.

7. Add 0.1 ml (per well) from each tube to three coated wells and one non-coated well. Incubate at the same temperature as in step 5 and for the time determined in *Protocol 6*.

8. Follow the ELISA procedure in *Protocol 1* from step 6. Since the antibody concentration was chosen in a range where the absorbance is proportional to the antibody concentration, the absorbance read for each well will correspond to the free antibody concentration.

5.5 Calculation

(a) A_0 is the absorbance measured for the well containing the total antibody concentration $[Ab_t]$ in the absence of antigen.

(b) $A1, A2,.....A10$ are the absorbances measured for the wells containing the total antigen concentrations Ag_{t1}, Ag_{t2},......Ag_{t10}, the total antibody concentration Ab_t, and the free antibody concentration $[Ab_1]$, $[Ab_2]$,......$[Ab_{10}]$ at equilibrium.

(c) $[Ab]$, the free antibody concentration in each well, is related to the absorbance A measured in the ELISA by the following equation:

$$[Ab] = [Ab_t] \, (A/A_0)$$

therefore:

$$[Ab_1] = [Ab_t] \, (A_1/A_0); [Ab_2] = [Ab_t] \, (A_2/A_0);$$
$$.....[Ab_{10}] = [Ab_t] \, (A_{10}/A_0).$$

(d) (x), the antigen–antibody complex concentration, is related to the total antibody site concentration $[Ab_t]$ and the free antibody site concentration $[Ab]$, (see Equation 1 in Section 5.1):

$$[x] = [Ab_t]-[Ab] = [Ab_t]-Ab_t(A/A_0) = [Ab_t](A_0-A)/A_0$$

therefore:

$$[x_1] = [Ab_t](A_0-A_1)/A_0; [x_2] = [Ab_t](A_0-A_2)/A_0;$$
$$....[x_{10}] = [Ab_t](A_0-A_{10})/A_0.$$

340

(e) [Ag], the free antigen concentration, is related to the total antigen site [Ag$_t$] and the antibody–antigen complex concentration [x] (see Equation 2 in Section 5.1):

$$[Ag] = [Ag_t]-[x]$$

therefore:

$$[Ag_1] = [Ag_{t1}]-[x_1]; [Ag_2] = [Ag_{t2}]-[x_2];....[Ag_{10}] = [Ag_{t10}]-[x_{10}].$$

(f) Since as indicated above:

$$[x] = [Ab_t](A_0-A)/A_0,$$

$$[Ag_1] = [Ag_{t1}]-[Ab_t](A_0-A_1)/A_0; [Ag_2] = [Ag_{t2}]-[Ab_t](A_0-A_2)/A_0;$$
$$......[Ag_{10}] = [Ag_{t10}]-[Ab_t](A_0-A_{10})/A_0.$$

(g) For the Scatchard equation calculate the values of [x] and [Ag] using the absorbance obtained for each triplicate. The Scatchard equation (see Equation 4 in Section 5.1) can be also written:

$$[x]/[Ab_t][Ag] = (1-[x]/[Ab_t])/K_D \qquad [6]$$

[x]/[Ab$_t$](the fraction of bound antibody) is usually referred as v.

(h) Plot v/[Ag] versus v. The straight line obtained (see *Figure 3*) has a slope equal to $1/K_D = (K_A)$.

(i) For Klotz representation (see Equation 5) calculate 1/[Ag] and [Ab$_t$]/[x] using the absorbance obtained for each triplicate.

(j) Plot [Ab$_t$]/[x] versus 1/[Ag]. The straight line obtained has a slope equal to $K_D = (1/K_A)$.

Moreover, by using programs now readily available for any standard microcomputer, non-linear regression methods can extract the K_D directly from the saturation curve in solution, e.g. plotting (A_0-A/A_0) versus [Ag].

5.6 Determination of K_D with impure antibody

The high sensitivity of the ELISA permits the total antibody concentration [Ab$_t$] to be very low compared to the total antigen concentration [Ag$_{t1}$], [Ag$_{t2}$],....[Ag$_{t10}$], e.g. [Ag$_t$] = 10 [Ab$_t$]. In such a case, the free antigen concentration can be approximated by the total antigen concentration, since the antibody–antigen complex concentration becomes negligible compared to the total antigen concentration. For example, for the total antigen concentration [Ag$_{t1}$] the equation:

$$[Ag_1] = [Ag_{t1}]-[Ab_t](A_0-A_1)/A_0$$

becomes:

$$[Ag_1] = [Ag_{t1}],$$

if [Ab$_t$] is sufficiently smaller than [Ag$_{t1}$].

Using the approximation [Ag$_1$] = [Ag$_{t1}$] and replacing [x$_1$]/[Ab$_t$] by its

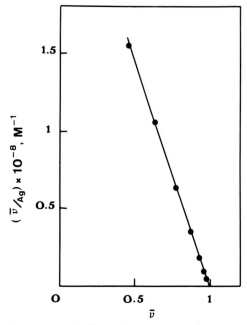

Figure 3. Scatchard plot of the binding of tryptophan synthase holo β2 subunit to mAb 46–9 measured by ELISA at 20°C. v is the fraction of bound antibody and Ag the concentration of free antigen at equilibrium. The total concentration Ab_t of antibody sites was 3×10^{-10} M. The value of K_D is 3.6×10^{-9} M.

expression given by the absorbances: $[x_1]/[Ab_t] = (A_0-A_1)/A_0$, the Scatchard equation (see Equation 6) can be written:

$$(A_0-A_1)/A_0 \times 1/[Ag_{t1}] = 1/K_D \times (1-(A_0-A_1)/A_0) \qquad [7]$$

and the Klotz equation (see Equation 5) can be written:

$$A_0/(A_0-A_1) = K_D/[Ag_{t1}] + 1 \qquad [8]$$

These two last equations no longer contain the total antibody concentration $[Ab_t]$. Therefore, when $[Ag_1]$ can be approximated by $[Ag_{t1}]$, both Equations 7 and 8 allow determination of the dissociation constant K_D even if the antibody site concentration is not known, such as in ascitic fluids, impure mAb, or non-titrated Fab preparations (see the Klotz plot of *Figure 4*). Remember that this is true only if the antigen is in large excess over the antibody (e.g. tenfold).

6. Affinity measurement in solution by a radioimmunoassay-based method

The rationale of the competition ELISA described has been adapted to a radioimmunoassay (RIA)-based method where the antigen is available as a

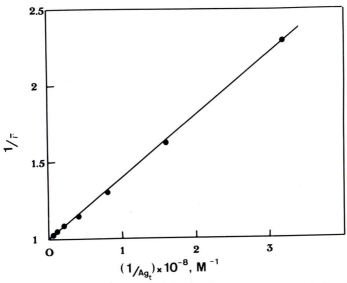

Figure 4. Klotz plot of the binding of tryptophan synthase holo β2 subunit to mAb 46–9 at 20 °C measured by ELISA with impure antibodies. ν is the fraction of bound antibody and Ag_t the total antigen concentration. The concentration of impure antibody was 10^{-4} mg/ml. The value of K_D is 3.8×10^{-9} M.

radiolabelled molecule only in minute amounts and not necessarily pure. It has been designed to measure the affinity of polypeptide chains synthesized in cell-free systems in the presence of radioactive amino acids. Chemically radio-labelled antigen can also be used, provided that the conformation of the protein is not altered upon labelling. The principle of this method is the following: the radiolabelled antigen at a constant concentration and the antibody at various known concentrations in large excess over the antigen concentration are incubated in solution until equilibrium is reached. The concentration of the free antigen (that has not reacted with the antibody) is then determined by using immunobeads, i.e. dextran beads (instead of the ELISA coated plate) to which the same antibody is covalently attached. The free antigen molecules are specifically trapped by the immobilized antibody, and the radioactivity of the immunotrapped antigen fraction is then determined (22,23).

The same requirements needed to determine the free antibody concentration at equilibrium in the competition ELISA must also be fulfilled to determine the free antigen concentration in this method:

(a) The concentration of antigen trapped on the immunobeads must be proportional to the concentration of antigen free in solution.

(b) Only a small percentage of free antigen molecules in the equilibrium solution must bind to the antibody immobilized on the beads to prevent any readjustment of the concentrations at equilibrium.

That these two requirements are fulfilled must be verified by comparing the total radioactivity of antigen in solution to the radioactivity bound to the beads during the RIA (23). Assuming that the antigen is the only radioactive molecule present in the incubation mixture, a simple radioactivity measurement in a scintillation counter can be done. In a more complex situation where the antigen is present in a heterogeneous mixture of radioactive molecules, a separation step such as gel electrophoresis has to be done, followed by the quantification of the amount of radioactive antigen on the gel using an appropriate radioactivity scanner.

7. General considerations about the competition methods

The methods described above offer many advantages, so long as they are used under conditions where equilibrium in solution is reached and not perturbed by the solid phase immunoassay (see Section 5.3). Only minute amounts of mAb and antigen are needed. Only one of the components (antigen or mAb) needs to be purified or titrated. It provides the real saturation curve in solution. In the ELISA-based method, the species at equilibrium in solution are not labelled, which ensures that the observed affinity is characteristic of native mAb and antigen. In the RIA-based method, the use of a protein directly labelled by incorporation of radioactive amino acids also ensures that the antigen should be native.

With the indirect competition ELISA or RIA, the two main problems one may encounter are related to the sensitivity of the solid phase assay (which sets a higher limit to the affinity values that can be measured) and to the complexity of the experimental curves obtained with multivalent antibodies or antigens (which complicates the extraction of the affinity constant from the experimental data).

The lower limits to the K_D values that can be determined by ELISA using classical chromogenic sustrates are about 10^{-10} M. However, commercially available fluorogenic substrates for ELISA increase the assay sensitivity by a factor of about 100, and render it possible to determine K_D values of 10^{-11} to 10^{-12} M. Higher affinities could certainly be approached by RIA-based indirect competition methods, using very high specific activity radiolabelled antigens.

The multivalence of antibodies and antigens, and sometimes the heterogeneity of the antibodies, complicate the determination of affinity. This has become the object of much attention (19,24). The simple data analysis proposed by Friguet *et al.* (14) provides satisfactory results even when divalent immunoglobulins are used, as long as one extracts the affinity only from the part of the saturation curve obtained at high saturation of the mAb by the antigen.

In summary, the competition ELISA method described above is the method of choice to determine the real equilibrium dissociation constants of mAb–Ag complexes in solution because of its simplicity and its modest requirements in terms of equipment, as well as quantity and purity of antigen or mAb. Some examples from the literature (25–29) illustrate the power and the versatility of this method. Its rationale adapted to the RIA-based method permits measurement of the affinity of mAb for radioactive polypeptide chains present in minute amounts in crude heterogeneous mixtures.

These methods can be transposed to the study of protein–protein interactions (e.g. see ref. 30) that do not involve antibodies.

8. Antibodies used as conformational probes

To detect and analyse conformational changes occurring upon ligand binding or association of the domains of the β2 subunit of *E. coli* tryptophan synthase, the affinity constants of mAb for the different forms of the protein were determined. The method described in Section 5.6 was used, permitting determination of the affinity of the antigen–antibody complex in solution without the need of a highly purified antibody preparation. Binding of the coenzyme pyridoxal phosphate and the substrate L-serine to the protein was found to modify the affinity constants of four out of five mAb, demonstrating extended conformational rearrangements of the protein upon ligand binding (16).

The effects of domain assembly on the conformation of the two domains of β2 were analysed using six mAb (17). The β2 subunit is made of two independently folding domains, F_1 and F_2, which can be isolated after a mild proteolytic treatment of β2 by trypsin. Thus, affinities were determined of the mAb for the isolated F_1 or F_2 domains and for the associated domains in the native apo β2 subunit (*Table 2*). The affinities of Fabs for the β2 subunit are higher than for the isolated domains. This indicates that the two domains F_1 and F_2 undergo a conformational change upon their assembly within β2. This conformational change involves a large area of the F_1 domain since affinities of five anti-F_1 antibodies, which recognize five different epitopes of β2, are affected. It is interesting to point out that these five different epitopes of F_1 show different changes in K_A (see ratios in *Table 2*). This suggests that the different areas of the domain are not affected to the same extent by the conformational changes.

Analysing the kinetics of association of mAbs with their protein antigen has provided information about the rate and equilibrium constants of some conformational fluctuations of the protein (31). Using mAbs specific for the native protein, the appearance of immunoreactive folding intermediates has been monitored quantitatively during *in vitro* renaturation of β2 (32,33) and of cytochrome *c* (34), or during biosynthesis of β chains (35) and the tailspike protein of phage P_{22} (36), providing information about the folding mechanisms.

Table 2. Affinity constants (K_a) and affinity ratios of mAb for apo β2 and the isolated domains, F_1 and F_2, of the β2 subunit of *E. coli* tryptophan synthase

Antibodies	K_a ($M^{-1} \times 10^{-8}$) of Fab for:			Ratio[a]
	Apo β2	F_1	F_2	K_a apo β2/K_a domain
15–1: anti-F_1	5	0.5	–	10
19–1: anti-F_1	5	1	–	5
31–2: anti-F_1	3	0.6	–	5
46–9: anti-F_1	1.5	0.25	–	6
68–1: anti-F_1	2.5	0.8	–	3
93–6: anti-F_2	1.5	–	0.03	50

[a] The results were reproducible within ± 20%.

In conclusion, we wish to point out, once again, how critical it is to design adequately the hybridoma screening, which will often determine the success or failure of the whole project. One illustration is the isolation (as reported above) of mAbs binding only the denatured antigen because the ELISA screening test using a coated antigen failed to discriminate between the native and denatured proteins. Similarly, one has to avoid studying the antigen–antibody interaction with one partner immobilized on a solid phase, since part of the protein may be hidden or denatured by the immobilization, and because the kinetics and thermodynamics of the interaction in heterogeneous phase are vastly different from what they are in solution.

An important remark is that the results obtained are sometimes difficult to interpret unambiguously. For example, the affinity changes reported in *Table 2* may result from three causes: a structural change of each domain; an increase in their rigidity which would render them less 'adaptable' to the antibody sites; or each epitope investigated being shared by both domains. Only the comparison of five distinct mAbs allowed us to favour the first hypothesis.

In spite of these drawbacks, mAbs are exquisite conformational probes, due to their unique sensitivity to minute changes in the state of the protein, the very small quantities of antigen needed, and the ability to perform large series of measurements, and therefore to screen a large array of factors (pH, ligands, temperature, etc...) liable to affect the protein conformation.

References

1. Crumpton, M. J. (1966). *Biochem. J.*, **100**, 223.
2. Galfré, G. and Milstein, C. (1981). In *Methods in enzymology* (ed. J. J. Langone and H. Van Vunakis), Vol. 73, p. 3. Academic Press, New York.
3. Zola, H. and Brooks, D. (1982). In *Monoclonal hybridoma antibodies: techniques and applications* (ed. J. G. R. Hurrel), p. 1. C.R.C. Press Inc.

4. Campbell, A. M. (1984). In *Laboratory techniques in biochemistry and molecular biology* (ed. R. H. Burdon and P. H. Knippenberg), Vol. 13, p. 101. Amsterdam, New York, Oxford Elsevier.
5. Legrain, P., Boegtle, D., Buttin, G., and Cazenave, P. A. (1981). *Eur. J. Immunol.*, **11**, 678.
6. Friguet, B., Djavadi-Ohaniance, L., and Goldberg, M. E. (1984). *Mol. Immunol.*, **12**, 673.
7. Suter, M. and Butler, J. E. (1986). *Immunol. Lett.*, **13**, 313.
8. Butler, J. E., Ni, L., Nessler, R., Joshi, K. S., Suter, M., Rosenberg, B., *et al.* (1992). *J. Immunol. Methods*, **150**, 77.
9. Smith, A. D. and Wilson, J. E. (1986). *J. Immunol. Methods*, **94**, 31.
10. Chaffotte, A. F., Djavadi-Ohaniance, L., and Goldberg, M. E. (1985). *Biochimie*, **67**, 75.
11. Harlow, E. and Lane, D. (1988). *Antibodies: a laboratory manual.* Cold Spring Harbor Laboratory.
12. Friguet, B., Djavadi-Ohaniance, L., Pagès, J., Bussard, A., and Goldberg, M. (1983). *J. Immunol. Methods*, **60**, 351.
13. Kenett, D. (1988). *J. Immunol. Methods*, **106**, 203.
14. Friguet, B., Chaffotte, A. F., Djavadi-Ohaniance, L., and Goldberg, M. E. (1985). *J. Immunol. Methods*, **77**, 305.
15. Larvor, M. P., Djavadi-Ohaniance, L., Friguet, B., Baleux, F., and Goldberg, M. E. (1991). *Mol. Immunol.*, **28**, 523.
16. Djavadi-Ohaniance, L., Friguet, B., and Goldberg, M. E. (1986). *Biochemistry*, **25**, 2502.
17. Friguet, B., Djavadi-Ohaniance, L., and Goldberg, M. E. (1986). *Eur. J. Biochem.*, **160**, 593.
18. Wilhelm, P., Friguet, B., Djavadi-Ohaniance, L., Pilz, I., and Goldberg, M. E. (1987). *Eur. J. Biochem.*, **164**, 103.
19. Stevens, F. J. (1987). *Mol. Immunol.*, **24**, 1055.
20. Djavadi-Ohaniance, L. and Friguet, B. (1991). In *The immunochemistry of solid-phase immunoassay* (ed. J. E. Butler), Chapter 10. C.R.C. Press Inc.
21. Butler, J. E. (1992). In *Structure of antigens* (ed. M. H. V. Van Regenmortel), Vol. 1, p. 209. C.R.C. Press Inc.
22. Friguet, B., Fedorov, A., Serganov, A. A., Navon, A., and Goldberg, M. E.(1993). *Anal. Biochem.*, **210**, 344.
23. Djavadi-Ohaniance, L., Goldberg, M. E., and Friguet, B. (1996). In *Antibody engineering: a practical approach* (ed. J. McCafferty), p. 77. IRL Press, Oxford.
24. Winzor, D. J., Bowles, M. R., Pentel, P. R., Schoof, D. D., and Pond, S. M. (1991). *Mol. Immunol.*, **28**, 995.
25. Randen, I., Brown, D., Thomson, K. M., Hugues-Jones, N., Pascual, V., Victor, K., *et al.* (1992). *J. Immunol.*, **148**, 3296.
26. Berneman, A., Ternynck, T., and Avrameas, S. (1992). *Eur. J. Immunol.*, **22**, 625.
27. Carter, P., Presta, I., Gorman, C. M., Ridgway, J. B. B., Henner, D., Wong, W. L. T., *et al.* (1992). *Proc. Natl. Acad. Sci. USA*, **89**, 4285.
28. Yeh, P., Landais, D., Lemaître, M., Maury, I., Crenne, J. Y., Becquart, J., *et al.* (1992). *Proc. Natl. Acad. Sci. USA*, **89**, 1904.
29. Lavoie, T. B., Drohan, W. N., and Smith-Gill, S. J. (1992). *J. Immunol.*, **148**, 503.
30. Nelson, R. M. and Long, G. L. (1992). *J. Biol. Chem.*, **267**, 8140.

31. Chaffotte, A. F. and Goldberg, M. E. (1987). *J. Mol. Biol.*, **197**, 131.
32. Blond-Elguindi, S. and Goldberg, M. E. (1990). *Biochemistry*, **29**, 2409.
33. Murry-Brelier, A. and Goldberg, M. E. (1988). *Biochemistry*, **27**, 7633.
34. Allen, M. J., Jemmerson, R., and Nall, B. T. (1994). *Biochemistry*, **33**, 3967.
35. Tokatlidis, K., Friguet, B., Deville-Bonne, D., Baleux, F., Fedorov, A. N., Navon, A., *et al.* (1995). *Phil. Trans. R. Soc. London*, **348**, 89.
36. Friguet, B., Djavadi-Ohaniance, L., King, J., and Goldberg, M. E (1994). *J. Biol. Chem.*, **269**, 15945.

14

Stabilization of protein structure by solvents

SERGE N. TIMASHEFF and TSUTOMU ARAKAWA

1. Introduction

It has been common practice for more than half a century for biochemists and biologists, when they have isolated an activity (e.g. an enzyme) or an organelle, to store the isolated material in concentrated (~ 1 M) glycerol, sucrose, or a similar substance, in order to preserve this activity. Certain salts, such as ammonium sulfate, have also been known for a long time to stabilize biological activity, when added at high concentration (~ 1 M). Yet the mechanism of such protein structure stabilization was totally unknown. At present, the manufacture of proteins, or modified proteins, by biosynthetic cloning technology has greatly revived interest in methods that are available for folding proteins into proper biologically active structures and for maintaining them in such a structure for prolonged periods. This chapter will be devoted to the question of the physical basis of the stabilization of native protein structures in aqueous solution by the addition of co-solvents at high (~ 1 M) concentration. Such understanding should make it possible to use such co-solvents rationally and effectively.

2. Principles of structure stabilization by solvent components

Why do some substances, when added at high concentration, stabilize protein structures? Let us look at the destabilization reaction, that is, at protein denaturation, expressed as a simple two-state equilibrium between the native (N) and denatured (or unfolded or destabilized) (D) states. And let this equilibrium be shifted toward the native, or stable, form of the protein by the solvent additive S. Then, the reaction can be expressed by:

$$N \underset{\text{stabilizer}}{\overset{K(S)}{\rightleftharpoons}} D, \qquad K = \frac{[D]}{[N]} \qquad [1]$$

By definition, then, the equilibrium constant of this reaction is a function of the concentration of the stabilizer S.

The effect of the stabilizer on the equilibrium constant can be expressed quantitatively by the Wyman linkage relation (1). It states that, upon changing only the co-solvent concentration, the change in the equilibrium constant is given by the difference in the number of co-solvent molecules bound by the denatured and native protein molecules:

$$\frac{d \log K}{d \log a_S} = \Delta \nu_S = \nu^D - \nu^N \tag{2}$$

where a_S is the thermodynamic activity of the additive, S, related to its free concentration by the activity coefficient; $\Delta \nu_S$ is the difference between the number of moles of solution component S (stabilizer) bound per mole of protein (as measured by dialysis equilibrium or a similar technique) between the denatured (ν^D) and native (ν^N) states of the protein.

In the case of stabilization, the reaction is shifted to the left, so the equilibrium constant must decrease with an increase in concentration of the additive. In this case, (d log K/d log a_S) is negative, so $\Delta \nu_S$ must also be negative. In other words, there must be less binding of the stabilizer to the denatured form of the protein than to the native one.

Measurements have been carried out by a dialysis equilibrium technique of the binding of a number of stabilizing solvents to a variety of proteins in the native state. Typical data abstracted from these studies are presented in *Table 1*. One striking result stands out: all the binding stoichiometries are negative. Reading these results as one normally does binding data, we find that, for example, in 1 M sucrose, tubulin binds minus 38 mol of sugar per mol of protein, in 1 M $MgSO_4$, bovine serum albumin (BSA) binds minus 27 mol of salt per mol of protein, etc.

What is the meaning of negative binding? This can be understood best if we consider the nature of dialysis equilibrium (2). *Figure 1* shows two possible end results of such an experiment. On the left-hand side we see the familiar result: at the end of the experiment, the solution inside the bag has a higher concentration of the ligand than the bulk solvent. On the right-hand side we see the opposite result—namely, at the end of the experiment, the solution inside the bag has a lower concentration of the ligand than the bulk solvent. Now, operationally, binding, ν, is defined as:

$$\nu = \frac{[L^{in}] - [L^{out}]}{[Protein]} \tag{3}$$

where $[L^{in}]$ and $[L^{out}]$ are the concentrations of ligand inside and outside the bag, respectively. In fact, an equilibrium binding experiment measures not the total amount of ligand bound to the protein, but the relative affinities

Table 1. Preferential interactions of stabilizing and precipitating co-solvents with proteins

Co-solvent	Protein	Apparent binding of co-solvent to protein			Preferential hydration of protein	Reference
		g/g	mol/mol	Ratio[a]	g H_2O/g protein	
1 M Sucrose	Tubulin	-0.106	-38.0	0.69	0.243	8
4.1 M Glycerol	Tubulin	-0.127	-149	Neg	0.154	2
1 M Sucrose	RNase A	-0.190	-7.6	0.53	0.437	8
2 M Glucose	BSA (pH 6.0)	-0.099	-37.4	0.72	0.212	9
3.4 M Glycerol	BSA (pH 5.8)	-0.113	-83.4	Neg	0.212	19
2 M Glycine	BSA (pH 6.1)	-0.069	-62.2	0.88	0.416	10
2 M Betaine	BSA (pH 6.1)	-0.125	-72.5	—	0.428	10
2 M Na glutamate	BSA (pH 7.0)	-0.171	-68.8	0.72	0.417	30
2 M Lys–HCl	BSA (pH 6.0)	-0.123	-45.8	0.49	0.248	30
1 M Lys–HCl	BSA (pH 6.0)	-0.047	-17.4	0.44	0.222	30
1 M Arg–HCl	BSA (pH 5.7)	-0.028	-9.09	0.35	0.114	15
1 M Na glutamate	BSA (pH 7.0)	-0.088	-35.5	0.84	0.477	30
1 M NaCl	BSA (pH 5.6)	-0.0145	-16.8	0.64	0.243	12
1 M Na_2SO_4	BSA (pH 5.6)	-0.074	-35.4	0.78	0.524	12
1 M NaO_2CCH_3	BSA (pH 5.6)	-0.027	-22.4	—	0.312	12
1 M $MgSO_4$	BSA (pH 4.5)	-0.047	-26.5	0.55	0.388	12
1 M $MgCl_2$	BSA (pH 4.5)	-0.0040	-2.8	0.13	0.041	12
2 M $MgCl_2$	BSA (pH 3.0)	-0.0158	-11.1	0.53	0.162	13
1 M Gdm_2SO_4	BSA (pH 4.5)	-0.052	-16.4	0.38	0.211	14
2 M Gdm_2SO_4	BSA (pH 4.5)	-0.184	-58.0	0.69	0.316	14
1 M $GdmO_2CCH_3$	BSA (pH 5.6)	-0.010	-5.8	—	0.077	14
1 M GdmCl	BSA (pH 4.5)	0.025	17.5	Neg	-0.239	14
3.9 M MPD (50%)	RNase A (pH 5.8)	-0.943	-109	Neg	1.031	18
3.9 M MPD (50%)	RNase A (pH 2.0)	-0.531	-61.6	Neg	0.602	18
0.4 M (30%) PEG_{1000}	β-Lg (pH 2.0)	-0.165	-3.04	Neg	0.411	6
1 M TMAO	RNase T1	-0.045	-6.6	Neg	0.553	22

[a] Ratio of the experimentally determined value of the effect of the co-solvent on the chemical potential of the protein to that calculated from the effect of the co-solvent on the surface tension of water.

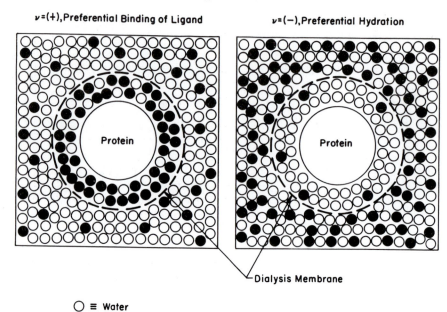

Figure 1. Schematic representation of solvent component distribution at dialysis equilibrium. Left, preferential binding; right, preferential hydration.

of the protein for ligand and for water. This is demonstrated analytically by:

$$\nu = (\partial m_S/\partial m_P)_{T,\mu_W,\mu_S} = \nu_S = \frac{m_S}{55.56}\nu_W \qquad [4]$$

where ν_S and ν_W are the total numbers of moles of ligand (stabilizer in the present case) and water interacting with each mole of protein, m_i is the molal concentration of component i (mol/1000 g of water), μ_i is its chemical potential, and T is the Kelvin temperature. The number 55.56 is the molal concentration of pure water. The subscripts W, P, and S refer to water, protein, and solvent additive, such as the stabilizing ligand.

The experimental binding parameter, ν, therefore, is preferential binding, that is it is the expression of the excess of binding of ligand over water relative to the bulk solvent composition. Therefore, a negative value of ν signifies that there is an excess of water in the domain of the protein, that is the protein is preferentially hydrated, or, as it is often expressed, there is preferential exclusion of the ligand from the protein domain. Preferential hydration can be calculated from preferential binding (3):

$$(\partial m_W/\partial m_P)_{T,\mu_W,\mu_S} = -\frac{55.56}{m_S}(\partial m_S/\partial m_P)_{T,\mu_W,\mu_S} = -\frac{55.56\nu}{m_S} \qquad [5]$$

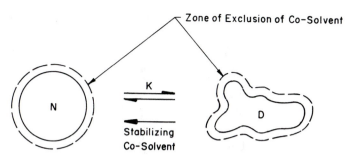

Zone of Exclusion of Co-Solvent

Figure 2. Schematic representation of a protein denaturation reaction in the presence of a stabilizing co-solvent. The zone of exclusion of the stabilizing co-solvents becomes greater as the protein–solvent interface increases during denaturation, due to an increase in the asymmetry of the protein.

How does preferential exclusion of ligand (v negative) lead to the stabilization of the native structure of a protein? Being a pure thermodynamic quantity, binding simply reflects the perturbation of the chemical potential of the protein by the ligand (4):

$$v = (\partial m_S/\partial m_P)_{T,\mu_w,\mu_S} = -(\partial \mu_P/\partial m_S)_{T,P,m_P}/(\partial \mu_S/\partial m_S)_{T,P,m_P} \qquad [6]$$

where P is pressure. Equation 6 shows that a negative value of v means that $(\partial \mu_P/\partial m_S)_{T,P,m_P}$ is positive, that is, addition of ligand (stabilizer) increases the chemical potential of the protein and, thus, the free energy of the system. This is a thermodynamically unfavourable situation. If, in the course of the denaturation reaction, the chemical nature of the interactions between protein and stabilizer does not change, the situation should become even more unfavourable thermodynamically in the unfolded state of the protein. This is illustrated by the schematic representation of a denaturation reaction in *Figure 2*. Here the protein is shown immersed in an aqueous solution of stabilizer, say sucrose. Preferential exclusion of the stabilizer can be expressed, as a model, by a zone of exclusion. Since denaturation leads to protein unfolding and an increase in structural asymmetry, the protein–solvent surface of contact increases, increasing the zone of exclusion. Then $(\partial \mu_P/\partial m_S)^D_{T,P,m_P} > (\partial \mu_P/\partial m_S)^N_{T,P,m_P}$, the situation becomes even more unfavourable thermodynamically and, by the Le Chatelier principle, the reaction is pushed toward the left, toward the native state. The net result is stabilization of the native structure.

3. Sources of the exclusion

All the systems examined up to the present have obeyed without exception what appears to be a general rule: those solvents which, at high concentration, stabilize the native structures of proteins are preferentially excluded from the domain of the protein. Is there a universal cause of the exclusion?

The answer is no. Since the choice of stabilizer may depend on the mechanism by which it is excluded from proteins, it seemed desirable to describe some of the known mechanisms of exclusion. These can be classified generally into two categories:

(a) Those in which the interactions with proteins are determined strictly by the properties of the solvent; that is the proteins are basically inert.

(b) Those in which the chemical nature of the protein surface determines the interactions (attractive or repulsive) between molecules of the protein and the co-solvent.

Obviously, there can also be intermediate cases in which both mechanisms operate simultaneously.

Taking the first category, two principal mechanisms can be identified. The simplest mechanism is that of steric exclusion, first proposed by Kauzmann (5). The basis of this phenomenon, depicted schematically in *Figure 3*, is the difference in size of co-solvent and water molecules. If the co-solvent cannot penetrate the protein structure, a shell that is impenetrable to co-solvent is formed around the protein molecule. Its thickness is determined by the distance of closest approach of protein and co-solvent. Water molecules, being smaller, can penetrate within this shell. As a result, a zone enriched in water is formed around the protein molecule. Effectively, this is preferential hydration and, by Equation 5, preferential exclusion of the co-solvent. This phenomenon is the source of the preferential hydration of proteins in the water polyethylene glycol (PEG) systems (6).

A second mechanism of the first category is based on the perturbation of the surface free energy (surface tension) of water by the co-solvent (7). Since the surface of contact between protein and solvent constitutes an interface, there must be in this surface an interfacial (surface) tension. Additives perturb the cohesive force of water and, hence, its surface tension. As shown by Gibbs in 1873, this results in an excess, or deficiency, of the additive in the

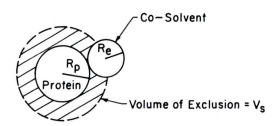

Figure 3. Schematic representation of the steric exclusion mechanism of preferential hydration. R_p is the radius of the protein; R_e is the radius of exclusion (co-solvent); V_s is the volume in which there is an excess of water (6). (Reproduced by permission from the American Chemical Society.)

surface layer. Adapted to our current notation, the Gibbs adsorption isotherm can be expressed as (8):

$$(\partial m_S/\partial m_P)^{calc}_{T,\mu_W,\mu_S} = -\frac{A_P}{RT}(\partial\sigma/\partial \ln a_S)_{T,P,m_P} \qquad [7]$$

where A_P is the molar surface area of the protein and σ is surface tension. The superscript 'calc' indicates that the value of the preferential interaction is a calculated quantity, rather than the one measured experimentally by dialysis equilibrium. It is clear that if a substance increases the surface tension of water, its excess in the surface layer will be negative; that is it will be preferentially excluded from the interface. If the interface is with a protein molecule, the result is preferential hydration of the protein. The contribution of this mechanism to preferential interactions can be evaluated by taking the ratio of the experimental value of $(\partial\mu_P/\partial m_S)_{T,P,m_P}$ to that calculated from the surface tension effect through a combination of Equations 6 and 7. Values of this ratio greater than 0.5 indicate the dominance of this effect. Examination of *Table 1*, column 5 shows that this is the mechanism that gives rise to the observed preferential exclusion of sugars (8,9), amino acids (10,11), and structure-stabilizing salts (12–14), and it appears to be the most widespread mechanism in protein stabilizing solvent systems (15).

In the second category, the exclusion is determined by chemical interactions between the solvent components and the protein structure. 2-Methyl-2,4-pentanediol (MPD) is a powerful protein crystallizing agent (16,17) and is strongly excluded from the protein surface (18). At pH 5.8 in 50% MPD, the preferential hydration of RNase A is 1.03 g of water per gram of protein. The source of this preferential hydration is the strong repulsion of MPD by electric charges (positive or negative). The surface of a globular protein at neutral pH can be regarded as a mosaic of charges, and the charge density may be very high even though the net charge may be close to zero. MPD is strongly repelled from the charges and migrates into the bulk solvent, leaving a layer enriched in water. Glycerol, on the other hand, seems to be excluded from protein surfaces by the solvophobic effect (19); that is contact between the non-polar regions of proteins and the glycerol–water mixture is entropically even more unfavourable than contact with water. As a result, glycerol molecules migrate away from such regions into the bulk solvent, leaving behind a layer enriched in water. Conversely, glycerol has an affinity for the polar regions of proteins (19). The net effect is a balance between the two types of interactions, with preferential exclusion generally predominating.

4. Balance between co-solvent exclusion and binding

The net interaction between stabilizing co-solvents and native globular proteins is that of preferential exclusion. This does not mean that co-solvent molecules are totally excluded from contact with protein molecules. In fact, they can penetrate into the zone of preferential hydration and bind to proteins at specific sites. The net observation is that of the balance between binding and exclusion, which, for stabilization, favours the latter. To illustrate this point, let us take the case of β-lactoglobulin in aqueous $MgCl_2$ (13). From the surface tension increment of $MgCl_2$, this salt at a concentration of 1 M should show a preferential hydration of 0.38 g water/g protein. The measured value at pH 5.1 is 0.027 g water/g protein. At pH 3.0, the value becomes 0.22 g water/g protein, which is much closer to that predicted. This difference can be explained by the binding of Mg^{2+} ions to the protein. This binding is much stronger at the iso-electric pH, 5.1 (11 moles of $MgCl_2$ per mol of protein) than at acid pH (3 mol of $MgCl_2$ per mol of protein), reflecting the balance between the affinity of Mg^{2+} ions for the protein and the large positive charge of the protein at acid pH which leads to strong electrostatic repulsion of the divalent cation.

In the case of salts, the contribution to binding and exclusion of each ion must be taken into account, and it is possible to classify them as either stabilizers (salting-out) or destabilizers (salting-in). This is clearly demonstrated in the study of Robinson and Jencks (20) who have measured the solubility of a model polypeptide in a series of salts. Both the cations and the anions exert a role. Thus, if we compare the sodium halides, we find that NaCl salts-out, NaBr has no effect, and NaI salts-in. In the series of cation chlorides, NaCl salts-out, LiCl has no effect, and $CaCl_2$ salts-in. This leads to the conclusion that the Na^+ ion salts-out (stabilizes), the Li^+ ion salts-in (destabilizes), the Cl^- ion salts-out (stabilizes) and the Br^- ion salts-in (destabilizes), and the effects of the ions are additive. As would be expected, when Li^+ and Br^- are introduced together as LiBr, they salt-in, while in their chloride and sodium salts, respectively, they were neutralized by the salting-out ions.

The same effect, when expressed as exclusion or binding, is well illustrated in a comparison of $MgSO_4$ and $MgCl_2$ (21). The measured preferential hydrations of β-lactoglobulin in 1 M $MgSO_4$ are 0.43 g water/g protein at pH 5.1 and 0.55 g water/g protein at pH 3.0. These values are much higher than those of $MgCl_2$ cited above and indicate essentially no binding of Mg^{2+} to the protein when the anion is SO_4^{2-}. This observation, in fact, reflects the relative effectiveness of the ions for binding or exclusion; SO_4^{2-} has a strong exclusion capacity, while that of Cl^- is much weaker; Mg^{2+} binds to proteins. The difference between the results obtained in $MgCl_2$ and $MgSO_4$ simply reflects the fact that SO_4^{2-} exclusion (a repulsive interaction) totally overwhelms the tendency of Mg^{2+} to bind (an attractive interaction), while Cl^- does not have that capability. The relative effectiveness of anions and cations in inducing the preferential hydration of native globular proteins is shown in *Table 2* for the sodium, divalent cation, and guanidinium salts of chloride, acetate, and sulfate (14). It is evident that this preferential hydration effectiveness increases in the orders $Cl^- < CH_3CO_2^- < SO_4^{2-}$ and $Gdm^+ < (Mg^{2+}, Ca^{2+}, Ba^{2+}) < Na^+$. Thus, Na_2SO_4 is much more strongly excluded than NaO_2CCH_3

Table 2. Extent of preferential hydration induced by salts[a]

| Cation/anion | Preferential hydration[b] (g water/g protein) | | |
	Cl^-	$CH_3CO_2^-$	SO_4^{2-}
Na^+	0.243	0.312	0.524
Mg^{2+}, Ca^{2+}, Ba^{2+}	0.041 to -0.046	0.109 to 0.180	0.388[c]
Gdm^+	-0.239	0.077	0.211

[a] The salt concentration was 1 M; the protein was BSA.
[b] Preferential hydration expressed in g water/g protein when the concentration is measured on the molal scale.
[c] Value for $MgSO_4$.

or $MgSO_4$. It should, therefore, be the better structure stabilizer. In the series of guanidinium salts, Gdm_2SO_4 can be expected to be a mild structure stabilizer, while GdmCl is a denaturant, as is well known. Comparison of the various salts identifies a set with intermediate capacity for inducing preferential hydration: it consists of NaCl, $Mg(O_2CCH_3)_2$, and Gdm_2SO_4. They should all act as mild precipitants of native proteins, and could be weak stabilizers. Urea, in fact, is similar in effectiveness as a denaturant to $GdmO_2CCH_3$ and much weaker than GdmCl.

The principle of additivity applies also to mixtures of co-solvents (22). This is well illustrated by the effects of urea and trimethylamine N-oxide (TMAO) on the stability of ribonuclease T1 (RNase T1), shown in *Figure 4*. In *Figure 4a*,

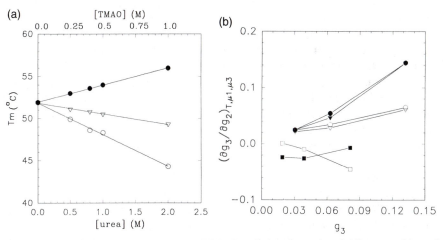

Figure 4. Effect of mixed solvents. (a) Midpoint of the thermal unfolding transition of RNase T1 in urea (o), TMAO (•), urea and TMAO at a molar ratio of 2 : 1 (▽). (b) Co-solvent concentration dependence of the preferential binding of co-solvent to protein: RNase T1 in urea (o); RCM-T1 in urea (•); RNase T1 in TMAO (□); RCM-T1 in TMAO (■); RNase T1 in solutions of urea and TMAO with a molar ratio of 2 : 1 (▽); RCM-T1 in solutions of urea and TMAO with a molar ratio of 2:1 (▼). The g_3 values for the ternary solvents indicate the concentration of urea in grams of urea per gram of water.

it is shown that the transition temperature (T_m) of RNase T1 is lowered by urea, but raised by TMAO. In a mixture of the two, T_m assumes values that are the arithmetic sum of the effects of the individual co-solvents. An explanation of this is found in *Figure 4b*. Here the preferential interactions of native RNase T1 and its unfolded form, reduced carbomethylated ribonuclease T1 (RCM-T1) (23), with urea, TMAO, and urea in the presence of TMAO are shown. It is clear that the preferential binding of urea to the protein is not affected by TMAO. The converse is also inferred, although measurements were impossible for technical reasons.

5. Co-solvent interactions in the denaturation reaction

When dealing with a reaction, such as the denaturation equilibrium of Equation 1, the protein co-solvent interactions must be known for both end states, since it is the difference between the preferential interactions that determines the effect of the co-solvent on the equilibrium. Combining Equations 2 and 4, we find that:

$$\frac{d \ln K}{d \ln a_S} = \Delta v = (v_S^D - v_S^N) - (m_S/55.56)(v_W^D - v_W^N) \tag{8}$$

Therefore, in which direction the unfolding reaction will be affected by any given co-solvent is not known a priori. When the interactions are defined by the properties of the solvent system alone, the protein being essentially an inert component, their extent should be determined by the surface of contact with solvent. This surface increases upon denaturation, so preferential exclusion should increase, rendering Δv of Equation 8 negative. This leads to protein structure stabilization.

When the interactions are determined by the chemical nature of the protein surface, the situation can become considerably more complicated, since denaturation can drastically change this property. Large non-polar regions, as well as peptide bonds, become exposed to solvent, potentially changing the nature of interactions between the co-solvent and the protein. A striking example of this situation is MPD. It has a good affinity for non-polar amino acid residues (24), and it also lowers the surface tension of water. Yet it is strongly excluded from native proteins, as shown above (18). This is the cause of its action as a strong protein precipitant. On the other hand, MPD facilitates protein unfolding (25). For example, at pH 5.8, 40% (v/v) MPD lowers the thermal unfolding transition temperature of RNase A by 15°C, even though the native protein is preferentially hydrated to the extent of 0.81 g water/g RNase A. Why is this so?

The denaturing action of MPD reflects the fact that its interactions with proteins are determined by the chemical nature of the solvent exposed surface. This is illustrated in *Figure 5*. As discussed above, MPD is strongly

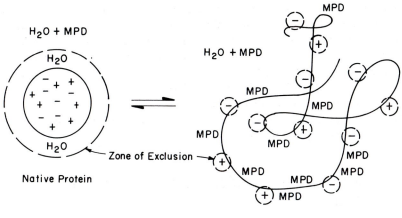

Figure 5. Pattern of protein–solvent interactions in the denaturation of RNase A in the water–MPD system. MPD exclusion due to the high density of charges on the native protein is replaced by MPD exclusion from individual charged sites and MPD binding to solvent exposed non-polar regions in the unfolded state of the protein.

repelled from the densely charged surface of a native protein molecule. The repulsion of charges for MPD overcomes both its attraction for exposed non-polar regions and its tendency to accumulate in the surface layer, due to its lowering of the surface tension of water. When the protein unfolds, its charge density becomes greatly reduced, permitting MPD molecules to penetrate to the protein surface. There they can bind to the newly exposed non-polar residues, while the local repulsion from individual charges is weakened. In this manner, the unfolded (or denatured) structure becomes stabilized. Since the interactions are more favourable with the denatured state of the protein, the balance is tipped toward denaturation and the net effect becomes one of protein destabilization. A similar situation is true for PEG; it is a non-polar substance (26) that lowers the unfolding transition temperatures of proteins (6), even though it is strongly excluded from the native proteins and is a good precipitant for them.

6. Practical considerations

The choice of stabilizing co-solvents must, to a great extent, be determined by the exact nature of the system and the particular aims of the experiment. For example, the purpose may be to stabilize for storage a purified protein or an assembled organelle, or to stabilize a protein during freezing or thawing, including lyophilization. It may be desirable to stabilize the native structure more or less strongly, depending upon whether molecular flexibility is necessary for it to exhibit its activity, or if the purpose is to arrest intermediate

Table 3. Nature of co-solvent interactions with proteins

Co-solvent	Mechanism of exclusion	Mechanism of binding
Class I—stabilizers		
Sugars	Surface tension increase	Weak binding
Some amino acids (glycine, alanine, glutamic and aspartic acids)	Surface tension increase	Weak binding
Salting-out salts (Na$_2$SO$_4$, NaCl, MgSO$_4$)	Surface tension increase	Weak binding
Glycerol	Solvophobicity	Affinity for polar regions
Class II—action dependent on conditions		
Weakly acting salts (MgCl$_2$, NaCl, MgSO$_4$)	Surface tension increase	To charged groups or peptide bonds
Arginine–HCl, lysine–HCl	Surface tension increase	To peptide bonds and negative charges
Valine (possibly other non-polar amino acids)	Surface tension increase	To hydrophobic regions
Methylamines	Unknown	To hydrophobic regions
Class III—destabilizers (under unfolding conditions)		
PEG	Steric exclusion	To hydrophobic regions
MPD	Repulsion from charges	To hydrophobic regions

species on the pathway of folding or activation. These various purposes may require stronger or weaker stabilizers, or a variation of stabilizing strength during the course of the process. In general, one should use stabilizer concentrations of 1–3 M, work at room temperature or below, and keep the pH not far from the isoelectric point of the protein.

In choosing a stabilizing co-solvent, one should consider the nature of its interactions with proteins. These are summarized for a number of compounds in *Table 3*. The compounds to consider first are those of class 1, where the protein surface is essentially inert to the co-solvent. These comprise: sugars, glycerol, amino acids, and some salts. All, except for glycerol, owe their preferential exclusion and, hence, protein stabilization, to the surface tension effect. Glycerol is excluded due to the solvophobic effect. None seem to bind to proteins either in the native or denatured state, although glycerol has an affinity for polar groups on proteins.

In the most general application, that of long-range stabilization, either in solution or in the frozen state, the compounds of choice seem to be sucrose and glycerol. In fact, they are those that had been identified empirically by biochemists over decades of experience. A good example is the stabilization of tubulin which, by nature, is a very unstable protein. In *Figure 6a* and *b* are shown, respectively, the effects of 1 M sucrose on tubulin stability in solution (27), and during freezing and thawing (28). In solution, the protein in dilute buffer rapidly undergoes a decay in its ability to bind the drug colchicine. Addition of 1 M sucrose totally eliminated this effect, rendering the protein fully stable. The freezing and thawing necessary in the storage of the protein resulted in the formation of aggregates of denatured material when tubulin was kept in dilute buffer. When the process was carried out in the presence of 1 M sucrose, this protein deterioration disappeared, indicating that 1 M sucrose protects this protein from damage by its preferential exclusion. In fact, it has been shown recently (29) that the agents that stabilize proteins in solutions can also be used for stabilization of proteins in the frozen state. The mechanism of stabilization is most probably the same as that described here for proteins in solution.

The salts and amino acids present a potential complication due to their charged state; electrostatic interactions are possible with charged sites on proteins. Amino acids, furthermore, are limited in the pH range over which they can be used due to the ionization of their carboxyl and amino groups. Furthermore, being organic molecules, they may interact in different modes with proteins. For example, arginine hydrochloride can interact with proteins via the formation of hydrogen bonds by its guanidine group. Valine can interact with hydrophobic groups on proteins (11). On the other hand, sodium glutamate should be a good stabilizer (30).

In selecting a salt for purposes of stabilization, refer to *Table 2*. Sulfate salts seem to be the natural choice because of the strong exclusion of SO_4^{2-} from proteins; the nature of the cation would be dictated by the particular

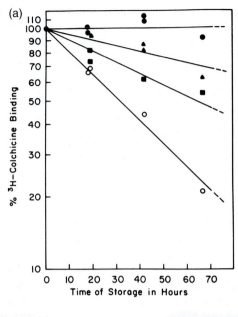

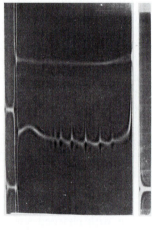
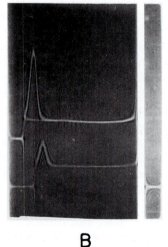

(b)   **A**   **B**

circumstances. The strongest stabilization will be certainly given by Na_2SO_4, which should be a desirable medium for long-range storage of proteins. Yet, if molecular flexibility is desired, the choice might be $MgSO_4$ or one of the intermediate stabilizers, such as $NaCl$, $Mg(O_2CCH_3)_2$, or Gdm_2SO_4, all of which have similar preferential hydration capacities (*Table 2*). In arresting intermediate states of proteins, a weak stabilizer like $GdmO_2CCH_3$ might be the best choice, changing to Gdm_2SO_4, and finally Na_2SO_4 to bring the protein gradually to progressively more compact structures.

Figure 6. Stabilization of calf brain tubulin by sucrose. (a) Loss of colchicine binding ability by tubulin during storage in PMG buffer (10 mM sodium phosphate, 5 mM MgCl$_2$, 0.1 mM GTP pH 7.0). The binding of 1 mole of [^{3}H] colchicine per α-β tubulin dimer, measured at 37°C, corresponds to native tubulin. The protein was stored at 4°C in PMG buffer containing varying amounts of sucrose: ○, no sucrose; ■, 0.2 M; ▲, 0.5 M; ●, 1.0 M sucrose. (Reproduced with permission, ref. 27.) (b) Stabilization of tubulin by 1 M sucrose in the freezing–lyophilization treatment: ultracentrifuge patterns. Tubulin was placed into a PMG buffer. Pattern A is that of a 3 mg/ml protein solution that had been lyophilized from that buffer and redissolved in it. The rapidly moving peaks are large non-specific aggregates. In pattern B, aliquots of the tubulin were frozen in the same buffer that had been made 1 M in sucrose, and then thawed. The upper pattern is for a protein concentration of 6 mg/ml, the lower pattern for 2 mg/ml. The single peak is native 5.8S α-β tubulin dimer; there is no evidence of aggregation or denaturation. (Reproduced with permission, ref. 28.)

Some strongly excluded co-solvents that belong to class 3, such as MPD or PEG, might seem at first sight to be good choices for stabilizing protein structure. These, in fact, do not act as expected because of their non-polar characters and their ability to bind to non-polar regions on proteins (31). Yet, these may be additives of choice for precipitating and crystallizing active proteins, if the work is done below the temperature range of unfolding. In fact, one might wish to combine their unfolding activity at higher temperatures with their precipitating action at room temperature by properly manipulating a temperature decrease in their presence. In the case of multidomain proteins, this may lead to gradual stabilization of particular domains of proteins, while others are kept flexible.

Acknowledgements

This is Publication No. 1642 from the Graduate Department of Biochemistry, Brandeis University, Waltham, MA 02254. This work was supported in part by the National Institutes of Health Grant No. GM-14603.

References

1. Wyman, J. (1964). *Adv. Protein Chem.*, **19**, 223.
2. Na, G. C. and Timasheff, S. N. (1981). *J. Mol. Biol.*, **151**, 165.
3. Timasheff, S. N. and Kronman, M. J. (1959). *Arch. Biochem. Biophys.*, **83**, 60.
4. Casassa, E. F. and Eisenberg, H. (1964). *Adv. Protein Chem.*, **19**, 287.
5. Kauzmann, W. as quoted in: H. K. Schachman and M. A. Lauffer (1949). *J. Am. Chem. Soc.*, **71**, 536.
6. Arakawa, T. and Timasheff, S. N. (1985). *Biochemistry*, **24**, 6756.
7. Sinanoglu, O. and Abdulnur, S. (1964). *Photochem. Photobiol.*, **3**, 333.
8. Lee, J. C. and Timasheff, S. N. (1981). *J. Biol. Chem.*, **256**, 7193.
9. Arakawa, T. and Timasheff, S. N. (1982). *Biochemistry*, **21**, 6536.
10. Arakawa, T. and Timasheff, S. N. (1983). *Arch. Biochem. Biophys.*, **224**, 169.

11. Arakawa, T. and Timasheff, S. N. (1985). *Biophys. J.*, **47**, 411.
12. Arakawa, T. and Timasheff, S. N. (1982). *Biochemistry*, **21**, 6545.
13. Arakawa, T. and Timasheff, S. N. (1984). *Biochemistry*, **23**, 5912.
14. Arakawa, T. and Timasheff, S. N. (1984). *Biochemistry*, **23**, 5924.
15. Kita, Y., Arakawa, T., Lin, T.-Y., and Timasheff, S. N. (1994). *Biochemistry*, **33**, 15178.
16. King, M. V., Magdoff, B. S., Adelman, M. B., and Harker, D. (1956). *Acta Crystallogr.*, **9**, 460.
17. Cotton, F. A., Hazen, E. E., and Richardson, D. C. (1966). *J. Biol. Chem.*, **241**, 4389.
18. Pittz, E. P. and Timasheff, S. N. (1978). *Biochemistry*, **17**, 615.
19. Gekko, K. and Timasheff, S. N. (1981). *Biochemistry*, **20**, 4667.
20. Robinson, D. R. and Jencks, W. P. (1965). *J. Am. Chem. Soc.*, **87**, 2470.
21. Arakawa, T., Bhat, R., and Timasheff, S. N. (1990). *Biochemistry*, **29**, 1914.
22. Lin, T.-Y. and Timasheff, S. N. (1994). *Biochemistry*, **33**, 12695.
23. Pace, C. N. and Creighton, T. E. (1986). *J. Mol. Biol.*, **188**, 477.
24. Pittz, E. P. and Bello, J. (1971). *Arch. Biochem. Biophys.*, **146**, 513.
25. Arakawa, T., Bhat, R., and Timasheff, S. N. (1990). *Biochemistry*, **29**, 1924.
26. Hammes, G. G. and Schimmel, P. R. (1967). *J. Am. Chem. Soc.*, **89**, 442.
27. Frigon, R. P. and Lee, J. C. (1972). *Arch. Biochem. Biophys.*, **153**, 587.
28. Lee, J. C., Frigon, R. P., and Timasheff, S. N. (1975). *Ann. N.Y. Acad. Sci.*, **253**, 284.
29. Carpenter, J. F. and Crowe, J. H. (1988). *Cryobiology*, **25**, 244.
30. Arakawa, T. and Timasheff, S. N. (1984). *J. Biol. Chem.*, **259**, 4979.
31. Lee, L. L.-Y. and Lee. J. C. (1987). *Biochemistry*, **26**, 7813.

A1

List of suppliers

Albright and Wilson

Albright and Wilson, P.O. Box 26229, Richmond, VA 23260-6229, USA.

Aldrich Chemical

Aldrich Chemical, 1001 West St. Paul Avenue, P.O. Box 2060, Milwaukee, WI 53201, USA.

Aldrich Chemical, The Old Brickyard, New Road, Gillingham, Dorset SP8 4JL, UK.

Aldrich Chemical, Postfach 1170, W-7920 Heidenheim, Germany.

Amersham

Amersham International plc., Lincoln Place, Green End, Aylesbury, Buckinghamshire HP20 2TP, UK.

Amersham International, 2636 South Clearbrook Drive, Arlington Heights, IL 60005, USA.

Amicon

Amicon, 72 Cherry Hill Drive, Beverly, MA 01915, USA.

Amicon, Upper Mill, Stonehouse, Gloucestershire GL10 2BJ, UK.

Amicon, Postfach 1103, D-58401 Witten, Germany.

Anderman

Anderman and Co. Ltd., 145 London Road, Kingston-Upon-Thames, Surrey KT17 7NH, UK.

Applied Biosystems

Applied Biosystems, 850 Lincoln Center Drive, Foster City, CA 94404, USA.

Applied Biosystems, 7 Kingsland Grange, Woolston, Warrington WA1 4SR, UK.

Aquebogue Machine and Repair Shop, Box 2055, Main Road, Aquebogue, NY 11931, USA.

Beckman Instruments

Beckman Instruments Inc., P.O. Box 3100, 2500 Harbor Boulevard, Fullerton, CA 92634, USA.

Beckman Instruments UK Ltd., Progress Road, Sands Industrial Estate, High Wycombe, Buckinghamshire HP12 4JL, UK.

Beckman Instruments, Frankfurter Ring 115, Postfach 1416, D-80807 München, Germany.

List of suppliers

Becton Dickinson

Becton Dickinson and Co., Between Towns Road, Cowley, Oxford OX4 3LY, UK.

Becton Dickinson and Co., 2 Bridgewater Lane, Lincoln Park, NJ 07035, USA.

Bio

Bio 101 Inc., c/o Statech Scientific Ltd, 61–63 Dudley Stree, Luton, Bedfordshire LU2 0HP, UK.

Bio 101 Inc., PO Box 2284, La Jolla, CA 92038–2284, USA.

Biometra

Biometra, 550 N. Reo Street, Tampa, FL 33609-1013, USA.

Biometra, Whatman House, St. Leonards Road, Maidstone, Kent ME16 0LS, UK.

Biometra, Rudolf-Wissel-Strasse 30, D-37079 Göttingen, Germany.

Bio-Rad Laboratories

Bio-Rad Laboratories Ltd, Division Headquarters, 3300 Regatta Boulevard, Richmond, CA 94804, USA.

Bio-Rad Laboratories Ltd., Bio-Rad House, Maylands Avenue, Hemel Hempstead HP2 7TD, UK.

Bio-Rad Laboratories, Heidemannstr. 164, D-80939 München, Germany.

Biosys

Biosys, 1057 East Meadow Circle, Palo Alto, CA 94303, USA.

Biosys, 21 Quai du Clos des Roses, F-60200 Compiegne, France.

Boehringer Mannheim

Boehringer Mannheim Biochemica, GmbH, Sandhofer Str. 116, Postfach 310120, D-68298 Mannheim, Germany.

Boehringer Mannheim Corporation, Biochemical Products, 9115 Hague Road, P.O. Box 504 Indianopolis, IN 46250–0414, USA.

Boehringer Mannheim UK, 1057 (Diagnostics and Biochemicals) Ltd., Bell Lane, Lewes, East Sussex BN17 1LG, UK.

British Drug Houses (BDH) Ltd., Poole, Droset, UK.

Calbiochem-Novabiochem

Calbiochem-Novabiochem, 10394 Pacific Center Court, San Diego, CA 92121 USA.

Calbiochem-Novabiochem, 3 Heathcoat Building, Highfields Science Park, University Boulevard, Nottingham NG7 2QJ, UK.

Calbiochem-Novabiochem, Lisztweg 1, P.O. Box 1167, D-65812 Bad Soden, Germany.

Difco Laboratories

Difco Laboratories Ltd., P.O. Box 14B, Central Avenue, West Molesey, Surrey KT8 2SE, UK.

Difco Laboratories, P.O. Box 331058, Detroit, MI 48232–7058, USA.

Dionex

Dionex, 4 Albany Court, Camberley, Surrey GU15 2PL, UK.

Dionex, Vicolo del casale di S. Nicola 2, I-00123 Roma, Italy.

Du Pont

Du Pont (UK) Ltd., Industrial Products Division, Wedgwood Way, Stevenage, Herts SG1 4Q, UK.

Du Pont Co. (Biotechnology Systems Division), P.O. Box 80024, Wilmington, DE 19880–002, USA.

Dynatech Laboratories

Dynatech Laboratories, 14340 Sullyfield Circle, Chantilly, VA 22021-1617, USA.

Dynatech Laboratories, Daux Road, Billingshurst, West Sussex RH14 9SJ, UK.

Dynatech Laboratories, Justinus Kerner Strasse 32, D-73770 Denkendorf, Germany.

Eppendorf, Postfach 650670, Barkhausenweg 1, D-22339 Hamburg, Germany.

European Collection of Animal Cell Culture, Division of Biologics, PHLS Centre for Applied Microbiology and Research, Porton Down, Salisbury, Wilts SP4 0JG, UK.

Falcon (Falcon is a registered trademark of Becton Dickinson and Co.).

Fisher Scientific Co., 711 Forbest Avenue, Pittsburgh, OA 15219–4785, USA.

Flow Laboratories, Woodcock Hill, Harefield Road, Rickmansworth, Herts. WD3 1PQ, UK.

Fluka

Fluka-Chemie AG, Industriestrasse 25, CH-9470, Buchs, Switzerland.

Fluka-Chemicals Ltd., Peakdale Road, Glossop, Derbyshire SK13 9XE, UK.

Fluka-Chemicals, 980 South Second Street, Ronkonkoma, NY 11779-7238, USA.

FMC BioProducts

FMC BioProducts, 191 Thomaston Street, Rockland, ME 04841, USA.

FMC BioProducts, Risingevej 1, DK-2665 Vallensbaek Strand, Denmark.

Gibco BRL

Gibco BRL (Life Technologies Inc.), 3175 Staler Road, Grand Island, NY 14072-0068, USA.

Gibco BRL (Life Technologies Ltd.), Trident House, Renfrew Road, Paisley PA3 4EF, UK.

Gilson Medical Electronics

Gilson Medical Electronics, 3000 West Beltline Highway, P.O. Box 620027, Middleton, WI 53562-0027, USA.

Gilson Medical Electronics, Boite Postale 45, 72 rue Gambetta, F-95400 Villiers le Bel, France.

Hamilton

Hamilton, P.O. Box 26, CH-7402 Bonaduz, Switzerland.

Hamilton, 4970 Energy Way, Reno, NV 89520, USA.

Harvard Apparatus, Fircroft Way, Ednebridge, Kent TN8 6HE, UK.

Hellma
Hellma, Postfach 1163, D-79371 Müllheim, Germany.
Hellma, P.O. Box 544, Borough Hall Station, Jamaica, NY 11424, USA.
Hewlett Packard
Hewlett Packard, 3495 Deer Creek Road, Palo, Alto, CA 94304, USA.
Hewlett Packard, Cain Road, Bracknell, Berkshire RG12 1HN, UK.
Hewlett Packard, 150 Route du Nant-d'Avril, CH-1217 Meyrin-Geneva 2, Switzerland.
Hitachi
Hitachi, 6 Kanda Surugadai-4-chome, Chiyoda-ku, Tokyo 101, Japan.
Hitachi, 3100 North First St., San Jose, CA 95134, USA.
Hitachi, Berlinerstrasse 91, D-40880 Ratingen, Germany.
Hoefer Scientific Instruments
Hoefer Scientific Instruments, 654 Minnesota Street, P.O. Box 77387, San Francisco, CA 94107-0387, USA.
Hoefer Scientific Instruments, Unit 12, Croft Road Workshops, Newcastle, Staffordshire ST5 0TW, UK.
Horwell, Arnold R., 73 Maygrove Road, West Hampstead, London NW6 2BP, UK.
Hybaid
Hybaid Ltd., 111–113 Waldegrave Road, Teddington, Middlesex TW11 8LL, UK.
Hybaid, National Labnet Corporation, P.O. Box 841, Woodbridge, N.J. 07095, USA.
HyClone Laboratories, 1725 South HyClone Road, Loga, UT 84321, USA.
IBF
IBF, 35 Avenue Jean-Jaures, F-92295 Villeneuve-la Garenne cedex, France.
IBF, 7151 Columbia Gateway Drive, Columbia, MD 24106, USA.
ICN Pharmaceuticals
ICN Pharmaceuticals, 3300 Hyland Avenue, Costa Mesa, CA 92626, USA.
ICN Pharmaceuticals, Unit 18, Thame Park Business Centre, Wenman Road, Thame, Oxfordshire OX9 3XA, UK.
ICN Pharmaceuticals, Thüringer Str. 15, D-37269 Eschwege, Germany.
International Biotechnologies Inc., 25 Science Park, New Haven, Connecticut 06535, USA.
Invitrogen Corporation
Invitrogen Corporation, 3985 B Sorrenton Valley Building, San Diego, C.A. 92121, USA.
Invitrogen Corporation c/o British Biotechnology Products Ltd., 4–10 The Quadrant, Barton Lane, Abingdon, OX14 3YS, UK.
ISCO
ISCO, 4700 Superior, P.O. Box 5347, Lincoln, NE 68504, USA.
ISCO, Bruschstrasse 17, CH-8708 Mannedorf, Switzerland.
Kodak Eastman Fine Chemicals, 343 State Street, Rochester, NY, USA.

Kratos Analytical

Kratos Analytical, 535 East Crescent Drive, Ramsey, NJ 07446, USA.

Kratos Analytical, Barton Dock Road, Urmston, Manchester M31 2LD, UK.

Kratos Analytical, 88 rue Phillipe de Girard, F-75018 Paris, France.

LC Packings, Baarsjesweg 154, 1057 HM Amsterdam, The Netherlands.

Life Technologies Inc., 8451 Helgerman Court, Gaithersburg, MN 20877, USA.

Malvern Instruments

Malvern Instruments, Spring Lane South, Malvern, Worcestershire WR14 1AQ, UK.

Malvern Instruments, 10 Southville Road, Southborough, MA 01772, USA.

Malvern Instruments, Seestrasse 7, Postfach 1142, D-82206 Herrsching, Germany.

Merck

Merck, Frankfurter Strasse, 250, Postfach 4119, D-64271 Darmstadt, Germany.

Merck Industries Inc., 5 Skyline Drive, Nawthorne, NY 10532, USA.

Merck, Merck House, Poole, Dorset BH15 1TD, UK.

Millipore

Millipore Corp./Biosearch, P.O. Box 255, 80 Ashby Road, Bedford, MA 01730, USA.

Millipore (UK) Ltd., The Boulevard, Blackmoor Lane, Watford, Herts WD1 8YW, UK.

Millipore, Postfach 5647, D-65731 Eschborn, Germany.

Molecular Dynamics

Molecular Dynamics, 880 East Arques Avenue, Sunnyvale, CA 94086, USA.

Molecular Dynamics, 4 Chaucer Business Park, Kemsing, Sevenoaks, Kent TN15 6PL, UK.

Molecular Dynamics, Elisabethstrasse 1030105, D-47799 Krefeld, Germany.

Molecular Probes

Molecular Probes, P.O. Box 22010, 4849 Pitchford Avenue, Eugene, OR 97402-9144, USA.

Molecular Probes, Poort Gebouw, Rijnsburgerweg 10, NL-2333 AA Leiden, The Netherlands.

New England Biolabs (NBL),

New England Biolabs (NBL), 32 Tozer Road, Beverley, MA 01915-5510, USA.

New England Biolabs (NBL), c/o CP Labs Ltd., P.O. Box 22, Bishops Stortford, Herts CM23 3DH, UK.

Nikon Corporation, Fuji Building, 2–3 Marunouchi 3-chome, Chiyoda-ku, Tokyo, Japan.

Nordic Immunological Laboratories

Nordic Immunological Laboratories, Langestraat 55-61, NL-5000 AA Tilburg, The Netherlands.

Nordic Immunological Laboratories, Dairy House, Moneyrow Green, Holyport, Maidenhead, Berkshire, UK.

Nordic Immunological Laboratories, Drawer 2517, Capistrano Beach, CA 92663, USA.

Nunc

Nunc, P 280, DK4000 Roskilde, Denmark.

Nunc, Hagenauer Str. 21a, D-65203 Wiesbaden-Biebrich, Germany.

Nunc, 2000 North Aurora Road, Naperville, IL 60563, USA.

Paul Marienfield, P.O. Box 1523, D-97965 Bad Mergentheim, Germany.

Perkin-Elmer

Perkin ElmerCetus (The Perkin-Elmer Corporation), 761 Main Avenue, Norwalk, CT 06859, USA.

Perkin Elmer Ltd., Post Office Lane, Beaconsfield, Bucks. HP9 1QA, UK.

Perkin Elmer Ltd., Postfach 101164, Askaniaweg 1, D-88647 Ueberlingen, Germany.

PerSeptive Biosystems, 38 Sidney Street, Cambridge, MA 02139, USA.

Pharmacia Biotech Europe Procordia EuroCentre, Rue de la Fuse-e 62, B-1130 Brussels, Belgium.

Pharmacia Biosystems

Pharmacia Biosystems Ltd., (Biotechnology Division), Davy Avenue, Knowlhill, Milton Keynes MK5 8PH, UK.

Pharmacia LKB Biotechnology AB, Björngatan 30, S-75182 Uppsala, Sweden.

Pierce Chemical

Pierce Chemical, 3747 N. Meridian Road, P.O. Box 117, Rockford, IL 61105, USA.

Pierce Chemical, Roentgenstrasse 18, P.O. Box 1512, NL-3260 BA Oud-Beijerland, The Netherlands.

Pierce Chemical, 36 Clifton Road, Cambridge CB1 4ZR, UK.

Polymicro Technologies, Phoenix, AZ, USA.

Promega

Promega Corporation, 2800 Woods Hollow Road, Madison, WI 53711–5399, USA.

Promega Ltd., Delta House, Enterprise Road, Chilworth Research Centre, Southampton SO1 7NS, UK.

Promega, Niels Bohrweg 11-13, NL-2333 CA Leiden, The Netherlands.

Protein Solutions

Protein Solutions, Unit 5, The Hillside Centre, Upper Green Street, High Wycombe, Buckinghamshire, UK.

Protein Solutions, 2300 Commonwealth Drive, Suite 102, Charlottesville, VA 22901, USA.

Qiagen

Qiagen Inc., 9600 Eton Avenue, Chatsworth, CA 91311, USA.

Qiagen Inc., c/o Hybaid, 111–113 Waldegrave Road, Teddington, Middlesex TW11 8LL, UK.

Qiagen, Max-Volmer-Str. 4, D-40724 Hilden, Germany.

Rheodyne, P.O. Box 996, Cotati, CA 94928, USA.

Sarstedt

Sarstedt, 68 Boston Road, Leicester LE4 1AW, UK.

Sarstedt, P.O. Box 468, Newton, NC 28658-9900, USA.

Sarstedt, Rommelsdorf, D-51588 Nuembrecht, Germany.

Savant Instruments, 110–103 Bi-County Blvd., Farmingdale, NY 11735, USA.

Schleicher and Schuell

Schleicher and Schuell Inc., 10 Optical Avenue, P.O. Box 2012, Keene, NH 03431, USA.

Schleicher and Schuell Inc., Postfach 4, Hahnestrasse 3, D-37856, Dassel, Germany.

Schleicher and Schuell Inc., 145 London Road, Kingston-upon-Thames, Surrey KT17 7NH, UK.

Schötte-Geräte, Im Langgewann 5, Postfach 1130, D-65701 Hofheim/Ts, Germany.

Schwarz-Mann, P.O. Box 28050, Cleveland, OH 44128, USA.

Separations Group, 17434 Mojave Street, P.O. Box 867, Hesperia, CA 92345, USA.

Serva Feinbiochemica

Serva Feinbiochemica, Carl-Benz Str. 7, P.O. Box 10 52 60, D-69115 Heidelberg, Germany.

Serva Feinbiochemica, 200 Shames Drive, Westburgy, NY 11590, USA.

Shadel, 1684 Hudson Avenue, San Francisco, CA 94124, USA.

Shandon Scientific Ltd., Chadwick Road, Astmoor, Runcorn, Cheshire WA7 1PR, UK.

Sigma Chemical Company

Sigma Chemical Company, 3050 Spruce Street, P.O. Box 14508, St. Louis, MO 63178-9916.

Sigma Chemical Company (UK), Fancy Road, Poole, Dorset BH12 4QH, UK.

Sigma Chemical Company (Germany), Grünwalder Weg 30, D-82041 Deisenhofen, Germany.

SLT Labinstruments

SLT Labinstruments, Unterbergstrasse 1A, A-5082 Grödig, Austria.

SLT Labinstruments, P.O. Box 13953, Research Triangle Park, NC 27709, USA.

Sorvall DuPont Company, Biotechnology Division, P.O. Box 80022, Wilmington, DE 19880-0022, USA.

Southern Biotechnology Associates, 160A Oxmoor Blvd., Birmingham, AL 35209, USA.

Stratagene

Stratagene Inc., 11011 North Torrey Pines Road, La Jolla, CA 92037, USA.

Stratagene Ltd., Unit 140, Cambridge Innovation Centre, Milton Road, Cambridge CB4 4FG, UK.

Stratagene Inc., Postfach 105466, D-69044, Heidelberg, Germany.

Takara Biochemical

Takara Biochemical, Shijo-Takakura, Shimogyo-ku, Kyoto 600, Japan.

Takara Biochemical, 719 Allston Way, Berkeley, CA 94710, USA.

United States Biochemical, P.O. Box 22400, Cleveland, OH 44122, USA.

Upchurch Scientific, 2969 North Goldie Road, P.O. Box 1529, Oak Harbor, WA 98034, USA.

Valco Instruments, P.O. Box 55603, Houston, TX 77255, USA.

Wako Chemicals

Wako Chemicals, 1,2 Doshomachi 3-Chome, Chou-ku, Osaka 541, Japan.

Wako Chemicals, 1600 Bellwood Road, Richmond, VA 23237, USA.

Wako Chemicals, Nissanstr. 2, D-41468 Neuss, Germany.

Waters Chromatography

Waters Chromatography, 34 Maple Street, Milfod, MA 01757, USA.

Waters Chromatography, 324 Chester Road, Hartford, Northwich, Cheshire CW8 2AH.

Waters Chromatography, 6 rue JP Timbaud, F-78180 Montigny les Breton-neux, France.

Wellcome Reagents, Langley Court, Beckenham, Kent BR3 3BS, UK.

Whatman

Whatman, Springfield Mill, Maidstone, Kent ME14 2LE, UK.

Whatman, 5 rue Pougin de la Maisonneuve, F-45200 Montargis, France.

Whatman, 9 Bridewell Place, Clifton, NJ 07014, USA.

Wyatt Technology, 802 East Cota St., P.O. Box 3003, Santa Barbara, CA 93130, USA.

Yellow Springs Instruments, P.O. Box 279, 1725 Brannum Lane, Yellow Springs, OH 45387, USA.

Index

Page numbers in bold indicate major entries

Index

ATP 109, 125, 145–6
ATZ *see* anilinothiazolinone
autocorrelation **224–9**
autolysis 121, 123, 126–7, 168
autoradiography 25, 84, **133**, 138, **141–43**, 145, 158, 187, 200–202
avidin 327
azide 83, 168

bacteria, expression of proteins **65–72**
bacteriophage **67–72**
band width **270–1**
baseline 203, 256–7, 263–4, 272–3, 290–94, 311
basic protein 1, **195–7**
basis spectra **295–6**
beta (β)-galactosidase **14–20, 325–6, 338–9**
beta (β)-sheet 59, **295–6**
BDB *see* bisdiazobenzidine
bead model 246–7
Beer-Lambert **253, 262**, 294
benzamidomethanol 179
benzidine 63
betaine 351
bicarbonate 34, **46**, 48–50, 121
binding
    antibody **323–48**
    Coomassie blue 199
    ligand **361–63**
    SDS **3–4**
    solvent **350–63**
biotin 80–82, 85–6, 87, 327
birefringent 291, 293
bisacrylamide 2, 7, **17–19, 193–7**
bisdiazobenzidine (BDB) **63**
biuret 259
blocking buffer **80–1, 142**
blotting 3, 12, 25, 47, **79–82**, 126, **137–44**
borate buffer **61, 268–9**
BPTI (bovine pancreatic trypsin inhibitor) **14–20**, 73, 74, 83, 121, **153, 155**, 176, 179, **188–90**, 200
bromphenol blue 10, 14, 72, 84, 128, 194–5, 199, 212
brownian motion 224
BSA (bovine serum albumin) 60–2, 68, 70, 81, 227, 337, 339, **350–1**
Bu₃P *see* tributylphosphine
buffers
    B **74**
    C **74**
    electrophoresis **4–5, 7–11, 17–21**, 144, **191–95, 207–8**
    ionization 308
    MS 33
    protease **121–23, 168**
    spectral properties **268–70,** 281, 292–3

volatile 121, 142
butanol 7–9
^{13}C 54–5
^{14}C 201

C-terminus 58, 63, 106, 121, 188–9
cacodylate buffer 268–9
CAD protein 117–8, **125, 133, 138–9**
calcium 22, 46–7, 206
calcium binding 122–4
calibration proteins **13–20, 221–24**
calmodulin **205–6**, 215
calorimetry 309, 315–6
camphorsulfonic acid **288–10**
capillary zone electrophoresis (CZE) 40, **132–4**
carbamate 104
carboxymethylation **152–8**, 357
carbamoyl phosphate synthetase 118
carbodiimide 60–1, 146, 189
carbohydrate 35, **99–103**, 235
carbonic anhydrase **14–20**, 227
carboxyl group 60, 268, 308, 361
carboxamidomethyl Cys 55
carboxymethyl Cys 55, **152–8**, 162
carboxypeptidase 97, 118
carrier protein **60–64**, 70
casein 81
casting box **206–12**
CD (circular dichroism) 29, 261–62, **287–96**, 300
cDNA (complementary DNA) 91, 93, 94
cell extract **73–9**, 80, 87
cellophane 201–202
Cerenkov radiation 132, 142–3, 146
chaotrope 33, 46, 121
CHAPS 34
charge
    electrophoresis 2, **131, 152–62**, 187–8
    hydration and 355–6, **359–61**
    MS 40–2
    nonideality and 235, 240
chelator 82, 124, 128, 206
chemical cleavage **118–20, 170**
chemical potential 353
chemiluminescence 78, 80–82, **138–41**
chiral reagent, chromatography 99
chloride ion 4–5, 7–8, **356–7**
chlorination 92, 98
chromate 268
chromatin 73, 76–8
chromatography *see* liquid chromatography
chromogenic 339
chymotrypsin
    digestion **99–103**, 106, **121–24**, 126, 128, 167–8

374

Index

Index

M